POLISH LOGIC
1920–1939

Oxford University Press, Ely House, London W. 1

GLASGOW NEW YORK TORONTO MELBOURNE WELLINGTON
CAPE TOWN SALISBURY IBADAN NAIROBI LUSAKA ADDIS ABABA
BOMBAY CALCUTTA MADRAS KARACHI LAHORE DACCA
KUALA LUMPUR HONG KONG TOKYO

POLISH LOGIC
1920—1939

PAPERS BY
Ajdukiewicz, Chwistek, Jaśkowski, Jordan
Leśniewski, Łukasiewicz, Słupecki
Sobociński, and Wajsberg

───

With an Introduction by
TADEUSZ KOTARBIŃSKI

Edited by
STORRS McCALL

Translated by
B. Gruchman, H. Hiż, Z. Jordan
E. C. Luschei, S. McCall, W. Teichmann
H. Weber, and P. Woodruff

OXFORD
AT THE CLARENDON PRESS
1967

PRINTED IN GREAT BRITAIN

*This book is dedicated to the memory
of those Polish logicians who
were killed in the Second World War*

EDITOR'S PREFACE

THIS book of translations makes available to the English reader eighteen papers by eminent Polish logicians, all but one of which were first published in the period between the two wars. That time saw an extraordinary flowering of logical thought in Poland—a fecundity so extensive as to have left its mark in every branch of contemporary logical development. Yet the original papers stemming from the period have remained for the most part inaccessible to the philosophical reader-at-large. Some have even been on the verge of extinction: only one offprint copy, for example, of Łukasiewicz's *Der Äquivalenzenkalkül* is known to have survived the war. All but two of the papers in the volume appear in English for the first time.

One comment should be made concerning the choosing of items to be translated, in which task the editor wishes to acknowledge the help of Professor A. N. Prior, Dr C. Lejewski, and Dr E. C. Luschei. This is the omission of writings by Tarski. It was felt that the appearance of Tarski's inter-war papers in the volume *Logic, Semantics, Metamathematics* (Oxford 1956) did full justice to this important author.

The editor has received much help from many people during the eight years it has taken to produce this book, and would like to thank the following. Professor Kotarbiński and Dr Sobociński provided constant encouragement, and I am particularly indebted to Professor Kotarbiński for writing the introduction, organizing a team of proof-readers in Poland, and for many other kindnesses. It is safe to say that the book would have died in its infancy without Dr Jordan's generous offer to translate from the Polish. Gene Luschei has gone to endless trouble over the Leśniewski items, both translating and checking, as has Henry Hiż over Wajsberg's 1931 paper, and Peter Woodruff over several papers. Help of a most tangible nature came from the research funds of McGill University and the University of Pittsburgh. All people mentioned above with the exception of

EDITOR'S PREFACE

Hiż and Lejewski read at least part of the proofs, as did Jay Rosenberg, Michael Dunn, Tony Quinton, David Wiggins, and my wife Ann. Esther Barfield, Betty Laubach, and Janet Hutchison are the best typists of Polish symbolism in the world. Additional thanks are due to W. H. Mayer and John Wilkinson for letting me see copies of the University of Chicago translations of papers by Ajdukiewicz and Łukasiewicz, and also to Bogdan Gruchman, Wendy Teichmann, Horst Weber, and Jonah Weinberg for giving generously of their time.

STORRS McCALL

Makerere University College
Kampala, Uganda
February 16th, 1967

CONTENTS

INTRODUCTION

NOTES ON THE DEVELOPMENT OF FORMAL LOGIC IN POLAND IN THE YEARS 1900–39

TADEUSZ KOTARBIŃSKI

SINCE the early years of this century Polish thinkers have contributed to the vigorous current of formal logic in the West. Their contributions, however, were almost all published at home, most of them in Polish, only a few appearing in German. The intention of this publication is to present in English—and in this way to make available to the largest possible body of readers—a review of the most significant works by Polish logicians in the period preceding the Second World War. To make for an easier understanding of the mutual relationships and the specific contents of these works some general notes on the people prominent in the period and on their activity will not be out of place.

The leading figure was unquestionably Jan Łukasiewicz (1878–1956) the senior member of the group—a man who was not only the colleague of his juniors, but also their venerated mentor. A philosopher by training, Łukasiewicz spent most of his university years in his home city of Lwów, in the part of Poland at that time—before the establishment in 1918 of the independent Polish state—under Austrian rule. In that part of Poland were the only two Polish universities active at the time, those of Cracow and Lwów. Both contributed greatly to the development of formal logic, but the role of Lwów was the more important. An outstanding teacher, Kazimierz Twardowski, had been active there ever since 1895 in his chair of philosophy. In the course of his lectures, which were encyclopaedic in character, he also spoke on the new line of study then being developed: algebra-like logic. This attracted Łukasiewicz,

who soon became its spokesman, enthusiast, and co-creator.
Łukasiewicz's scholarly training was rather versatile and included, in addition to that given in the methodical school of
Twardowski, substantial knowledge of Greek and Latin, as
well as mathematical studies, particularly in the theory of
probability.[1] With this background he embarked early on a most
ambitious venture: to open up in logic vistas comparable to
those opened in geometry by the introduction of non-Euclidean
systems. What if one questioned some of the fundamental
assumptions of the narrow traditional logic of Aristotle?
Such questioning directed Łukasiewicz's scholarly pursuits.
A start had to be made with a reconstruction of authentic
Aristotelianism in logic, deformed as this had been by philolo-
gists and philosophers not sufficiently versed in a knowledge
that can be acquired only through extensive study of mathematical logic. Thus a work on the principle of contradiction in
Aristotle's writings[2] was the first of Łukasiewicz's books; and
his last, many years later, was on Aristotle's syllogistic analysed
from the standpoint of modern logic.[3] On the way, so to speak,
he made a new departure in the interpretation of Stoic logic
and its relation to Aristotle's syllogistic.[4] Stoic logic was by
no means an application of the latter, but, on the contrary, a
system logically more fundamental, an anticipation of the
present-day propositional calculus, while Aristotle's syllogistic belonged to the domain of the calculus of predicates.

---

[1] It is worth mentioning that in his works on the theory of probability
Łukasiewicz defines probability as a characteristic property of linguistic
structures and not of events, preceding Keynes by eight and Reichenbach by
twenty-two years. Cf. *Die logischen Grundlagen der Wahrscheinlichkeitsrechnung*, Cracow, 1913.

[2] *O zasadzie sprzeczności u Arystotelesa* (*On the Principle of Contradiction in
Aristotle's Writings*), Cracow, 1910. In 1918 Łukasiewicz said that even in 1910
he had been thinking about the formulation of a non-Aristotelian logic. Cf.
L. Borkowski and J. Słupecki, 'The Logical Works of Łukasiewicz', *Studia
Logica* 8, Poznań, 1958.

[3] *Aristotle's Syllogistic from the Standpoint of Modern Formal Logic*, Oxford,
1st edition 1951, 2nd edition 1957.

[4] 'Z historii logiki zdań' (On the History of the Logic of Propositions),
Przegląd Filozoficzny (*Philosophical Review*) 37, 1934, pp. 117–37; 'Zur Geschichte der Aussagenlogik', *Erkenntnis* 5, 1935–6, pp. 111–31 (translated as
paper 4 of this volume).

It was indeed a shame that though people had been aware of this even towards the end of the Middle Ages, they had lost this rational orientation in more recent times. And it was only the development of the formal theory of deduction and the formal systems of algebraic logic that made possible a return to truths mistakenly given up.

Łukasiewicz's logical studies were concentrated in the second place primarily on propositional calculus. This led, among other things, to the construction of an axiom system for this calculus maximally simplified when compared with Frege's, with the signs of implication and negation as the sole primary terms; to a simplification of the axiom system of Whitehead and Russell's *Principia Mathematica*; to the construction of an alternation-negation axiom system; to a simplification of Nicod's disjunctive axiom; to the construction of the axiom of equivalence; to a system of propositional calculus with quantifiers, based on the implicative axiom system of Tarski and Bernays;[1] and to the discovery of a bracketless and wholly unpunctuated method of formula notation.[2]

Thus prepared on all fronts, Łukasiewicz attacked his central problem, sketching, and elaborating as one of the first logicians ever, a system of three-valued propositional calculus,[3] and shortly afterwards also one of multi-valued propositional calculus,[4] that is, a logic not two-valued, i.e. non-Chrysippian. Indeed, this is how one should label—rather than naming it non-Aristotelian logic—any system which permits other logical values for a sentence besides truth or falsehood. It was none other than Chrysippus, the protagonist of Stoicism, who stubbornly fought for the exclusiveness of these two logical values.

---

[1] Cf. Borkowski and Słupecki, op. cit.

[2] 'O znaczeniu i potrzebach logiki matematycznej' (On the Importance and Needs of Mathematical Logic) *Nauka Polska* (*Polish Science*) 10 (1929), pp. 604–20.

[3] 'O logice trójwartościowej' (On Three-Valued Logic), *Ruch Filozoficzny* (*Philosophical Movement*) 5 (1920), pp. 170–1 (paper 1 of this volume). The first was N. A. Vasil'ev; cf. George L. Kline, 'N. A. Vasil'ev and the development of many-valued logics', in *Contributions to Logic and Methodology in honor of J. M. Bocheński*, Amsterdam, 1965.

[4] 'Interpretacja liczbowa teorii zdań' (Numerical Interpretation of the Theory of Propositions), ibid. 7 (1922–3), pp. 92–93.

Aristotle was hesitant on this point: he doubted the possibility of classifying as true or false, before the act, any forecast about the future occurrence or non-occurrence of some action dependent on someone's future decision. Before the decision, both eventualities are probably possible, even if not to the same degree. Thus it can be seen at once that the Aristotelian doubts open up prospects for the creation of systems of multi-valued logic permitting a multiplicity of different logical values, and linking this variety of values with modifications of the modal functors expressing possibility and necessity. Łukasiewicz's investigations went in this direction, and problems of this nature absorbed him most in the closing years of his life.[1]

This interest of his was certainly connected with his propensity to indeterminism. That in turn was in harmony with his general views, which were far from materialistic, and with his religious attitude, though he pursued his logical studies as one pursues professionally any of the mathematical sciences, in a technological spirit independent of any general philosophical preconceptions. Philosophy in the sense of speculation, as we know it from textbooks on the history of philosophy (e.g. Kant's theory of cognition), he held in contempt, seeing in it an expression of the very sins of method against which everyone must rebel who has ever passed through the exigent school of modern mathematical logic.[2]

Among Łukasiewicz's disciples and later colleagues Stanisław Leśniewski (1886–1939), also a philosopher by training, was distinguished by an extraordinary originality of mind, which went with a capacity for accurate and most powerful argumentation. Devoured by a passion for an absolute exactness of statement, he looked for inspiration to Hans Cornelius, who was

---

[1] 'A System of Modal Logic', *The Journal of Computing Systems*, vol. i, no. 3 (1953), pp. 111–49; *Aristotle's Syllogistic*, 2nd edition, cf. above.

[2] For a general picture of Łukasiewicz's logical studies see the above-mentioned article by Borkowski and Słupecki, as well as the following: B. Sobociński, 'In Memoriam Jan Łukasiewicz', *Philosophical Studies* 6 (December 1956), Maynooth, Ireland; T. Kotarbiński, 'Jan Łukasiewicz's Works on the History of Logic', *Studia Logica* 8, Poznań, 1958; 'Bibliografia prac Jana Łukasiewicza' (A Bibliography of the Works of Jan Łukasiewicz), collective publication, ibid. 5, 1957.

at the same time close to both empirio-criticism and Kantianism,
to the writings on general grammar and the philosophy of language of the Brentanist Marty, and to the semantic chapters of
Mill's *System of Logic*; observing with mistrust attempts at
formulating theses on anything else than language before the
construction of a system of principles for the rational use of
language. After studies abroad he came to Lwów, where he
listened both to Twardowski and to Łukasiewicz, then lecturing
on philosophy in Lwów University. As far as I can remember,
Łukasiewicz was at that time speaking in very critical terms of
the issues and doctrines of the theory of cognition, and popular-
izing Schröder's algebra of logic and the theory of relationships;
he also held exercises on the theory of probability for philosophy
students; read Aristotle in the original at seminars and com-
mented on him, interpreting, among others, passages pertaining
to the paradoxical arguments by subscribers to the view that
only the real was possible and only the necessary was real. His
students often returned in private conversation to Russell's
paradox of a class of classes not members of themselves and to
other antinomies, a host of which had been discovered by the
contemporary analysis of the foundations of mathematics. In
this climate, Leśniewski moved from the position as it were of
a searcher after an appropriate linguo-technical introduction
to philosophy to that of a searcher for replies to problems concerning the general theory of objects, which after all must
constitute the basis for any more special researches, and consequently also for linguo-technical investigations. And this
general theory of objects must be pursued with the help of the
most appropriate and the most adequate logistic symbols—
provided these are applied with the necessary exactness, consistency, and self-control. It seems that Leśniewski was most
powerfully impressed by the supplement on 'The Principle of
Contradiction and Symbolic Logic' attached to the above-
mentioned work by Łukasiewicz on the principle of contradiction in Aristotle's writings, which contained an exposition of
the algebra of logic based mainly on Couturat's 1905 textbook.

What happened later was highly paradoxical. Leśniewski embarked on the construction of a system of logical foundations of mathematics that would be free of all antinomies, fully rational, and cognitionally adequate, though he was himself a philosopher by training, and by no means a mathematician. He was, moreover, a man who knew very little about numerical mathematical calculations and so-called higher mathematics in the usual meaning of the words, nor did he wish to learn more. But the issue for him was clear: the problems on which he was engaged were obviously logically independent of the theorems, be it of differential calculus, analytical geometry, or other branches of higher mathematics, while those same branches of mathematics were logically dependent on the former basic problems. The point of contact between the two worlds—that of the logician and that of the mathematician—was in the domain of the basic concepts and theorems of the theory of sets, which began to prosper very much in Poland and some representatives of which were most willing to co-operate with the logicians on matters of mutual interest. In Warsaw these contacts became in fact quite lively, and in the early years of independent Poland after the First World War Łukasiewicz and Leśniewski met as professors in the University of Warsaw, where they co-operated in harmony, and, thanks to that harmony, stimulated most beneficially the minds of their students.

What did Leśniewski achieve in the course of his extraordinarily lively career as scholar and teacher? There is hardly any disagreement on that among people conversant with his work. He constructed a truly original deductive system of the calculus of names,[1] a radically new deductive system with variables, the values of which consisted of interpropositional connectives (such as 'and' and 'if') or, in other cases, of modifying propositional particles (such as 'not'),[2] as well as a truly original deductive system of the theory of the relation of the part to the whole, with part understood as a fragment of the whole (in the sense in which, for instance, a hand is a part of

---

[1] J. Słupecki, 'St. Leśniewski's Calculus of Names', *Studia Logica* 3, Warsaw, 1955.   [2] Słupecki, 'St. Leśniewski's Protothetics', ibid. 1, 1953.

the whole human body).[1] The best developed of these systems is the first, which we refrain from calling a theory of predicates, since such a term implies that some expressions are suitable to the role of predicates and predicates only, and not, for instance, to that of subjects of singular propositions. In fact, the basic originality of Leśniewski's system (which he himself rather preferred to call an 'ontology') consists in that in it expressions suitable as subjects of singular propositions are suitable also as predicates of propositions, and vice versa; in a word, the subjects of singular propositions and the predicates of propositions belong in this system to the same semantic category, the category of names. That is why we choose rather to speak here about a system of the calculus of names. This is connected with Leśniewski's original conception of semantic categories. According to it, there are but two such fundamental categories: the category of names and the category of propositions. The other categories are filled by the so-called functors, the category of each being determined by whether it creates from the terms it joins a name or a proposition, and also whether these terms are themselves propositions or names, as well as how many there are. Thus, for instance, the copula 'is' in the singular proposition 'A is B' belongs to the category of proposition-forming functors attached to two name terms, the particle 'not' in the proposition 'not p' to the category of proposition-forming functors attached to one propositional term, the connective 'and' in the proposition 'every A is B and C' to the category of name-forming functors attached to two name terms, etc. Leśniewski's ontology makes possible the investigation of dependences, involving semantic categories of any degree of complication. It is accompanied by a system of directives, the most exactly defined of any I know, for the construction of definitions, while the system of ontological axioms is brought to such a degree of simplification by its author that it consists of a single short inscription containing the only primary specific

---

[1] Leśniewski, *Podstawy ogólnej teorii mnogości I* (*Foundations of General Set Theory I*), Moscow, 1916; Leśniewski, 'O podstawach matematyki' (On the Foundations of Mathematics), *Przegląd Filozoficzny* 30–34, 1927–31; Słupecki, 'Towards a Generalized Mereology of Leśniewski', *Studia Logica* 8, 1958.

term. This term consists precisely of the copula 'is', linking the subject of the singular proposition with the predicate. In Greek this would be ἐστί, the infinitive being εἶναι, or 'to be'. The term 'ontology', as a name for the whole system, is derived from the participle of the word εἶναι, which is ὤν, οὖσα, ὄν, genitive ὄντος, οὔσης, ὄντος.

Logically, ontology is, of course, preceded by the normal propositional calculus. Leśniewski's innovation in the field of the foundations of the propositional calculus consisted in basing it on what he called prototothetic, a system the variables of which take as their values functors of the propositional calculus. Let me give an example of a protothetical thesis. First, however, let me recall that, besides the particle 'not', three other particles can be used which have the character of proposition-forming functors of one propositional argument. These differ from one another depending on what logical value is acquired by the proposition with the particle in question for different logical values of the same proposition without the particle. Let them be the functors 'as', 'ver', and 'fal', besides the functor 'neg' (since this is how we shall denote the negation 'not'). Let their distribution of values conform to the following table:

p	as p	neg p	ver p	fal p
1	1	0	1	0
0	0	1	1	0

where '1' is the symbol of a true proposition and '0' the symbol of a false proposition.

With the above notations, and if, in addition, 'F' is the variable taking as its values the functors 'as', 'neg', 'ver', and 'fal', we obtain, for instance, the following equivalence,[1] indicating the possibility of defining conjunction with the help of 'F' and the quantifier:

$$\Pi p \, \Pi q \{ pq = \Pi F[p = (Fp = Fq)] \}.$$

Leśniewski's third system was named 'mereology', i.e. the theory of the relation of the fragment to the whole, which assumes both the propositional calculus with protothetic and

---

[1] Cf. T. Czeżowski, *Logika* (*Logic*), Warsaw, 1949, p. 91.

ontology as systems logically precedent. Its formulation is linked up closely in the history of Leśniewski's studies with his efforts at a solution of Russell's paradox of classes. The latter ceases to hold if the relation of element to class and the relation of sub-class to class is formulated precisely as a relation of the component fragment to the whole. Because in that case it cannot be inferred from the fact that x is an element of the M class that x is M, which passage is one of the steps in the construction of the paradox. Such a conclusion would not in fact be correct, since x's being a fragment of a whole composed of M's does not imply that the whole has no other fragments. While it is true, for example, that the heart is a fragment of a whole composed of cells, the body of a creature with a heart being composed of cells, the heart itself is not a cell.

Leśniewski offered solutions to all the then-known antinomies of logic, semantics, and set theory, in some of them showing (a pioneering exercise at that time) how indispensable it was, if nonsense was to be avoided, to distinguish between the scope of a given name in a given language and the scope of an identically sounding (or identically shaped) name in another language, the latter being used to talk about the expressions of the former. Some of the solutions to antinomies given by Leśniewski were questioned; his solution of Russell's paradox was among those attacked. The argument put forward against that solution was that it suggested a different interpretation of class from that assumed in the formulation of the paradox. Views differ also on the importance of his achievements in relation to problems relevant to modern science.[1]

I am most curious concerning what Leśniewski's own answer to his critics would be, since I cannot recall even a single logical discussion in which he was not shown to be right. This at least is how that extraordinary man impressed himself on the memory of his contemporaries.

---

[1] A. Grzegorczyk, 'The Systems of Leśniewski in Relation to Contemporary Logical Research', *Studia Logica* 3, 1955. Grzegorczyk has some objections. These are answered by Eugene C. Luschei in *The Logical Systems of Leśniewski*, North-Holland, Amsterdam, 1962, the only documentary and exhaustive monograph on the whole of Leśniewski's opus.

Leśniewski's attitude to philosophy was most interesting. He was rather contemptuous of the usual ways of philosophical thought, but himself refrained from pursuing problems of the theory of cognition or general philosophical questions, e.g. the problem of the relation of mind to matter. When asked for the reason, he used to answer: 'These are problems too difficult for me.'

In any list of the *coryphaei* of what used to be called the Warsaw school of logic, the name of Alfred Tarski is usually mentioned in the same breath as those of Łukasiewicz and Leśniewski. Quite correctly. If, then, I am not now going to write in any greater detail about him, this is for the sole reason that this outstanding logician and mathematician, who has now lived in the United States for more than twenty years, is rather well known both there and in the world at large from his numerous publications in the English language. Let me just recall that he was a disciple of both Łukasiewicz and Leśniewski, and also of the Warsaw professors of the theory of sets, and that some of his most important writings date from his creative years in Warsaw. Suffice it to say that his paper on the notion of truth in formalized languages, which was the foundation for the vast building of logico-mathematical semantics erected later, first appeared in Polish in 1933, and that its German version was also published in Poland a few years later.[1] In his Warsaw period Tarski, with Wajsberg (see below), also played a prominent role in developing the multi-valued logic of Łukasiewicz.[2] In contrast to Łukasiewicz and Leśniewski, he was not a philosopher who devoted himself to formal logic; first and foremost he was a mathematical logician, but one who had had a thorough philosophical training.

A philosopher by training, on the other hand, like Łukasiewicz and Leśniewski, was that other pillar of the then Warsaw group, Kazimierz Ajdukiewicz (1890–1963), soon afterwards professor

---

[1] A. Tarski, *Pojęcie prawdy w językach nauk dedukcyjnych* (*The Notion of Truth in the Languages of the Deductive Sciences*), Warsaw, 1933; 'Der Wahrheitsbegriff in den formalisierten Sprachen', *Studia Philosophica* (Lwów, 1936), pp. 261–405, reprint 1935.

[2] Borkowski and Słupecki, op. cit.

in his home city of Lwów. Another student of both Twardowski and Łukasiewicz, he was in later years closely associated with neo-positivism, as well as a spokesman for his own conception of conventionalism. The reader will get an idea of his intellectual style from his paper on syntactic connexion included in the present collection. In it he gives a clear exposition of the new ideas he evolved from Leśniewski's conception of semantic categories. But the role of Ajdukiewicz in the group of Polish logicians cannot be described merely from his publications. Besides his other merits, he was a supreme connoisseur of modern logic and will not easily find his peer, in Poland at least, in his ability to give a critical analysis of modern logical literature. In writing this, I am partly drawing on my knowledge of his activity since the last war. But the quality of Ajdukiewicz's intellect and pen was clearly apparent even in the period covered by this introduction. It was fully displayed, for example, in his critical scrutiny of my own university textbook.[1]

If I mention this book, and thus also myself, this is because I too was at that time among the Warsaw group of professors of logic. With a philosophical background, not very well conversant with the foundations of mathematics, and mainly concerned with problems of the methodology of practical abilities and questions relating to the general education of prospective teachers, I did not contribute to the construction of aprioristic deductive systems. To be quite frank, formal logic never appealed to me strongly. I did believe, though, that the teaching of formal logic according to traditional stereotypes was an anachronism and that in instruction use ought to be made of the methods and results of mathematical logic. Consequently, in my textbook I popularized both Russell's calculus of propositions and Leśniewski's calculus of names. At the same time, however, when lecturing on these different calculi, I spread on my own responsibility a certain methodological idea, which

---

[1] Kotarbiński, *Elementy teorii poznania, logiki formalnej i metodologii nauk* (*Elements of the Theory of Cognition, Formal Logic and Scientific Methodology*), Lwów, 1929, 2nd revised edition, Warsaw, 1961. English translation, *Gnosiology—The Scientific Approach to the Theory of Knowledge*, London 1966. Ajdukiewicz's review appears as an appendix to the 1961 edition.

I then called reism and later somatism, and which I now prefer to call concretism. If there is anything sound in it, this can be reduced to the proposal that one should strive for statements, the names in which would be names only of spatio-temporal objects, physically definable. For all the abuses of this tendency committed in the above-mentioned *Elementy*, I bear the undivided responsibility. It seems to me that Łukasiewicz would very much object to my aversion to Platonism, to my freethinking, and to the monistically materialist aspect of concretism. I mention this as an example to show how wide was the range of the general philosophical attitudes of the various members of a group, who were still all agreed on the importance of mathematical logic as the only adequately scientific modern form of formal logic.

The teachers of the Warsaw centre had the greatest satisfaction in seeing how a number of young people were becoming their partners in logico-mathematical investigations. Some of them became known as authors of valuable ideas even in the period here under discussion. Of those who fell victim to the Nazi invasion let me recall Mordchaj Wajsberg, whom I have already mentioned, Miss Janina Hosiasson, the author of studies on induction, and Adolf Lindenbaum, who collaborated with Tarski in working out paradoxical topological structures. But even younger people, who did not really make their mark until after the war, were then first setting out on the path of science, sometimes with striking effect. Among them were Andrzej Mostowski, later author of numerous specialized works and a very well-received textbook on mathematical logic (1948); Henryk Greniewski, later author of an interesting compendium, especially of the theory of induction (1955); Stanisław Jaśkowski, Jerzy Słupecki, Czesław Lejewski (now resident in England), Bolesław Sobociński (now in the United States) and others. Not in Warsaw at that time was I. M. Bocheński, later author of works on the history of formal logic, including the monumental *Formale Logik* (1956).

In the period about which I have been writing, mathematical logic was also studied at the Jagellonian University in Cracow, though on a lesser scale. There two men were most active, one rather well known in the West, the other almost unknown. The former was Leon Chwistek (1884–1944). The simplification to which he subjected Russell's ramified theory of types is widely known. In a desire to avoid Richard's antinomy, he developed a theory of types free of Russell's premiss in the form of the so-called axiom of reducibility.[1] Chwistek published the results of his investigations in English as well as in Polish. He later tried to work out some very general formalized system of semantics. Expressed in a manner so concise as to make it almost mysterious, this system awaits a competent interpreter.[2]

The other man, unknown to the West, was Jan Śleszyński, a mathematician of the older generation. Much indeed can be learned from the rich collection of his papers on various subjects in the realm of formal logic, and of mathematical logic and its history in particular—a collection published in two volumes under the rather non-informative title *Teoria dowodu*.[3] Introduction to mathematical logic, complete proof, mathematical proof, exposition of the theory of propositions, the Boolean calculus, Grassmann's logic, Schröder's algebra, Porecki's seven laws, Peano's doctrine, Burali-Forti's doctrine—these are some of the themes pursued in this work, from which I personally have learned a great deal and thanks to which I have got a clear idea of many an unclear thing. Unfortunately, for lack of available sources, I am not in a position to give any personal details about Śleszyński, whose contribution to the advancement of mathematical logic in Poland was highly appreciated by Łukasiewicz.[4]

[1] Cf. Bocheński, *Formale Logik*, Freiburg-München, 1956, translated by I. Thomas as *A History of Formal Logic* (Notre Dame, 1961), pp. 388 and 399–400; Chwistek, 'Antynomje logiki formalnej' (Antinomies of Formal Logic), *Przegląd Filozoficzny* 24 (1921), pp. 164–71 (paper 17 of this volume); Chwistek, 'The Theory of Constructive Types', *Rocznik Polskiego Towarzystwa Matematycznego (Year-Book of the Polish Mathematical Society)* 2, (1924), pp. 9–48; 3 (1925), pp. 82–141; Chwistek, 'Zasady czystej teorii typów' (The Principles of a Simple Theory of Types), *Przegląd Filozoficzny* 25 (1922).

[2] Chwistek, *Granice nauki (The Limits of Science)*, Warsaw-Lwów, 1935; Chwistek, *The Limits of Science*, London, 1947, New York, 1948.

[3] *Teoria dowodu (The Theory of Proof)*, edited by S. K. Zaremba from the university lectures by Professor Jan Śleszyński, Cracow, vol. 1 1925, vol. 2 1929.   [4] Borkowski and Słupecki, op. cit., p. 56.

Apart from Warsaw and Cracow, research on mathematical logic was also conducted in other intellectual centres in the country. Let me mention two more philosophers who engaged in such investigations, but who have published didactic outlines of their researches only since the last war. I have in mind now Tadeusz Czeżowski, at that time active in Wilno, whose textbook of mathematical logic meant primarily for philosophers appeared in 1949, and Adam Wiegner, then professor in the University of Poznań, who later wrote another textbook of mathematical logic, published in 1948.[1]

TADEUSZ KOTARBIŃSKI
President of the
Polish Academy of Sciences
Warsaw, 28 February 1961   *1957–1962*

[1] Some bibliographic data on Polish logical writings can be found, among others, in the present writer's article 'La Logique en Pologne', contained in vol. i of *Philosophy in the Mid-Century* (Florence, 1958), a publication of the International Philosophical Institute edited by R. Klibansky. Some historical information is contained in the author's outline of the history of logic in Poland, *La Logique en Pologne: son originalité et les influences étrangères*, which appeared as No. 7 of the collection of lectures published by the Rome Library of the Polish Academy of Sciences (1959). [Ed. note: a bibliography is also appended to paper 18 of this volume.]

I. ON THE NOTION OF POSSIBILITY
II. ON THREE-VALUED LOGIC†

JAN ŁUKASIEWICZ

I

BEFORE the 206th scientific meeting of the Polish Philosophical Society (Lwów), held on 5 June 1920, Professor Jan Łukasiewicz delivered a talk entitled 'On the notion of possibility'. He based his analysis of the notion of possibility on an examination of the logical relations between the following six sentences: S is P, S is not P, S can be P, S cannot be P, S can be non P, S cannot be non (i.e. must be) P. There may be three opinions concerning the logical relations between these sentences:

(1) If it is true that S must be P, then it is true that S is P, and if it is true that S cannot be P, then it is true that S is not P. When one assumes that no other relation independent of these two holds between the given sentences, one takes the point of view of traditional logic.

(2) Besides the relations listed under (1) the following relations hold: If it is true that S is P, then it is true that S must be P, and if it is true that S is not P, then it is true that S cannot be P. Those who consider as impossible anything that contains a contradiction and as necessary anything whose denial contains a contradiction are forced to this opinion (Leibniz). This view corresponds to ontological determinism.

(3) Besides the relations listed under (1), the following relations hold: if it is true that S can be P, then it is true that S can be non P, and vice versa, if it is true that S can be non P, then it is true that S can be P. This view corresponds to

† These papers, entitled respectively 'O pojęciu możliwości' and 'O logice trojwartosciowej', appeared originally in *Ruch Filozoficzny* 5 (1920), pp. 169–71. Translated by H. Hiż.

ontological indeterminism and is held by those who accept the notion of 'pure' or 'two-sided' possibility, according to which only that can happen which does not have to happen (Aristotle).

All three kinds of relations, according to the speaker, have a considerable obviousness, although taken together they lead to many contradictions. The following table shows these contradictions (0 means falsity, 1 means truth):

S is P	0	1	–
S is not P	1	0	–
S can be P	0	–	1
S cannot be P	1	–	0
S can be non P	–	0	1
S cannot be non P	–	1	0

This table is constructed in such a way that all relations listed under (1), (2), and (3) are satisfied. However, neither 0 nor 1 can be written in the empty places marked by the dash. In either case a contradiction arises. (E.g. in the first column, if 'S can be non P' is 0, then 'S cannot be non P' is 1, and then, because of the relation (1), 'S is P' must also be 1. But it is 0 in the table. If, on the other hand, 'S can be non P' is 1, then, because of the relation (3), 'S can be P' must also be 1. But it is 0 in the table.) It is possible to eliminate the contradictions by assuming that, besides truth and falsity, there is a third logical value of sentences. The speaker calls the third value 'possibility' and denotes it by the numeral 2. He therefore recognizes, in addition to true and false sentences, sentences which are neither true nor false but only 'possible'. Filling up in the table the empty places by 2's we obtain the correct and consistent totality. But then we are not in the Aristotelian logic; rather we face a new, non-Aristotelian logic which can be called 'three-valued logic'. (Reported by the speaker.)

II

Before the 207th scientific meeting held on 19 June 1920, Professor Jan Łukasiewicz delivered a talk entitled 'On three-valued logic'. The Aristotelian logic assumes that every sentence is either true or false. Therefore it recognizes only two kinds of logical values: truth and falsity. Denoting truth by 1, falsity by 0, identity by $=$, and consequence by $<$, it is possible to deduce all the laws of Aristotelian logic from the following principles and definitions:

I. The principles of identity of falsity, of identity of truth, and of difference between truth and falsity: $(0 = 0) = 1$, $(1 = 1) = 1$, $(0 = 1) = (1 = 0) = 0$.

II. The principles of consequence:

$$(0 < 0) = (0 < 1) = (1 < 1) = 1, \quad (1 < 0) = 0.$$

III. The definitions of denial, of addition, and of multiplication: $a' = (a < 0)$, $a+b = [(a < b) < b]$, $ab = (a'+b')'$.

In the definitions, a and b are variables accepting only two values, 0 or 1. All logical laws expressed with variables can be verified by substituting for letters the symbols 0 and 1; e.g. $(a = 1) = a$ is true because $(0 = 1) = 0$ and $(1 = 1) = 1$.

The three-valued logic is a system of non-Aristotelian logic, for it assumes that besides true and false sentences there are also sentences which are neither true nor false, thus accepting the existence of a third logical value. We can interpret the third value as 'possibility' and we can denote it by 2. In order to construct the system of three-valued logic it is necessary that the principles concerning 0 and 1 should be supplemented by the principles concerning 2. One can do this in many ways. The system which the speaker accepts in the present state of his research and which deviates the least from the 'two-valued' logic is as follows:

I. The principles of identity:

$$(0 = 2) = (2 = 0) = (1 = 2) = (2 = 1) = 2,$$

$$(2 = 2) = 1.$$

II. The principles of consequence:

$$(0 < 2) = (2 < 1) = (2 < 2) = 1,$$

$$(2 < 0) = (1 < 2) = 2.$$

The principles listed above concerning 0 and 1, as well as the definitions of denial, addition, and multiplication remain unchanged in the three-valued logic except that the variables a and b can accept three values 0, 1, and 2.

The laws of three-valued logic are partly different from the laws of two-valued logic. Some laws of Aristotelian logic are only 'possible' in three-valued logic, e.g. the principle of the syllogism in the usual formulation $(a < b)(b < c) < (a < c)$ {however, the principle of the syllogism formulated as

$$(a < b) < [(b < c) < (a < c)]$$

is true}, the principle of contradiction $aa' = 0$, of the excluded middle $a + a' = 1$, etc. Some laws of two-valued logic are false in three-valued logic, among others the law $(a = a') = 0$, because when $a = 2$ the sentence $a = a'$ is true. From this fact results the absence of antinomies in three-valued logic.

The speaker thinks that three-valued logic has, above all, theoretical significance as the first attempt to create a non-Aristotelian logic. What practical significance, if any, the new system of logic will have may be known only after a careful examination, in relation to the new logical laws, of logical phenomena, especially those occurring in the deductive sciences, when it may become possible to compare with experience the consequences of the indeterministic view which is the metaphysical basis of the new logic.

2

ON DETERMINISM†

JAN ŁUKASIEWICZ

THIS article is a revision of an address which I delivered as
Rector of Warsaw University at the Inauguration of the aca-
demic year 1922–3. As was my habit, I spoke without notes.
I wrote down my address later on, but never published it.

In the course of the next twenty-four years I frequently returned to the editing of my lecture, improving its form and
content. The main ideas, and in particular the critical examination of the arguments in favour of determinism, remained,
however, unchanged.

At the time when I gave my address those facts and theories
in the field of atomic physics which subsequently led to the
undermining of determinism were still unknown. In order not
to deviate too much from, and not to interfere with, the original
content of the address, I have not amplified my article with
arguments drawn from this branch of knowledge.

Dublin, November 1946

᠅

1. It is an old academic custom that the Rector should open
a new session with an inaugural address. In such a lecture he
should state his scientific creed and give a synthesis of his
investigations.

A synthesis of philosophical investigations is expressed in a
philosophical system, in a comprehensive view of the world and
life. I am unable to give such a system, for I do not believe

† This paper, entitled 'O Determiniźmie', was published for the first time
in *Z Zagadnień Logiki i Filozofii*, an anthology of Łukasiewicz's works edited
by J. Słupecki, Warsaw, 1961. Translated by Z. Jordan.

that today one can establish a philosophical system satisfying
the requirements of scientific method.

I belong, with a few fellow workers, to a still tiny group of
philosophers and mathematicians who have chosen mathematical logic as the subject or the basis of their investigations. This
discipline was initiated by Leibniz, the great mathematician and
philosopher, but his efforts had fallen into oblivion when, about
the middle of the nineteenth century, George Boole became its
second founder. Gottlob Frege in Germany, Charles Peirce in
the United States, and Bertrand Russell in England have been
the most prominent representatives of mathematical logic in our
own times.

In Poland the cultivation of mathematical logic has produced
more plentiful and fruitful results than in many other countries.
We have constructed logical systems which greatly surpass not
only traditional logic but also the systems of mathematical logic
formulated until now. We have understood, perhaps better than
others, what a deductive system is and how such systems should
be built. We have been the first to grasp the connexion of
mathematical logic with the ancient systems of formal logic.
Above all, we have achieved standards of scientific precision
that are much superior to the requirements accepted so far.

Compared with these new standards of precision, the exactness of mathematics, previously regarded as an unequalled
model, has not held its own. The degree of precision sufficient
for the mathematician does not satisfy us any longer. We require
that every branch of mathematics should be a correctly constructed deductive system. We want to know the axioms
on which each system is based, and the rules of inference of
which it makes use. We demand that proofs should be carried
out in accordance with these rules of inference, that they should
be complete and capable of being mechanically checked. We
are no longer satisfied with ordinary mathematical deductions,
which usually start somewhere 'in the middle', reveal frequent
gaps, and constantly appeal to intuition. If mathematics has
not withstood the test of the new standard of precision, how are
other disciplines, less exact than mathematics, to stand up to

it? How is philosophy, in which fantastic speculations often stifle systematic investigations, to survive?

When we approach the great philosophical systems of Plato or Aristotle, Descartes or Spinoza, Kant or Hegel, with the criteria of precision set up by mathematical logic, these systems fall to pieces as if they were houses of cards. Their basic concepts are not clear, their most important theses are incomprehensible, their reasoning and proofs are inexact, and the logical theories which often underlie them are practically all erroneous. Philosophy must be reconstructed from its very foundations; it should take its inspiration from scientific method and be based on the new logic. No single individual can dream of accomplishing this task. This is a work for generations and for intellects much more powerful than those yet born.

2. This is my scientific creed. Since I cannot give a philosophical system, today I shall try to discuss a certain problem which no philosophical synthesis can ignore and which is closely connected with my logical investigations. I should like to confess in advance that I am unable to examine this problem, in all its details, with the scientific precision that I demand from myself. What I give is only a very imperfect essay, of which perhaps somebody will one day take advantage to establish, on the basis of these preliminary examinations, a more exact and mature synthesis.

I want to speak of determinism. I understand by determinism something more than that belief which rejects the freedom of the will. I shall first explain what I mean by an example.

John met Paul in the Old Town Square in Warsaw yesterday noon. The fact of yesterday's meeting no longer exists today. Yet that fact of yesterday is not a mere illusion today, but some part of the reality which both John and Paul have to take into account. They both remember their yesterday's meeting. The effects or traces of that meeting somehow exist in them today. Each of them could take an oath in a court of law that he saw the other in the Old Town Square in Warsaw yesterday noon.

On the basis of these data I say, 'it is true at every instant of today that John met Paul in the Old Town Square in Warsaw

yesterday noon'. I do not intend to maintain by this that the
sentence 'John met Paul in the Old Town Square in Warsaw
yesterday noon' is true at every instant of today, for such a
sentence, if nobody utters it or thinks of it, may not exist at
all. I make use of the expression 'it is true at instant t that p'—
in which 'instant' means an unextended time point and 'p' any
statement of fact—as equivalent to 'it is the case at instant t
that p'. For the present I am unable to give a further analysis
of the latter expression.

We believe that what has happened cannot be undone, *facta
infecta fieri non possunt*. What once was true remains true for
ever. All truth is eternal. These sentences seem to be intuitively
certain. We believe, therefore, that if an object A is b at instant
t, it is true at any instant later than t that A is b at instant t.
If John met Paul in the Old Town Square in Warsaw yesterday
noon, it is true at any instant later than yesterday noon that
John met Paul in the Old Town Square in Warsaw yesterday
noon.

The question occurs whether it was also true at any instant
earlier than yesterday noon that John would meet Paul in the
Old Town Square in Warsaw yesterday noon? Was it true the
day before yesterday and one year ago, at the moment of John's
birth and at any instant preceding his birth? Is everything
which will happen and be true at some future time true already
today, and has it been true from all eternity? Is every truth
eternal?

Intuition fails us in this case and the problem becomes controversial. The determinist answers the question in the affirmative
and the indeterminist in the negative. By determinism I understand the belief that if A is b at instant t it is true at any instant
earlier than t that A is b at instant t.

Nobody who adopts this belief can treat the future differently
from the past. If everything that is to occur and become true
at some future time is true already today, and has been true
from all eternity, the future is as much determined as the past
and differs from the past only in so far as it has not yet come
to pass. The determinist looks at the events taking place in

the world as if they were a film drama produced in some cinematographic studio in the universe. We are in the middle of the performance and do not know its ending, although each of us is not only a spectator but also an actor in the drama. But the ending is there, it exists from the beginning of the performance, for the whole picture is completed from eternity. In it all our parts, all our adventures and vicissitudes of life, all our decisions and deeds, both good and bad, are fixed in advance. Even the moment of our death, of yours and mine, is laid down beforehand. We are only puppets in the universal drama. There remains for us nothing else to do but watch the spectacle and patiently await its end.

This is a strange view and by no means obvious. However, there are two arguments of considerable persuasive power which have been known for a long time and which seem to support determinism. One of them, originating with Aristotle, is based on the logical principle of the excluded middle, and the other, which was known to the Stoics, on the physical principle of causality. I shall try to present these two arguments, however difficult and abstract they are, in a way as easy to understand as possible.

3. Two sentences of which one is the denial of the other are called *contradictory*. I shall illustrate this notion by an example taken from Aristotle. 'There will be a sea battle tomorrow' and 'There will not be a sea battle tomorrow' are contradictory sentences. Two famous principles derived from Aristotle, the principle of the excluded contradiction and the principle of the excluded middle, are concerned with contradictory sentences. The first of these states that two contradictory sentences are not true together, that is, that one of them must be false. In my subsequent inquiry I shall not deal with this important principle, which Aristotle and, following him, numerous other thinkers regarded as the deepest mainstay of our thinking. I am concerned here with the principle of the excluded middle. It lays down that two contradictory sentences are not false together, that is, that one of them must be true. Either there will be or there will not be a sea battle tomorrow. *Tertium non datur.*

There is nothing in between the arguments of this alternative, no third thing that, being true, would invalidate both its arguments. It may sometimes happen that two disputants, of whom one regards as white what the other considers black, are both mistaken, and the truth lies somewhere in between these two assertions. There is no contradiction, however, between regarding something as white and considering the same thing as black. Only the sentences stating that the same thing is and is not white would be contradictory. In such cases truth cannot lie in between or outside of these sentences, but must inhere in one of them.

To return to our everyday example, if the principle of the excluded middle holds, and if Peter says today 'John will be at home tomorrow noon' and Paul denies it by saying 'John will not be at home tomorrow noon', then one of them speaks the truth. We may not know today which one of them does so, but we shall learn by visiting John tomorrow noon. If we find John at home, Peter made a true statement, and if John is away, Paul spoke the truth today.

Therefore, either it is already true today that John will be at home tomorrow noon or it is true today that John will not be at home tomorrow noon. If someone utters the sentence 'p', and someone else utters its denial, 'not-p', then one of them makes a true statement not only today but at any instant t; for either 'p' or 'not-p' is true. It does not matter at all whether anyone actually expresses these sentences or even thinks of them; it seems to be in the very nature of the case that either it is true at instant t that 'p' or it is true at instant t that 'not-p'. This alternative seems to be intuitively true. As applied to our example, it takes the following form:

(a) *Either it is true at instant* t *that John will be at home tomorrow noon or it is true at instant* t *that John will not be at home tomorrow noon.*

Let us keep in mind this sentence as the first premiss of our reasoning.

The second premiss is not based on any logical principle and can be expressed in general form as the conditional 'if it is true at instant t that p, then p'. In this conditional, 'p' stands for any sentence, either affirmative or negative. If we substitute for 'p' the negative sentence 'John will not be at home tomorrow noon' we obtain

(b) *If it is true at instant* t *that John will not be at home to-morrow noon, then John will not be at home tomorrow noon.*

This premiss also seems to be intuitively true. If it is true at an arbitrary instant t, e.g. now, that John will not be at home tomorrow noon—for we know that he has just left for a distant destination and for a long time—there is no use calling upon John tomorrow noon. We are certain that we shall not find him at home.

We accept both premisses without proof as intuitively certain. The thesis of determinism is based upon these premisses. Its proof will be carried out rigorously in accordance with the so-called theory of deduction.

4. Thanks to mathematical logic we know today that the basic system of logic is not the small fragment of the logic of terms known as Aristotle's syllogistic, but the logic of propositions, incomparably more important than syllogistic. Aristotle made intuitive use of the logic of propositions, and only the Stoics, with Chrysippus at their head, formulated it systematically. In our own times the logic of propositions was constructed in an almost perfect axiomatic form by Gottlob Frege in 1879; it was discovered independently of Frege and enriched with new methods and theorems by Charles Peirce in 1895; and under the name of 'the theory of deduction' it was made the basis of mathematics and logic by Bertrand Russell in 1910. It was also Bertrand Russell who extended knowledge of it to the scientific community at large.

The theory of deduction should become as universally well known as elementary arithmetic, for it comprises the most important rules of inference used in science and life. It teaches us how to use correctly such common words as 'not', 'and', 'or', 'if-then'. In the course of the present exposition, which I begin with our second premiss, we shall become acquainted with three rules of inference included in the theory of deduction.

The second premiss is a conditional of the form 'if α, then not-β' in which 'α' stands for the sentence 'it is true at instant t that John will not be at home tomorrow noon' and 'β' for the sentence 'John will be at home tomorrow noon'. In the consequent of premiss (b) there occurs the denial of the sentence 'β', that is, the sentence 'not-β', 'John will not be at home tomorrow noon'. In accordance with the theory of deduction the premiss 'if α, then not-β' implies the conclusion 'if β, then not-α'. For if 'α' implies 'not-β' then 'α' and 'β' exclude each other, and therefore 'β' implies 'not-α'. According to this rule of inference, premiss (b) is transformed into the sentence

(c) *If John will be at home tomorrow noon, then it is not true at instant* t *that John will not be at home tomorrow noon.*

Let us now pass to the first premiss, to the alternation of the form 'γ or α', in which 'γ' signifies the sentence 'it is true at instant t that John will be at home tomorrow noon' and 'α' the same sentence as before, 'it is true at instant t that John will not be at home tomorrow noon'. It follows from the theory of deduction that the premiss 'γ or α' implies the conclusion 'if not-α, then γ'. For an alternative is true if and only if at least one of its arguments is true. If the second argument is false, the first one must be true. In accordance with this rule of inference premiss (a) is transformed into the sentence

(d) *If it is not true at instant* t *that John will not be at home tomorrow noon, then it is true at instant* t *that John will be at home tomorrow noon.*

Let us now compare sentences (c) and (d). They are both conditionals, and the consequent in (c) is equiform with the antecedent in (d); these two sentences have the form 'if β, then not-α' and 'if not-α, then γ'. According to the theory of deduction two such premisses imply the conclusion 'if β, then γ'. For if it is true that 'if the first, then the second' and 'if the second, then the third', then it is also true that 'if the first, then the third'. This is the law of the hypothetical syllogism, as known by Aristotle. If we remember that 'β' stands for the sentence 'John will be at home tomorrow noon', and 'γ' for the sentence 'it is true at instant t that John will be at home tomorrow noon', we obtain the conclusion

(e) *If John will be at home tomorrow noon, then it is true at instant t that John will be at home tomorrow noon.*

Instant t is an arbitrary instant; therefore, it is either earlier than or simultaneous with or later than tomorrow noon. It follows that if John is at home tomorrow noon, then it is true at an arbitrary or at any instant that John will be at home tomorrow noon. To put it in general form, it has been proved on the basis of a particular example that if A is b at instant t, then it is true at any instant, and therefore at any instant earlier than t, that A is b at instant t. The thesis of determinism has been proved by deducing it from the principle of the excluded middle.

5. The second argument in favour of determinism is based on the principle of causality. It is not easy to present this argument in a comprehensible way, for neither the word 'cause' nor the proposition known as the principle of causality have acquired an established meaning in science. They are only associated with a certain intuitive meaning which I should like to make explicit by giving a few explanations.

I say that the ringing of the bell at the entrance door to my apartment at this moment is a fact taking place now. I regard John's presence at home at instant t as a fact occurring at instant t. Every fact takes place somewhere at some time. Statements of fact are singular and include an indication of time and place.

Fact F occurring at instant s is called the *cause* of fact G occurring at instant t, and fact G the *effect* of fact F, if instant s is earlier than instant t, and if facts F and G are so connected with each other that by means of known laws obtaining between the respective states of affairs it is possible to infer the statement of fact G from the statement of fact F.† For instance,

---

† [Editorial note from the 1961 edition: This definition of the concept of cause differs from the definition accepted in Łukasiewicz's paper 'Analiza i

I consider the pressing of the button of an electric bell the cause of its ringing, because the bell is pressed at an instant earlier than the instant of its ringing, and I can deduce the statement of the second fact from the statement of the first one by means of the known laws of physics on which the construction of an electric bell is based.

The definition of cause implies that the causal relation is transitive. This means that for any facts F, G, and H, if F is the cause of G and G is the cause of H, then F is the cause of H.

I understand by the principle of causality the proposition that every fact G occurring at instant t has its cause in some fact F occurring at instant s earlier than t, and that at every instant later than s and earlier than t there occur facts which are both effects of fact F and causes of fact G.

These explanations are intended to make explicit the following intuitions. The fact which is the cause takes place earlier than the fact which is the effect. I first press the button of the bell and the bell rings later, even if it appears to us that both facts happen simultaneously. If there occurs a fact which is the cause of some other fact, then the latter fact, which is the effect of the former, follows the cause inevitably. Thus if I press the button, then the bell rings. It is possible to infer the effect from the cause. As the conclusion is true provided that its premisses are true, in a similar way the effect has to occur provided that its causes exist. Nothing happens without cause. The bell does not ring of itself; this only happens because of some earlier facts. In the set of facts succeeding each other, ordered by the causal relation, there are neither gaps nor jumps. From the instant when the button is pressed to the instant when the bell rings there constantly occur facts each of which is simultaneously an effect of the pressing of the button and a cause of the ringing of the bell. Moreover, every one of these facts occurring earlier is the cause of every one occurring later.

konstrukcja pojęcia przyczyny' (The analysis and construction of the concept of cause), *Przegląd Filozoficzny* 9 (1906), pp. 105–79. Both definitions lay down, however, that the relation of causality is transitive, and this point is of paramount importance in Łukasiewicz's subsequent investigations.]

6. The argument deducing the thesis of determinism from the principle of causality may become more intelligible after these explanations. Let us assume that a certain fact F occurs at instant t; for instance, that John is at home tomorrow noon. Fact F has its cause in some fact F_1, taking place at instant t_1 earlier than t. Again, fact F_1 has its cause in some fact F_2, taking place at instant t_2 earlier than t_1. Since according to the principle of causality every fact has its cause in some earlier fact, this procedure can be repeated over and over again. Therefore, we obtain an infinite sequence of facts which extends back indefinitely

$$... F_n, F_{n-1}, ..., F_2, F_1, F.$$

because the facts take place at ever earlier instants

$$... t_n, t_{n-1}, ..., t_2, t_1, t.$$

In this sequence every earlier fact is the cause of every later fact, for the causal relation is transitive. Moreover, if fact F_n occurring at instant t_n is the cause of fact F occurring at instant t, then, in accordance with the principle of causality, at every instant later than t_n and earlier than t there occur facts which are simultaneously effects of fact F_n and causes of fact F. Since these facts are infinitely many, we are unable to order all of them in the sequence and can designate only some, for instance F_{n-1}, F_2, or F_1.

While everything seems to be in order so far, the most important step in the determinist's argument comes only now. His reasoning would probably take the following course.

As the sequence of facts which occur earlier than and which are the causes of fact F is infinite, at every instant earlier than t, and therefore at every present and past instant, there occurs some fact that is the cause of F. If it is the case that John will be at home tomorrow noon, then the cause of this fact exists already today and also at every instant earlier than tomorrow noon. If the cause exists or existed, all the effects of this cause must inevitably exist. Therefore it is already true today and has been true from all eternity that John would be at home tomorrow noon. In general, if A is b at instant t, it is true at every instant earlier than t that A is b at instant t; for at every instant earlier than t there exist the causes of this fact. Thus the thesis of determinism may be proved by means of the principle of causality.

These are the two strongest arguments which can be used in support of determinism. Should we give up and accept them? Should we believe that everything in the world takes place of necessity and that every free and creative act is only an illusion? Or, on the contrary, should we reject the principle of causality along with the principle of the excluded middle?

7. Leibniz writes that there are two famous labyrinths in which our reason is often lost. One of them is the problem of freedom and necessity, and the other is concerned with continuity and infinity. While writing this Leibniz did not think it plausible that these two labyrinths should constitute one single whole and that freedom, if it exists at all, could be concealed in some nook of infinity.

Should the causes of all facts which could ever occur exist at every instant, there would be no freedom. Fortunately, the principle of causality does not compel us to accept this consequence. Infinity and continuity come to our rescue.

There is an error in the argument which derives the thesis of determinism from the principle of causality. For it is not the case that if John is at home tomorrow noon, then the infinite sequence of causes of this fact must reach the present and every past instant. This sequence may have its lower limit at an instant later than the present instant: one which, therefore, has not yet come to pass. This is clearly implied by the following considerations.

Let us consider time as a straight line and let us establish a one-to-one correspondence between a certain interval of time and the segment $(0, 1)$ of that line. Let us assume that the present instant corresponds to point 0, that a certain future fact occurs at instant (corresponding to point) 1, and that the causes of this fact occur at instants determined by real numbers greater than $\frac{1}{2}$. This sequence of causes is infinite and has no beginning, that is, no first cause. For this first cause would have to take place at the instant corresponding to the smallest real number greater than $\frac{1}{2}$, and no such real number exists; not even the smallest rational number greater than $\frac{1}{2}$ exists. In the set of real numbers, and similarly in the ordered set of rational numbers, there are no two numbers succeeding each other immediately, that is, being the immediate predecessor and successor of each other; between any two numbers there is always another one, and consequently there are infinitely many numbers between any two of them. Similarly there are no two instants succeeding each other immediately, that is, being the immediate predecessor and successor of each other; between any two instants there is another one, and consequently there are infinitely many instants between any two of them. In accordance with the principle of causality, every fact of the sequence under consideration has its cause in some earlier fact. Although it has a lower limit at instant $\frac{1}{2}$, which is later than the present instant 0 and has not yet been attained, the sequence is infinite. Furthermore, this sequence cannot exceed its lower limit and therefore cannot reach back to the present instant.

This reasoning shows that there might exist infinite causal sequences which have not yet begun and which belong entirely to the future. This view is not only logically possible but also seems to be more prudent than the belief that each, even the smallest, future event has its causes acting from the beginning of the universe. I do not doubt at all that some future facts have their causes already in existence today and have had them from eternity. By means of observations and the laws of motion of the heavenly bodies astronomers predict eclipses of the moon and sun with great precision many years in advance. But nobody is able to predict today that a fly which does not yet exist will buzz into my ear at noon on 7 September of next year. The belief that this future behaviour of that future fly has its causes already today and has had them from all eternity seems to be a fantasy, rather than a proposition supported by even a shadow of scientific validation.

Therefore the argument based on the principle of causality

falls to the ground. One can be strongly convinced that nothing happens without cause, and that every fact has its cause in some earlier fact, without being a determinist. There remains to be considered the argument based on the principle of the excluded middle.

8. Although the argument based on the principle of the excluded middle is independent of that derived from the principle of causality, the former indeed becomes fully intelligible if every fact has its causes existing from all eternity. I shall explain what I mean by an example taken from ordinary life. Let us assume that John will be at home tomorrow noon. If the causes of all facts exist from all eternity, we should recognize that at the present instant there exists the cause of John's presence at home tomorrow noon. Therefore it is true, i.e. it is the case at the present instant, that John will be at home tomorrow noon. The somewhat confused expression 'it is the case at instant t that p', in which 'p' stands for sentences about future events, which I have previously been unable to elucidate, now becomes perfectly intelligible. It is the case at the present instant that 'John will be at home tomorrow noon' implies first that at the present instant there exists a fact which is the cause of John's presence at home tomorrow noon, and secondly that this future effect is as much comprehended in that cause as a conclusion is included in its premisses. The cause of the future fact, which the sentence 'p' states and which exists at instant t, is an actual correlate of the sentence 'it is the case at instant t that p'.

Should we assume that John will not be at home tomorrow noon, we can follow the same course of reasoning. If we recognize that the causes of every fact exist from all eternity, we must also accept the fact that the cause of John's absence from home tomorrow noon exists already at the present instant. Therefore the sentence 'it is true, i.e. it is the case at the present instant, that John will not be at home tomorrow noon' has its actual correlate in the cause of the stated fact, and this cause exists at present.

As John will or will not be at home tomorrow noon, there exists either the cause of his presence at or of his absence from home tomorrow noon, provided that the causes of all facts exist from all eternity. Therefore, either it is true at the present instant that John will be at home tomorrow noon or it is true at the present instant that John will not be at home tomorrow noon. The argument based on the principle of the excluded middle has additional support in the argument derived from the principle of causality.

9. However, the second of these arguments has proved itself to be invalid. In accordance with the preceding investigations, we may assume that at the present instant there exists as yet neither the cause of John's presence nor the cause of John's absence from home tomorrow noon. Thus it might happen that the infinite sequence of causes, which bring about John's presence or absence from home tomorrow noon, has not yet begun and lies entirely in the future. To put it colloquially, we can say that the question whether John will or will not be at home tomorrow noon is not yet decided either way. How should we argue in this case?

We might adopt the following course. The sentence 'it is true at the present instant t that John will be at home tomorrow noon' has no actual correlate, for the cause of this fact does not exist at instant t; therefore nothing compels us to recognize this sentence as true. Thus it might happen that John would not be at home tomorrow noon. In the same way the sentence 'it is true at the present instant t that John will not be at home tomorrow noon' has no real correlate, for the cause of this fact does not exist at instant t; again, nothing compels us to recognize this sentence as true. Thus it might happen that John would be at home tomorrow noon. We may, therefore, reject both these sentences as false and accept their denials 'it is not true at instant t that John will be at home tomorrow noon', and 'it is not true at instant t that John will not be at home tomorrow noon'.† The previously established conditional (e), 'if John will be at home tomorrow noon, then it is true at instant t that John will be at home tomorrow noon' becomes invalid.

† [The first of these latter two sentences was omitted in the Polish edition and is inserted here by the Editor.]

For its antecedent turns out to be true if John is at home tomorrow noon, and its consequent becomes false if we choose an instant t, earlier than tomorrow noon, at which the cause of John's presence at home tomorrow noon does not yet exist. But with conditional (e) the thesis of determinism, 'if A is b at instant t, it is true at every instant earlier than t that A is b at instant t' also becomes invalid; for we can substitute values for variables A, b, and t such that the antecedent of this thesis becomes true and the consequent false.

If on the assumption that a certain future fact is not yet decided either way the thesis of determinism becomes false, the deduction of this thesis from the principle of the excluded middle must involve an error. Indeed, if we reject as false the sentence 'it is true at instant t that John will be at home tomorrow noon' as well as the sentence 'it is true at instant t that John will not be at home tomorrow noon', we must also reject alternative (a) which is composed of these sentences as its arguments and which has been the starting-point of the deduction. An alternative both of whose arguments are false is itself false. So also conditional (d), obtained by transforming premiss (a), 'if it is not true at instant t that John will not be at home tomorrow noon, then it is true at instant t that John will be at home tomorrow noon', turns out to be false, for we accept its antecedent and reject its consequent. It is no wonder that the inference produces a false conclusion if one of its premisses and one of its intervening theorems are false.

It should be pointed out that the rejection of alternative (a) is not a transgression of the principle of the excluded middle; for its arguments do not contradict each other. Only the sentences 'John will be at home tomorrow noon' and 'John will not be at home tomorrow noon' are contradictory, and the alternative composed of these sentences, 'either John will be at home tomorrow noon or John will not be at home tomorrow noon', must be true in accordance with the principle of the excluded middle. But the sentences 'it is true at instant t that John will be at home tomorrow noon' and 'it is true at instant t that John will not be at home tomorrow noon' are not contradictory, for the one is not the denial of the other, and their presentation as alternatives need not be true. Premiss (a) has been deduced from the principle of the excluded middle on the basis of purely intuitive investigations and not by applying a logical principle. However, intuitive investigations may be fallacious, and they seem to have deceived us in this case.

10. Although this solution appears to be logically valid, I do not regard it as entirely satisfactory, for it does not satisfy all my intuitions. I believe that there is a difference between the non-acceptance of the sentence 'it is true at the present instant that John will be at home tomorrow noon' because John's presence at or absence from home tomorrow is not yet decided, and the non-acceptance of this sentence because the cause of his absence tomorrow already exists at the present instant. I think that solely in the latter case have we the right to reject the sentence in question and say, 'it is not true at the present instant that John will be at home tomorrow noon'. In the former case we can neither accept nor reject the sentence but should suspend our judgement.

This attitude finds its justification both in life and in colloquial speech. If John's presence at or absence from home tomorrow noon is not yet decided, we say, '*it is possible* that John will be at home tomorrow noon, but also *it is possible* that John will not be at home tomorrow noon'. On the other hand, if the cause of John's absence from home tomorrow noon exists already at the present instant, we say, provided that we know this cause, '*it is not possible* that John will be at home tomorrow noon'. On the assumption of John's presence at or absence from home tomorrow noon not yet being decided, the sentence 'it is true at the present instant that John will be at home tomorrow noon' can be neither accepted nor rejected, that is, we cannot consider it either true or false. Consequently also the denial of this sentence, 'it is not true at the present instant that John will be at home tomorrow noon', can be neither accepted nor rejected, i.e. we cannot consider it as either true or false. The previous reasoning, which consisted in the rejection of the sentence under discussion and in the acceptance of its denial, is now inapplicable.

In particular conditional (*d*), which was previously rejected, for its antecedent was accepted and its consequent rejected, need not now be rejected, for it is not true any longer that its antecedent is accepted and its consequent rejected. Furthermore, since conditional (*d*) together with premiss (*c*), which does not seem to involve any doubts whatsoever, suffice to validate the thesis of determinism, it appears as though Aristotle's argument regains its persuasive power.

11. However, this is not the case. I think that only now do we achieve a solution which is in agreement both with our intuitions and with the views of Aristotle himself. For Aristotle formulated his argument in support of determinism solely for the purpose of its subsequent rejection as invalid. In the famous chapter 9 of *De Interpretatione* Aristotle seems to have reached the conclusion that the alternative 'either there will be a sea battle tomorrow or there will not be a sea battle tomorrow' is already true and necessary today, but it is neither true today that 'there will be a sea battle tomorrow' nor that 'there will not be a sea battle tomorrow'. These sentences concern future contingent events and as such they are neither true nor false today. This was the interpretation of Aristotle given by the Stoics, who, being determinists, disputed his view, and by the Epicureans, who defended indeterminism and Aristotle.

Aristotle's reasoning does not undermine so much the principle of the excluded middle as one of the basic principles of our entire logic, which he himself was the first to state, namely, that *every proposition is either true or false*. That is, it can assume one and only one of two truth-values: truth or falsity. I call this principle *the principle of bivalence*. In ancient times this principle was emphatically defended by the Stoics and opposed by the Epicureans, both parties being fully aware of the issues involved. Because it lies at the very foundations of logic, the principle under discussion cannot be proved. One can only believe it, and he alone who considers it self-evident believes it. To me, personally, the principle of bivalence does not appear to be self-evident. Therefore I am entitled not to recognize it, and to accept the view that besides truth and falsehood there exist other truth-values, including at least one more, the third truth-value.

What is this third truth-value? I have no suitable name for it. But after the preceding explanations it should not be difficult to understand what I have in mind. I maintain that there are propositions which are neither true nor false but *indeterminate*. All sentences about future facts which are not yet decided belong to this category. Such sentences are neither true at the present moment, for they have no real correlate, nor are they false, for their denials too have no real correlate. If we make use of philosophical terminology which is not particularly clear, we could say that ontologically there corresponds to these sentences neither being nor non-being but possibility. Indeterminate sentences, which ontologically have possibility as their correlate, take the third truth-value.

If this third value is introduced into logic we change its very foundations. A trivalent system of logic, whose first outline I was able to give in 1920, differs from ordinary bivalent logic, the only one known so far, as much as non-Euclidean systems of geometry differ from Euclidean geometry. In spite of this, trivalent logic is as consistent and free from contradictions as is bivalent logic. Whatever form, when worked out in detail, this new logic assumes, the thesis of determinism will be no part of it. For in the conditional in terms of which this thesis is expressed, 'if A is b at instant t, then it is true at every instant earlier than t that A is b at instant t', we can assign such values to variables 'A', 'b', and 't' that its antecedent changes into a true sentence and its consequent into an indeterminate one, that is, into a sentence having the third truth-value. This always happens when the cause of the fact that A is b at a future instant t does not yet exist today. A conditional whose antecedent is true and consequent indeterminate cannot be accepted as true; for truth can imply only truth. The logical argument which seems to support determinism falls decisively.

12. I am near the end of my investigations. In my view, the age-old arguments in support of determinism do not withstand the test of critical examination. This does not at all imply that

determinism is a false view; the falsehood of the arguments does not demonstrate the falsehood of the thesis. Taking advantage of my preceding critical examination, I should like to state only one thing, namely that determinism is not a view better justified than indeterminism.

Therefore, without exposing myself to the charge of thoughtlessness, I may declare myself for indeterminism. I may assume that not the whole future is determined in advance. If there are causal chains commencing only in the future, then only some future facts and events, those nearest to the present time, are causally determined at the present instant. On the basis of present knowledge even an omniscient mind could predict fewer and fewer facts the deeper into the future it tried to reach: the only thing actually determined is the ever broader framework within which facts occur, and within which there is more and more room for possibility. The universal drama is not a picture completed from eternity; the further away we move from the parts of the film which are being shown just now, the more gaps and blanks the picture includes. It is well that it should be so. We may believe that we are not merely passive spectators of the drama but also its active participants. Among the contingencies that await us we can choose the better course and avoid the worse. We can ourselves somehow shape the future of the world in accordance with our designs. I do not know how this is possible, but I believe that it is.

We should not treat the past differently from the future. If the only part of the future that is now real is that which is causally determined by the present instant, and if causal chains commencing in the future belong to the realm of possibility, then only those parts of the past are at present real which still continue to act by their effects today. Facts whose effects have disappeared altogether, and which even an omniscient mind could not infer from those now occurring, belong to the realm of possibility. One cannot say about them that they took place, but only that they were *possible*. It is well that it should be so. There are hard moments of suffering and still harder ones of guilt in everyone's life. We should be glad to be able to erase them not only from our memory but also from existence. We may believe that when all the effects of those fateful moments are exhausted, even should that happen only *after* our death, then their causes too will be effaced from the world of actuality and pass into the realm of possibility. Time calms our cares and brings us forgiveness.

3

PHILOSOPHICAL REMARKS ON MANY-VALUED SYSTEMS OF PROPOSITIONAL LOGIC†

JAN ŁUKASIEWICZ

CONTENTS

In the communication 'Untersuchungen über den Aussagen- kalkül' (Investigations into the Sentential Calculus) which ap- peared in this issue under Tarski's and my name, § 3 is devoted to the 'many-valued' systems of propositional logic established by myself. Referring the reader to this communication as far as logical questions are concerned, I here propose to clarify the origin and significance of those systems from a philosophical point of view.

§ 1. Modal propositions

The three-valued system of propositional logic owes its origin to certain inquiries I made into so-called 'modal propositions'

† This paper appeared originally under the title 'Philosophische Bemerkun- gen zu mehrwertigen Systemen des Aussagenkalküls' in *Comptes rendus des séances de la Société des Sciences et des Lettres de Varsovie*, Cl. iii, 23 (1930), pp. 51–77. Translated by H. Weber.

and the notions of possibility and necessity closely connected with them.[1]

By *modal propositions* I mean propositions that have been formed after the pattern of one of the following four expressions:

(1) It is possible that p in symbols: Mp
(2) It is not possible that p „ NMp
(3) It is possible that not-p „ MNp
(4) It is not possible that not-p „ $NMNp$

The letter 'p' designates here any proposition; 'N' is the symbol of *negation* ('Np' = 'not-p'); 'M' corresponds to the words 'it is possible that'. Instead of saying 'it is not possible that not-p', one can also use the phrase 'it is necessary that p'.

The expressions listed here are not identical with Kant's 'problematical' and 'apodeictic' judgements. Rather they correspond to the modal propositions of medieval logic originating in Aristotle and formed from the four 'modes': *possibile* (e.g. *Socratem currere est possibile*), *impossibile*, *contingens*, and *necessarium*. Besides these four modes, two more modes were cited by the logicians of the Middle Ages; namely, *verum* and *falsum*. However, these modes were given no further consideration, as the modal propositions corresponding to them, 'it is true that p' and 'it is false that p', were regarded as being equivalent to the propositions 'p' and 'Np'.[2]

The expression 'it is possible that' is not defined here; its sense is made clear by the theorems which hold for modal propositions.

§ 2. Theorems concerning modal propositions

In the history of logic we meet with three groups of theorems concerning modal propositions.

Among the *first* group I count those well-known theorems which have been handed down to us from classical logic and

---

[1] I read a paper on these inquiries at the meeting on 5 June 1920 of the Polish Philosophical Society at Lwów. The essential parts of this paper were published in the Polish periodical *Ruch Filozoficzny* 5 (Lwów, 1920), pp. 169–70. [Translated as paper 1 of this volume—Ed.]

[2] Cf. Prantl, *Geschichte der Logik im Abendlande*, vol. iii, p. 14, note 42; p. 117, note 542.

have been regarded by it as truths evident without demonstration:

(a) *Ab oportere ad esse valet consequentia.*

(b) *Ab esse ad posse valet consequentia.*

By contraposition we get from (b) a third proposition:

(c) *A non posse ad non esse valet consequentia.*

The latter proposition means: 'The inference from not-being-possible to not-being is valid'. For instance: It is not possible to divide a prime number by four; therefore no prime number is divisible by four. This example is plausible, and just as plausible is the following general theorem which we shall keep in mind as representative of the first group:

I. *If it is not possible that p, then not-p.*

Less well known, but no less intuitive, seems to be the following theorem of the *second* group quoted by Leibniz in the *Théodicée*:[1]

(d) *Unumquodque, quando est, oportet esse.*

'Whatever is, *when* it is, is necessary.' This theorem dates back to Aristotle, who, to be sure, holds that not everything which is is necessary and not everything which is not is impossible, but when something which is is, then it is also necessary; and when something which is not is not, then it is also impossible.[2]

The theorems just quoted are not easily interpreted. First I shall give some examples.

It is not necessary that I should be at home this evening. But *when* I am at home this evening, then on this assumption it is *necessary* that I should be at home this evening. A second example: It rarely happens that I have no money in my pocket, but *if* I have now (at a certain moment *t*) no money in my pocket, it is *not possible*, on this assumption, that I have money (at just the same moment *t*) in my pocket.

Note has to be taken of two things about these examples.

---

[1] *Philos. Schriften* (ed. Gerhardt), vol. 6, p. 131.

[2] *De interpr.* 9. 19ᵃ23: Τὸ μὲν οὖν εἶναι τὸ ὂν ὅταν ᾖ, καὶ τὸ μὴ ὂν μὴ εἶναι, ὅταν μὴ ᾖ, ἀνάγκη· οὐ μὴν οὔτε τὸ ὂν ἅπαν ἀνάγκη εἶναι οὔτε τὸ μὴ ὂν μὴ εἶναι.

First, the propositions: 'I am at home this evening' and 'I have (at the moment *t*) no money in my pocket' are supposed to be *true*, and on this supposition the necessity or impossibility respectively is inferred. Secondly, the word *quando* in (*d*), and the corresponding 'ὅταν' of Aristotle, is not a conditional, but a temporal particle. Yet the temporal merges into the conditional, if the determination of time in the temporally connected propositions is included in the content of the propositions.

The examples given are, moreover, evident enough to establish the following general theorem, which we shall keep in mind as representative of the second group:

II. *If it is supposed that not-p, then it is (on this supposition) not possible that p.*

The *third* group consists of only *one* theorem based on the Aristotelian concept of 'two-sided' possibility. According to Aristotle there are some things which are possible in both directions, i.e. which *can* be, but *need* not be. It is possible, for instance, that this cloak should be cut; but it is also possible that it should not be cut.[1]

Again, it is possible that the patient will die, but it is also possible that he will recover, and therefore not die. This concept of two-sided possibility is deeply rooted in everyday thinking and speech. The following theorem, to which we will return, seems therefore to be just as evident as the two preceding ones:

III. *For some p: it is possible that p and it is possible that not-p.*

§ 3. Consequences of the first two theorems concerning modal propositions

We shall now draw some inferences from theorems I and II cited above. For this purpose we shall first represent those theorems in the symbolism of propositional logic.

Let '*Cpq*' symbolize the implication: 'if *p*, then *q*', '*p*' and '*q*'

---

[1] *De interpr.* 9. 19ᵃ9: ὅλως ἔστιν ἐν τοῖς μὴ ἀεὶ ἐνεργοῦσι τὸ δυνατὸν εἶναι καὶ μὴ ὁμοίως· ἐν οἷς ἄμφω ἐνδέχεται, καὶ τὸ εἶναι καὶ τὸ μὴ εἶναι . . . οἷον ὅτι τουτὶ τὸ ἱμάτιον δυνατόν ἐστι διατμηθῆναι . . . ὁμοίως δὲ καὶ τὸ μὴ διατμηθῆναι δυνατόν.

denoting any proposition. It is evident that theorem I can be expressed in the form of an implication, which I call 'thesis' 1:[1]

1. *CNMpNp.*

Meaning: 'If it is not possible that *p*, then not-*p*.'

It is not equally evident, but can be proven, that theorem II can be represented as an implication which is the converse of 1. For if a proposition '*β*' is valid on the assumption '*α*', this means no more than that '*β*' is true *if* '*α*' is true. The implication 'if *α*, then *β*' therefore holds, if '*α*' is true. Since this implication must also hold if '*α*' is false, it holds in both cases. We thus arrive at the thesis:

2. *CNpNMp.*

This means: 'If not-*p*, then it is not possible that *p*.' Theorem II cannot be expressed in any other way in the two-valued propositional calculus.

From these theses and using the usual propositional calculus, we shall prove several consequences. All the following demonstrations are strictly formalized and carried out by means of two rules of inference: *substitution* and *detachment*. These well-known rules of inference will not be discussed here. I will only explain how formalized proofs are recorded in the symbolism which I introduced.

Before each thesis to be proved (to which consecutive numbers are assigned for purposes of identification) is an unnumbered line, which I call the 'derivational line'. Each derivational line consists of two parts separated by the sign '×'. The symbols before and after the separation sign denote the same expression, but in different ways. Before the separation sign, a substitution is indicated, which is to be carried out on a thesis already proven. In the first derivational line, for example, the expression '3 *q/Mp*' means that '*Mp*' should be substituted for '*q*' in 3. The resultant thesis, which is omitted in the proof for the sake of brevity, would be:

3'. *CCNMpNpCpMp.*

The expression '*C*1—7' after the separation sign refers to this

[1] Following Leśniewski, I understand by 'theses' axioms as well as theorems of a deductive system.

thesis 3′ and indicates that the rule of detachment can be applied to 3′. Thesis 3′ is asserted as a substitution instance of thesis 3; but since it is an implication whose antecedent is thesis 1, its consequent may be detached and asserted as thesis 7. In the second derivational line the number '8' denotes the thesis obtained from 7 by the substitution 'p/Np'. In the derivational line of thesis 10, the rule of detachment is used twice. After these explanations, I believe the reader will have no difficulty in understanding the demonstration below.

In addition to theses 1 and 2, which appear as axioms, four well-known auxiliary theses from the ordinary propositional calculus appear in the demonstration: three laws of transposition, numbered 3–5; and the principle of the hypothetical syllogism, thesis 6. All of these theses I place at the head of the demonstration as premisses.

1. $CNMpNp$
2. $CNpNMp$
3. $CCNqNpCpq$
4. $CCNpqCNqp$
5. $CCpNqCqNp$
6. $CCpqCCqrCpr$
   *
   $3\ q/Mp \times C1$—7
7. $CpMp$
   $7\ p/Np \times 8$
8. $CNpMNp$
   $4\ q/MNp \times C8$—9
9. $CNMNpp$
   $6\ p/NMNp,\ q/p,\ r/Mp \times C9$—$C7$—10
10. $CNMNpMp$
    $4\ p/MNp,\ q/Mp \times C10$—11
11. $CNMpMNp$
    *
    $3\ q/p,\ p/Mp \times C2$—12
12. $CMpp$
    $12\ p/Np \times 13$

13. $CMNpNp$

  $5\,p/MNp,\ q/p \times C13\text{---}14$

14. $CpNMNp$

  $6\,p/Mp,\ q/p,\ r/NMNp \times C12\text{---}C14\text{---}15$

15. $CMpNMNp$

  $5\,p/Mp,\ q/MNp \times C15\text{---}16$

16. $CMNpNMp.$

Theses 7–11 are consequences of 1; 12–16 result from 2. Thesis 1 says: 'if p, then it is possible that p'. Thesis 9 says: 'if it is not possible that not-p, then p'. The latter thesis corresponds to theorem (a) in classical logic, cited above, the first to theorem (b). Both are evident. In fact, all theses of the first group, 7–11, are evident.

Not so evident are the theses of the second group, 12–16. Thesis 12 reads: 'if it is possible that p, then p'. On the basis of this thesis we can infer: It is possible that the patient will die; hence he will die. This inference will be admitted only by those making no distinction between possibility and being. Theses of the second group, 12–16, are the converses of theses of the first group, 7–11. Whoever admits both groups of theses must assume the following propositions to be equivalent: 'p', 'it is possible that p', and 'it is not possible that not-p', or 'it is necessary that p'. Also the propositions 'not-p', 'it is possible that not-p', and 'it is not possible that p'. But then the concepts of possibility and necessity become dispensable. This unpleasant consequence results from the acceptance of our symbolic formulation of theorem II, which is evident in ordinary language and can be recognized as being true without reservation. Nevertheless it seems to me impossible to express proposition II in the symbolic language of the two-valued propositional calculus in any other way than by a simple implication which is the converse of thesis 1.

§ 4. Consequences of the third theorem on modal propositions

The symbolic formulation of the third theorem leads to another unwelcome result.

Theorem III can be expressed only by means of the symbolism of the extended propositional calculus. Let 'Σ' be the *existential quantifier*, and let 'Σp' denote the expression 'for some p'. Let 'Kpq' be the symbol of *conjunction*, 'p and q', where 'p' and 'q' denote any propositions. Theorem III can then be expressed symbolically as follows:

17. $\Sigma p K M p M N p$.

This means verbally: 'For some p: it is possible that p, and it is possible that not-p.'

The existential quantifier 'Σ' can be expressed by means of the *universal* quantifier 'Π'. If 'Πp' says: 'for every p', and if '$\alpha(p)$' represents any expression containing 'p', the following definition is evident:

$D1$. $\Sigma p \alpha(p) = N \Pi p N \alpha(p)$.

$D1$ states that the expressions: 'for some p, $\alpha(p)$ (holds)' and 'it is not true that for each p not-$\alpha(p)$ (holds)' mean the same thing. Thesis 17 then becomes the following thesis:

18. $N \Pi p N K M p M N p$.

There is, however, besides the extended propositional calculus, a still more general logical system created by Leśniewski, which he has termed 'prototehtic'.[1] The main difference between prototehtic and the extended propositional calculus is the occurrence in the latter of variable 'functors'[2] as well as constants.

Denoting a variable functor to which *one* proposition only is attached as argument by 'ϕ', we can prove the following proposition in prototehtic:

$$C K \phi p \phi N p \phi q.$$

In words: 'If ϕ of p and ϕ of not-p, then ϕ of q.'

Since this proposition is valid for all functors with one argument, it is also valid for the functor 'M'. We thus obtain:

19. $C K M p M N p M q$.

---

[1] S. Leśniewski, 'Grundzüge eines neuen Systems der Grundlagen der Mathematik', introduction and §§ 1–11, *Fund. Math.* 14, Warsaw, 1929.

[2] In the function 'Cpq', 'C' is the 'functor', and 'p' and 'q' the 'arguments'. The term 'functor' was introduced by Kotarbiński.

Theses 18 and 19, as well as two auxiliary theses from the ordinary propositional calculus, viz. the principle of transposition 4 mentioned above, and another rule of transposition, thesis 20, are premisses of the formalized proof given below. Besides substitution and detachment, the rule for the *introduction of a quantifier* is used in the proof. This rule runs thus: If in the consequent of an implication which is a thesis there occurs a free propositional variable 'p' which does not occur in the antecedent of that implication, the symbol 'Πp' may be put before the consequent. This rule of inference is denoted below by '$+\Pi$'. Beginning with the premisses, our demonstration then reads thus:

18. $N\Pi p N K M p M N p$
19. $C K M p M N p M q$
20. $C C p q C N q N p$
   $*$
   $20\,p/KMpMNp,\ q/Mq \times C19\text{---}21$
21. $C N M q N K M p M N p$
   $21 + \Pi \times 22$
22. $C N M q \Pi p N K M p M N p$
   $4\,p/Mp,\ q/\Pi p N K M p M N p \times C22q/p\text{---}C18\text{---}23$
23. Mp.

The result obtained, thesis 23, has to be admitted as being true. This thesis, which in words reads 'it is possible that p', holds for any p. We therefore have to admit as true the proposition 'it is possible that 2 is a prime number', as well as the proposition 'it is possible that 2 is not a prime number'. Freely speaking, we have been led to admit everything as possible by reason of theorem III. Yet if everything is possible, then nothing is impossible and nothing necessary. For if the proposition 'Mp' is admitted, we obtain from it by substitution the proposition 'MNp', and the expressions 'NMp' and '$NMNp$' have to be rejected as negations of those preceding.

These are consequences running contrary to all of our intuitions. Yet I see no possibility of expressing theorem III, in the symbolism of the extended propositional calculus, in any other form than that of thesis 17 or 18.

§ 5. Incompatibility of the theorems on modal propositions in the two-valued propositional calculus

The unpleasant consequences to which we were led by theorems II and III considered separately become wholly unacceptable when we consider both theorems together.

Indeed, when we combine thesis 12, resulting from the symbolic formulation of theorem II, with thesis 23:

12. $CMpp$
23. Mp

we immediately obtain:

$$12 \times C23\text{---}24$$

24. p.

If therefore theses 12 and 23 are valid, any proposition p is valid too. Hence we arrive at the inconsistent system of all propositions. Theorems II and III are incompatible when symbolically represented as theses 2 and 18.

We can obtain the same result without employing thesis 19, which presupposes a proposition from protothetic. In the following demonstration we use the theses 12, 13, and 20 alone, as well as certain auxiliary theses of the ordinary propositional calculus:

25. $CpCqp$
26. $NKpNp$
27. $CCpqCCrsCKprKqs$
      *
$$27\,p/Mp,\ q/p,\ r/MNp,\ s/Np \times C12\text{---}C13\text{---}28$$
28. $CKMpMNpKpNp$
$$20\,p/KMpMNp,\ q/KpNp \times C28\text{---}C26\text{---}29$$
29. $NKMpMNp$
$$25\,p/NKMpMNp \times C29\text{---}30$$
30. $CqNKMpMNp$
$$30+\Pi \times 31$$
31. $Cq\Pi pNKMpMNp$
$$31\,q/CpCqp \times C25\text{---}32$$
32. $\Pi pNKMpMNp$.

E

Theses 18 and 32 contradict each other. Therefore propositions II and III are incompatible.

The demonstration given above could be made intuitively plausible in the following manner: If according to proposition III the expressions '$M\alpha$' and '$MN\alpha$' were jointly true for a certain proposition 'α', then the propositions 'α' and '$N\alpha$' would also have to be true according to theses 12 and 13. Yet this is impossible, because 'α' and '$N\alpha$' contradict each other.

In view of this fact the problem of modal propositions could be solved in two ways, taking the two-valued propositional calculus as a basis. Theorem I and those theses of the first group connected with it (viz. theses 1 and 7–11) have to be accepted unconditionally; they were actually never called in question. Of theorems II and III only one can be selected. If we decide in favour of theorem II and those theses of the second group connected with it (viz. theses 2 and 12–16), then all modal propositions become equivalent to non-modal ones. The consequence of this is that it is not worth while to introduce modal propositions into logic. Also, the extremely intuitive concept of two-sided possibility must then be rejected as being inconsistent. If, on the other hand, we decide in favour of proposition III, we are compelled to admit the paradoxical consequence that everything is possible. On this condition again it is senseless to introduce modal propositions into logic; moreover, we would then have to do without the intuitively evident theorem II in order to avoid contradiction. None of these solutions can claim to be satisfactory.

A different result was not to be expected. This becomes especially clear when the system of the two-valued propositional calculus is defined by the so-called matrix method. On the basis of this method it is assumed that all propositional variables can take only two constant values, namely '0' or 'the false' and '1' or 'the true'. It is further laid down that:

$$C00 = C01 = C11 = 1, \quad C10 = 0, \quad N0 = 1, \quad \text{and} \quad N1 = 0.$$

These equations are recorded in the following table, which is the 'matrix' of the two-valued propositional calculus based on 'C' and 'N'.

C	0	1	N
0	1	1	1
1	0	1	0

In a two-valued system only four different functions of *one* argument can be formed. If 'ϕ' denotes a functor of one argument, then the following cases are possible: (1) $\phi 0 = 0$ and $\phi 1 = 0$; this function we denote by 'Fp' ('*falsum* of p'). (2) $\phi 0 = 0$ and $\phi 1 = 1$; ϕp is equivalent to p. (3) $\phi 0 = 1$ and $\phi 1 = 0$; this is the negation of p, 'Np'. (4) $\phi 0 = 1$ and $\phi 1 = 1$; this function we denote by 'Vp' ('*verum* of p').

'Mp' must be identical with one of these four cases. But each of theses 1, 2, and 18 excludes certain cases. By direct verification with '0' and '1' it can be ascertained that:

$$(A) \begin{cases} 1.\ CNMpNp & \text{holds only for } Mp = p \text{ or } Mp = Vp. \\ 2.\ CNpNMp & \text{,,} \quad \text{,,} \quad Mp = p \text{ or } Mp = Fp. \\ 18.\ N\Pi pNKMpMNp & \text{,,} \quad \text{,,} \quad Mp = Vp. \end{cases}$$

Thesis 18 is verified by the statement: $\Pi p\alpha(p) = K\alpha(0)\alpha(1)$. One then obtains:

$$\begin{aligned} N\Pi pNKMpMNp &= NKNKM0MN0NKM1MN1 \\ &= NKNKM0M1NKM1M0 \\ &= NKNKM0M1NKM0M1 \\ &= NNKM0M1 = KM0M1. \end{aligned}$$

The last conjunction obtains only on the condition that

$$M0 = M1 = 1.$$

The conditions (A) make it evident that theses 1 and 2 can be valid jointly only for $Mp = p$; just as theses 1 and 18 can be valid only for $Mp = Vp$. Theses 2 and 18 are incompatible, as there is no function for 'Mp' which would simultaneously verify both theses.

§ 6. Modal propositions and the three-valued propositional calculus

When I recognized the incompatibility of the traditional theorems on modal propositions in 1920,[1] I was occupied with establishing the system of the ordinary 'two-valued' propositional calculus by means of the matrix method.[2] I satisfied myself at that time that all theses of the ordinary propositional calculus could be proved on the assumption that their propositional variables could assume only two values, '0' or 'the false', and '1' or 'the true'.

To this assumption corresponds the basic theorem that *every proposition is either true or false*. For short I will term this the *law of bivalence*. Although this is occasionally called the law of the excluded middle, I prefer to reserve this name for the familiar principle of classical logic that two contradictory propositions cannot be false simultaneously.

The law of bivalence is the basis of our entire logic, yet it was already much disputed by the ancients. Known to Aristotle, although contested for propositions referring to future contingencies; peremptorily rejected by the Epicureans, the law of bivalence makes its first full appearance with Chrysippus and the Stoics as a principle of their dialectic, which represents the ancient propositional calculus.[3] The quarrel about the law of bivalence has a metaphysical background, the advocates of the law being decided determinists, while its opponents

---

[1] In the report cited in note 1 (p. 41) I had defined the concept of two-sided possibility more strictly by assuming that the propositions 'it is possible that *p*' and 'it is possible that not-*p*' must always hold simultaneously, which in conjunction with propositions of the two first groups leads to numerous contradictions. I had in mind here the Aristotelian concept of 'pure' possibility. It seems that Aristotle distinguished between two essentially different kinds of possibility: possibility in the proper sense or pure possibility, by which something is only possible if it is not necessary; and possibility in the improper sense, which is connected with necessity and results from it according to our thesis 10. Cf. H. Maier, *Die Syllogistik des Aristoteles*, part i (Tübingen, 1896), pp. 180, 181.

[2] The results of these inquiries have been published in my article 'Logika dwuwartościowa' (Two-valued Logic), which appeared in the Polish philosophical review *Przegląd Filozoficzny* 23 (Studies in honour of Professor Twardowski), (Lwów, 1921), pp. 189–205.

[3] Cf. the appendix: 'On the history of the law of bivalence', pp. 63 ff.

tend towards an indeterministic *Weltanschauung*.[1] Thus we have re-entered the area of the concepts of possibility and necessity.

The most fundamental law of logic seems after all to be not quite evident. Relying on venerable examples, which go back to Aristotle, I tried to refute the law of bivalence by pursuing the following line of thought.

I can assume without contradiction that my presence in Warsaw at a certain moment of next year, e.g. at noon on 21 December, is at the present time determined neither positively nor negatively. Hence it is *possible*, but not *necessary*, that I shall be present in Warsaw at the given time. On this assumption the proposition 'I shall be in Warsaw at noon on 21 December of next year', can at the present time be neither true nor false. For if it were true now, my future presence in Warsaw would have to be necessary, which is contradictory to the assumption. If it were false now, on the other hand, my future presence in Warsaw would have to be impossible, which is also contradictory to the assumption. Therefore the proposition considered is at the moment *neither true nor false* and must possess a third value, different from '0' or falsity and '1' or truth. This value we can designate by '$\frac{1}{2}$'. It represents 'the possible', and joins 'the true' and 'the false' as a third value.

The three-valued system of propositional logic owes its origin to this line of thought. Next the matrix had to be given by which this new system of logic could be defined. Immediately it was clear to me that if the proposition concerning my future presence in Warsaw took the value $\frac{1}{2}$, its negation must take the same value $\frac{1}{2}$. Thus I obtained the equation $N\frac{1}{2} = \frac{1}{2}$. For implication I still had to determine the five equations containing the value $\frac{1}{2}$, namely $C0\frac{1}{2}$, $C\frac{1}{2}0$, $C\frac{1}{2}\frac{1}{2}$, $C\frac{1}{2}1$, and $C1\frac{1}{2}$. Equations not containing the value $\frac{1}{2}$, I took over from the two-valued system of propositional logic, as well as the values for '$N0$' and '$N1$'. The desired equations I obtained on the basis of detailed considerations, which were more or less plausible to me. In this

---

[1] In the inaugural address which I delivered as Chancellor of the University of Warsaw in 1922, I tried to solve the problem of an indeterministic philosophy by three-valued logic. A revised version of this lecture will be published shortly in Polish. [Ed. note: this address appears as paper 2 of this volume.]

way I finally arrived at the formulation of a three-valued propositional calculus, defined by the matrix below. The system originated in 1920.[1]

C	0	$\frac{1}{2}$	1	N
0	1	1	1	1
$\frac{1}{2}$	$\frac{1}{2}$	1	1	$\frac{1}{2}$
1	0	$\frac{1}{2}$	1	0

§ 7. Definition of the concept of possibility

On the basis of this system I then tried to construct a definition of the concept of possibility which would allow me to establish all the intuitive traditional theorems for modal propositions without contradiction. I did this with regard to the concept of 'pure' possibility, and soon found a satisfactory definition.[2] Later on, however, I became convinced that the

[1] I reported on this system to the Polish Philosophical Society at Lwów on 19 June, 1920. The essential contents of this report have been published in *Ruch Filozoficzny* 5 (Lwów, 1920), p. 170. [Ed. note: See paper 1 of this volume.]

[2] The definition found was rather complicated and read thus:

$D*1.\ Mp = AEpNp\Pi qNCpKqNq.$

That is: The expression 'it is possible that p' means 'either p and not-p are equivalent to one another, or there is no pair of contradictory propositions implied by p'. 'A' is the sign of *alternation*; 'E' the sign of *equivalence*. In three-valued logic the following definitions hold:

$D*2.\ Apq = CCpqq$
$D*3.\ Kpq = NANpNq.$
$D*4.\ Epq = KCpqCqp.$

The definition of 'impossibility' is more evident:

$D*5.\ NMp = KNEpNp\Sigma qCpKqNq.$

That is, the expression 'it is not possible that p' means 'p and not-p are not equivalent to one another, and there is a pair of contradictory propositions implied by p'.

From $D*1$ the following equations are obtained for 'M': $M0 = 0$, $M\frac{1}{2} = 1$, $M1 = \frac{1}{2}$. By means of these equations and the matrix of the three-valued propositional calculus the following theses can be easily verified:

(1) $CpCpNMNp$
(2) $CNpCNpNMp$
(3) $CMpCMpMNp$
(4) $CMNpCMNpMp.$
(5) $CNMpCNMpNp.$
(6) $CNMNpCNMNpp.$

Thesis (5) allows us to obtain by two detachments, in accordance with theorem

wider concept of possibility *in general* was to be preferred to
the more narrow concept of *pure* possibility. In what follows,
therefore, I discuss a definition of the latter concept, which
satisfies all the requirements of theorems I–III.

The definition in question was discovered by Tarski in 1921
when he attended my seminars as a student at the University
of Warsaw. Tarski's definition is as follows:

$$D2. \quad Mp = CNpp.$$

Expressed verbally this says: 'it is possible that p' means 'if
not-p then p'.

One must grasp the intuitive meaning of this definition. The
expression '$CNpp$' is according to the three-valued matrix false
if and only if 'p' is false. Otherwise '$CNpp$' is true. Thus we
obtain the equations:

$$M0 = 0, \qquad M\tfrac{1}{2} = 1, \qquad M1 = 1.$$

Hence, if any proposition 'α' is false, the proposition 'it is
possible that α' is false too. And if 'α' is true, or if it takes the
third value, that of 'possibility', then the proposition 'it is possible that α' is true. This agrees very well with our intuitions.

In two-valued logic the expression '$CNpp$' is equivalent to
the expression 'p'; but not in three-valued logic. The thesis
'$CCNppp$', valid in the two-valued calculus, and appearing as
an axiom in my system of the ordinary propositional calculus,[1]
is not valid for $p = \tfrac{1}{2}$ in the three-valued system. Vailati has
written an interesting monograph on the thesis '$CCNppp$',[2] in

I and on the basis of the admitted proposition 'it is not possible that α' ('$NM\alpha$'),
the proposition 'not-α' ('$N\alpha$'). Conversely we get by two detachments the
proposition 'it is not possible that α' ('$NM\alpha$') from thesis (2), in accordance
with theorem II, on the basis of the admitted proposition 'not-α' ('$N\alpha$'). Furthermore, if one of the propositions 'it is possible that α' ('$M\alpha$'), and 'it is possible
that not-α' ('$MN\alpha$') is admitted, the other of these propositions has to be
admitted too, by theses (3) and (4). From the admitted propositions 'α' and 'it is
necessary that α' no inference can be made to the proposition 'it is possible
that α', since we are dealing here with 'pure' possibility, which is incompatible
with necessity. Cf. note 1, page 52.

[1] Cf. *Elementy logiki matematycznej (Elements of Mathematical Logic)*, a
lithographed edition of lectures given by me at the University of Warsaw in the
autumn of 1928–9, revised by M. Presburger (Warsaw, 1929), p. 45. [Ed. note:
an English translation (Warsaw, 1963) has appeared.]

[2] *Scritti di G. Vailati*, Leipzig–Firenze, 1911. *CXV. A proposito d'un passo
del Teeteto e di una dimostrazione di Euclide*, pp. 516–27.

which it is shown that Euclid made use of this thesis in demonstrating one of his theorems, without formulating it expressly.[1] It was Clavius, a commentator on Euclid from the second half of the sixteenth century, a Jesuit and the constructor of the Gregorian calendar, who first paid attention to this thesis.[2] Since that time it appears to have acquired a certain popularity among Jesuit scholars under the name *consequentia mirabilis*.[3] The notable Jesuit Gerolamo Saccheri in particular was so taken by the thesis '*CCNppp*' that he attempted to demonstrate Euclid's parallel postulate on the basis of it. The attempt failed, but Saccheri gained the title of being a precursor of non-Euclidean geometry.[4]

The thesis '*CCNppp*' states that if for a certain proposition, say 'α', the implication '*CNαα*' holds, then 'α' holds too. The implication 'if not-α, then α' does not, to be sure, mean the same as the expression 'α can be inferred from not-α', yet the more general concept of implication covers the more special case of inference. If therefore from a proposition 'not-α' the proposition 'α' can be inferred, then 'α' is true. It would, however, not be correct to assume with Saccheri that the fact 'from not-α is inferred α' stamps the proposition 'α' as a *prima veritas*.[5] On

---

[1] Cf. Vailati, op. cit., pp. 518 ff. It seems to have escaped Vailati that the above-mentioned thesis was already known to the Stoics, although not in its pure form. We read in Sextus Empiricus, *Adv. math.* viii. 292; εἰ τὸ πρῶτον, τὸ πρῶτον· εἰ οὐ τὸ πρῶτον, τὸ πρῶτον· ἤτοι τὸ πρῶτον ἢ οὐ τὸ πρῶτον· τὸ πρῶτον ἄρα. If in this schema the self-evident premisses εἰ τὸ πρῶτον, τὸ πρῶτον and ἤτοι τὸ πρῶτον ἢ οὐ τὸ πρῶτον are omitted, we obtain the consequence εἰ οὐ τὸ πρῶτον, τὸ πρῶτον· τὸ πρῶτον ἄρα, which corresponds to the thesis '*CCNppp*'.

[2] Cf. Vailati, op. cit., p. 521.

[3] I find the name *consequentia mirabilis* for this thesis in the writings of Polish Jesuits. Adam Krasnodebski, in his *Philosophia Aristotelis explicata* (Warsaw, 1676), *Dialecticae Prolegomenon* 21, writes, for instance, the following: *Artificium argumentandi per consequentiam mirabilem in hoc positum est (uti de re speculativa optime in Polonia meritus. R. P. Tho. Młodzianowski Tr. I de Poenit. disp. 1. quae. 1. difficul. 1 No. 20 refert), ut ex propositione quam tuetur respondens, ab argumentante eliciatur contradictoria.*

[4] Cf. Vailati, op. cit. CIX. *Di un'opera dimenticata del P. Gerolamo Saccheri* ('*Logica demonstrativa*' 1697), pp. 477–84.

[5] Cf. Vailati, op. cit., p. 526, where the following words of Saccheri are quoted: 'Nam hic maxime videtur esse cujusque primae veritatis veluti character ut non nisi exquisita aliqua redargutione ex suo ipso contradictorio assumpto ut vero illa ipsi sibi tandem restitui possit' (*Euclides ab omni naevo vindicatus*, p. 99).

the contrary, the thesis '*CCNppp*' strikes us as outrightly paradoxical, as is also indicated by its name, *consequentia mirabilis*. *This* alone is certain: if any proposition *can* be inferred from its contradictory opposite, it is certainly not false, hence not impossible either. It is *possible*, as Tarski's definition states. This definition will perhaps be even more obvious, if it is applied to the concept of necessity. For we obtain in accordance with *D2*:

D3. $NMNp = NCpNp$,

which says that 'it is necessary that *p*' means 'it is not true that if *p*, then not-*p*'. Freely speaking, we can then assert that a certain proposition '*α*' is necessary, if and only if it does not contain its own negation.

Without stressing the intuitive character of the above definition, we have to admit in any case that this definition meets all of the requirements of theorems I–III. Indeed, as Tarski has shown, it is the *only* positive definition in the three-valued system which meets these requirements. We will now proceed to demonstrate these last assertions.

§ 8 Consequences of the definition of the concept of possibility

From definition *D2* it follows that all theses of the first group are verified, i.e. thesis 1, corresponding to theorem I, and theses 7–11. For in three-valued propositional logic the thesis

T1. $CpCqp$

holds good. We thus obtain:

$T1\ q/Np \times T2$
T2. $CpCNpp$
$T2.D2 \times T3$
T3. $CpMp$.

In the derivational line belonging to thesis *T3* a rule of inference has been used which permits us to replace the right side of a definition by its left. Since all laws of transposition as well as the principle of the syllogism hold true in the three-valued calculus, we obtain all of the remaining theses of the first group from $T3$. All these theses are perfectly evident.

The theses of the second group are not valid. However, not all these theses are evident in any case. Two of them, of which one corresponds to theorem II, are in a certain sense valid, though not as simple implications. To be exact, by definition $D2$ the following propositions hold true in the three-valued calculus:

$$CpCpNMNp \quad \text{and} \quad CNpCNpNMp,$$

although the expressions

$$CpNMNp \quad \text{and} \quad CNpNMp$$

are not valid. This is caused by the fact that in the three-valued calculus the thesis '$CCpCpqCpq$' does not hold, and because of this the expressions '$C\alpha C\alpha\beta$' and '$C\alpha\beta$' are not equivalent to each other as they are in the ordinary two-valued calculus. The above-mentioned propositions can be demonstrated by means of the following auxiliary theses, which also hold true in three-valued propositional logic:

$T4.$ $CpCCpqq$

$T5.$ $CpCCNNpqq$

$T6.$ $CCpCqrCpCNrNq$

$T7.$ $CCpCqNrCpCrNq$

$\qquad *$

$\qquad T6\,p/Np,\ q/CNpp,\ r/p \times CT4\,p/Np,\ q/p—T8$

$T8.$ $CNpCNpNCNpp$

$\qquad T8.D2 \times T9$

$T9.$ $CNpCNpNMp$

$\qquad T7\,q/CNNpNp,\ r/p \times CT5\,q/Np—T10$

$T10.$ $CpCpNCNNpNp$

$\qquad T10.D2\,p/Np \times T11$

$T11.$ $CpCpNMNp.$

If the proposition 'not-α' is admitted, then by double detachment applied to thesis $T9$, the proposition 'it is not possible that α' is obtained. If the proposition 'α' is admitted, then, by $T11$ and double detachment, one arrives at the proposition: 'it is not possible that not-α' which means the same as 'it is necessary that α'. It can therefore be correctly inferred: 'I have no money in my pocket; hence it is not possible that I have money in my pocket.' Or again, 'I am at home in the evening; hence it is necessary that I am at home in the evening'. The intuitively evident theorem II has been shown to hold good, moreover, in such a way that the Aristotelian maxim is maintained, according to which not everything which is is necessary and not everything which is not is impossible. For the expressions 'α' and '$NMN\alpha$' as well as '$N\alpha$' and '$NM\alpha$' are not equivalent to each other. Nor can being be inferred from possibility, as long as 'Mp' means the same as '$CNpp$', since neither '$CMpp$' nor '$CMpCMpp$' holds true in the three-valued propositional calculus.

Finally, theorem III is verified in the form of the theses:

$T12.\ \Sigma pKMpMNp$

or

$T13.\ N\Pi pNKMpMNp,$

in which the following definitions are assumed:

$D4.\ Apq = CCpqq$

$D5.\ Kpq = NANpNq.$

Theses $T12$ and $T13$ are easily verified with the help of the matrix of the three-valued calculus and the equations given for 'M' in the preceding section. For $p = \frac{1}{2}$ we obtain:

$$KM\tfrac{1}{2}MN\tfrac{1}{2} = K1M\tfrac{1}{2} = K11 = 1.$$

There is, therefore, a value for p for which the expression '$KMpMNp$' is correct.

As a résumé of the above findings we are now able to establish the following theorem:

All the traditional theorems for modal propositions have been established free of contradiction in the three-valued propositional calculus, on the basis of the definition '$Mp = CNpp$'.

This result seems to me highly significant. For it appears that those of our intuitions which are connected with the concepts of possibility and necessity point to a system of logic which is fundamentally different from ordinary logic based on the law of bivalence.

It remains to prove that the definition given by Tarski is the only one in the three-valued calculus which meets the requirements of theorems I–III. This can be shown in the following manner. Since according to theorem I the proposition '$N\alpha$' follows from the proposition '$NM\alpha$', by the law of transposition '$M\alpha$' must follow from 'α'. Hence, if $\alpha = 1$, then $M\alpha = M1 = 1$. We thus obtain the equation $M1 = 1$. On the other hand, according to theorem II the proposition '$NM\alpha$' follows from the proposition '$N\alpha$'. Hence if $\alpha = 0$, or $N\alpha=1$, then $NM\alpha = NM0 = 1$. But $NM0$ can equal 1 only under the condition that $M0 = 0$. We thus obtain the second equation: $M0 = 0$. Finally also theorem III, '$\Sigma pKMpMNp$', must be true. But it is not true for $p = 0$ or $p = 1$, for in both cases one term of the conjunction is false; hence the conjunction itself must be false too. We therefore have to assume that $M\frac{1}{2} = 1$, since only then does the conjunction '$KMpMNp$' equal 1 for $p = \frac{1}{2}$. In this way the function 'Mp' is fully determined for the three-valued propositional calculus, and can be defined only by '$CNpp$' or by some other expression equivalent to it.

§ 9. Philosophical significance of many-valued systems of propositional logic

Besides the three-valued system of propositional logic, I discovered an entire class of closely related systems in 1922, which I defined by means of the matrix method in the following manner:

When 'p' and 'q' denote certain numbers of the interval (0, 1), then:

$$Cpq = 1 \qquad \text{for} \quad p \leqslant q,$$
$$Cpq = 1-p+q \quad \text{,,} \quad p > q,$$
$$Np = 1-p.$$

If only the limiting values 0 and 1 are chosen from the interval (0, 1), the above definition represents the matrix of the ordinary two-valued propositional calculus. If, in addition, the value $\frac{1}{2}$ is included, we obtain the matrix of the three-valued system. In a similar manner 4, 5,... n-valued systems can be formed.

It was clear to me from the outset, that among all the many-

valued systems only two can claim any philosophical significance: the three-valued and the infinite-valued ones. For if
values other than '0' and '1' are interpreted as 'the possible',
only two cases can reasonably be distinguished: either one assumes that there are no variations in degree of the possible and
consequently arrives at the three-valued system; or one assumes
the opposite, in which case it would be most natural to suppose
(as in the theory of probabilities) that there are infinitely many
degrees of possibility, which leads to the infinite-valued propositional calculus. I believe that the latter system is preferable to
all others. Unfortunately this system has not yet been investigated sufficiently; in particular the relation of the infinite-valued
system to the calculus of probabilities awaits further inquiry.[1]

If the definition of possibility established by Tarski is assumed
for the infinite-valued system, there result, as in the three-valued
system, all theses mentioned in the preceding section. The intuitively evident theorems I–III are therefore also verified in
the infinite-valued propositional calculus.

The three-valued system is a proper part of the two-valued,
just as the infinite-valued system is a proper part of the
three-valued one. This means that all theses of the three- and
infinite-valued systems (without quantifiers) hold true for the
two-valued system. There are, however, theses which are valid
in the two-valued calculus but not in the infinite-valued system.
But when it is a question of the best known propositional theses
—for instance those listed in *Principia Mathematica*[2]—the difference between the three-valued and the infinite-valued propositional calculus is minimal. To be sure, I cannot find a single
thesis in this work that would be valid in the three-valued
system without being also true in the infinite-valued one.

The most important theses of the two-valued calculus which
do not hold true for the three- and infinite-valued systems concern certain apagogic inference schemata that have been suspect

[1] My little book *Die logischen Gründlagen der Wahrscheinlichkeitsrechnung*,
Cracow, 1913, Akad. d. Wiss., tries to base the notion of probability on quite
a different idea.
[2] Cf. A. N. Whitehead and B. Russell, *Principia Mathematica* (Cambridge,
1910), vol. i, pp. 94–131.

from time immemorial. For example, the following theses do not hold true in many-valued systems: '*CCNppp*', '*CCpNpNp*', '*CCpqCCpNqNp*', '*CCpKqNqNp*', '*CCpEqNqNp*'. The first of these theses has been discussed above; the second differs from the first only by the introduction of the negation of *p* for *p*. The two other theses justify us in assuming a proposition '*Nα*' to be true, when from its opposite '*α*' two mutually contradictory propositions can be derived. The last thesis asserts that a proposition from which the equivalence of two contradictory propositions follows is incorrect. There are modes of inference in mathematics, among others the so-called 'diagonal method' in set theory, which are founded on such theses not accepted in the three- and infinite-valued systems of propositional logic. It would be interesting to inquire whether mathematical theorems based on the diagonal method could be demonstrated without propositional theses such as these.

Although many-valued systems of propositional logic are merely fragments of the ordinary propositional calculus, the situation changes entirely when these systems are extended by the addition of the universal quantifier. There are theses of the extended many-valued systems which are not valid in the two-valued system. *T*13 serves as an example of such a thesis. If the expression '*Mp*' in *T*13 is replaced in accordance with *D*2 by '*CNpp*', and '*MNp*' by '*CNNpNp*', we obtain the thesis:

*T*14. *NΠpNKCNppCNNpNp*,

which is false in the two-valued calculus. The three-valued system of propositional logic with quantifiers, which owing to the research of Tarski and Wajsberg can be represented axiomatically, is the simplest example of a consistent logical system which is as different from the ordinary two-valued system as any non-Euclidean geometry is from the Euclidean.

I think it may be said that the system mentioned is the *first* intuitively grounded system differing from the ordinary propositional calculus. It was the main purpose of this communication to prove that this intuitive basis lay in the theorems I–III, which are intuitively evident for modal propositions, but which are not jointly tenable in ordinary logic. It is true that Post has investigated many-valued systems of propositional logic from a purely formal point of view, yet he has not been able to interpret them logically.[1] The well-known attempts of Brouwer, who rejects the universal validity of the law of the excluded middle and also repudiates several theses of the ordinary propositional calculus, have so far not led to an intuitively based *system*. They are merely fragments of a system whose construction and significance are still entirely obscure.[2]

It would perhaps not be right to call the many-valued systems of propositional logic established by me 'non-Aristotelian' logic, as Aristotle was the first to have thought that the law of bivalence could not be true for certain propositions. Our new-found logic might be rather termed 'non-Chrysippean', since Chrysippus appears to have been the first logician to consciously set up and stubbornly defend the theorem that every proposition is either true or false. This Chrysippean theorem has to the present day formed the most basic foundation of our entire logic.

It is not easy to foresee what influence the discovery of non-Chrysippean systems of logic will exercise on philosophical speculation. However, it seems to me that the philosophical significance of the systems of logic treated here might be at least as great as the significance of non-Euclidean systems of geometry.

APPENDIX

ON THE HISTORY OF THE LAW OF BIVALENCE

The law of bivalence, i.e. the law according to which every proposition is either true or false, was familiar to Aristotle, who explicitly characterized a proposition, 'ἀπόφανσις', as discourse which is either true or false. We read in *De interpr.* 4. 17ᵃ2: ἀποφαντικὸς δὲ (scil. λόγος; λόγος ἀποφαντικός = ἀπόφανσις) οὐ πᾶς, ἀλλ' ἐν ᾧ τὸ ἀληθεύειν ἢ ψεύδεσθαι ὑπάρχει. Aristotle, however,

---

[1] See E. L. Post, 'Introduction to a general theory of elementary propositions', *Am. Journ. of Math.* 43 (1921), p. 182: '. . . the highest dimensioned intuitional proposition space is two.'

[2] Cf., e.g., L. E. J. Brouwer, 'Intuitionistische Zerlegung mathematischer Grundbegriffe', *Jahresber. d. Deutsch. Math.-Vereinigung* 33 (1925), pp. 251 ff.; 'Zur Begründung der intuitionistischen Mathematik.' I', *Math. Ann.* 93 (1925), pp. 244 ff.

does not accept the validity of this law for propositions dealing with
contingent future events. The famous chapter 9 of *De interpretatione* is
devoted to this matter. Aristotle believes that determinism would be the
inevitable consequence of the law of bivalence, a consequence he is unable
to accept. Hence he is forced to restrict the law. He does not, however,
do this decisively enough, and for this reason his way of putting the matter
is not quite clear. The most important passage reads as follows (*De interpr.*
9. 19ᵃ36): τούτων γὰρ (scil. τῶν μὴ ἀεὶ ὄντων ἢ μὴ ἀεὶ μὴ ὄντων) ἀνάγκη μὲν θάτερον
μόριον τῆς ἀντιφάσεως ἀληθὲς εἶναι ἢ ψεῦδος, οὐ μέντοι τόδε ἢ τόδε ἀλλ᾽ ὁπότερ᾽ ἔτυχε,
καὶ μᾶλλον μὲν ἀληθῆ τὴν ἑτέραν, οὐ μέντοι ἤδη ἀληθῆ ἢ ψευδῆ. Another
passage of *De interpretatione*, viz. 18ᵇ8: τὸ γὰρ ὁπότερ᾽ ἔτυχεν οὐδὲν μᾶλλον
οὕτως ἢ μὴ οὕτως ἔχει ἢ ἕξει, allowed the Stoics to maintain that Aristotle denied
the law of bivalence. Thus we find in Boethius, *ad Arist. de interpr.*, ed.
secunda, rec. Meiser, p. 208 (ed. Bas., p. 364), the passage: 'putaverunt
autem quidam, quorum Stoici quoque sunt, Aristotelem dicere in futuro
contingentes nec veras esse nec falsas'. The Peripatetics attempted to
defend Aristotle against this objection by puzzling out a 'distinction'
between the *definite verum* and the *indefinite verum*, non-existent in the
Stagirite's works. Thus Boethius says (*ad Arist. de interpr.*, ed. prima,
rec. Meiser, p. 125): 'manifestum esse non necesse esse omnes adfirma-
tiones et negationes *definite* veras esse (sed deest "definite" atque ideo
subaudiendum est)'. The sentence in parentheses has been taken
almost literally from Greek commentators. Cf. Ammonius, *in librum Arist.
de interpr.*, ed. Busse, p. 141, 20: προσυπακουομένου δηλονότι τοῦ "ἀφωρι-
σμένως".

There can be no doubt that the Epicureans, who embraced an indeterministic *Weltanschauung*, made Aristotle's idea their own. One of the
most important passages bearing witness to this has been transmitted
to us by Cicero, *de fato* 37: 'Necesse est enim in rebus contrariis duabus
(contraria autem hoc loco ea dico, quorum alterum ait quid, alterum
negat) ex his igitur necesse est, invito Epicuro, alterum verum esse, alterum falsum: ut "sauciabitur Philocteta", omnibus ante seculis verum fuit,
"non sauciabitur", falsum. Nisi forte volumus Epicureorum opinionem
sequi, qui tales enuntiationes *nec veras nec falsas* esse dicunt: aut, cum
id pudet, illud tamen dicunt, quod est impudentius, veras esse ex contrariis disiunctiones; sed, quae in his enuntiata essent, eorum neutrum
esse verum.' Cicero opposes this opinion and then continues: 'Tenebitur
ergo id quod a Chrysippo defenditur: omnem enuntiationem aut veram
aut falsam esse.' That not only the Epicureans shared the opinion of
Aristotle, follows from a passage of Simplicius, *in Arist. cat.*, ed. Kalbfleisch, p. 406 (f. 103ᴀ ed. Bas.): ῾Ο δὲ Νικόστρατος αἰτιᾶται κἀνταῦθα λέγων
μὴ ἴδιον εἶναι τῶν κατὰ ἀντίφασιν ἀντικειμένων τὸ διαιρεῖν τὸ ἀληθὲς καὶ τὸ ψεῦδος.
... αἱ γὰρ εἰς τὸν μέλλοντα χρόνον ἐγκεκλιμέναι προτάσεις οὔτε ἀληθεῖς εἰσιν
οὔτε ψευδεῖς διὰ τὴν τοῦ ἐνδεχομένου φύσιν· οὔτε γὰρ τὸ "ἔσται ναυμαχία" ἀληθὲς
οὔτε τὸ "οὐκ ἔσται", ἀλλ᾽ ὁπότερον ἔτυχεν. The latter example is borrowed from

Aristotle's *De interpr.* 9. 19ᵃ30. For Nikostratos see Prantl, vol. i, pp. 618–20.

In conscious opposition to this, the Stoics, as outspoken determinists, and especially Chrysippus, established the law of bivalence as the fundamental principle of their dialectic. As evidence the following quotations, taken from J. v. Arnim's *Stoicorum veterum fragmenta*, vol. ii, may be cited: (1) Page 62, fr. 193: *Diocles Magnes apud Diog. Laert.* vii. 65: ἀξίωμα δέ ἐστιν ὅ ἐστιν ἀληθὲς ἢ ψεῦδος.(2) Page 63, fr. 196: Cicero, *Acad. Pr.* ii. 95: 'Fundamentum dialecticae est, quidquid enuntietur (id autem appellant ἀξίωμα—) aut verum esse aut falsum.' (3) Page 275, fr. 952: Cicero, *defato* 20: 'Concludit enim Chrysippus hoc modo: "Si est motus sine causa, non omnis enuntiatio, quod ἀξίωμα dialectici appellant, aut vera aut falsa erit; causas enim efficientis quod non habebit, id nec verum nec falsum erit. Omnis autem enuntiatio aut vera aut falsa est. Motus ergo sine causa nullus est. 21. Quod si ita est, omnia, quae fiunt, causis fiunt antegressis. Id si ita est, fato omnia fiunt. Efficitur igitur fato fieri, quaecunque fiant." . . . Itaque contendit omnis nervos Chrysippus ut persuadeat omne ἀξίωμα aut verum esse aut falsum.'

I have compiled thus many quotations on purpose, for, although they illuminate one of the most important problems of logic, it nevertheless appears that many of them were either unknown to the historians of logic, or at least not sufficiently appreciated. The reason for this is in my opinion that the history of logic has thus far been treated by philosophers with insufficient training in logic. The older authors cannot be blamed for this, as a scientific logic has existed only for a few decades. *The history of logic must be written anew*, and by an historian who has a thorough command of modern mathematical logic. Valuable as Prantl's work is as a compilation of sources and materials, from a logical point of view it is practically worthless. To give only one illustration of this, Prantl, as well as all the later authors who have written about the logic of the Stoa, such as Zeller and Brochard, have entirely misunderstood this logic. For anybody familiar with mathematical logic it is self-evident *that the Stoic dialectic is the ancient form of modern propositional logic.*[1]

Propositional logic, which contains only propositional variables, is as distinct from the Aristotelian syllogistic, which operates only with name variables, as arithmetic is from geometry. The Stoic dialectic is not a development or supplementation of Aristotelian logic, but an achievement of equal rank with that of Aristotle. In view of this it seems only fair to demand of an historian of logic that he know something about logic. Nowadays it does not suffice to be merely a philosopher in order to voice one's opinion on logic.

[1] I have already expressed this idea, in 1923, in a paper read to the first congress of Polish philosophers in Lwów. A short summary of it appeared in *Przegląd Filozoficzny* 30 (Warsaw, 1927), p. 278.

4

ON THE HISTORY OF THE LOGIC OF PROPOSITIONS†

JAN ŁUKASIEWICZ

MODERN mathematical logic has taught us to distinguish within formal logic two basic disciplines, no less different from one another than arithmetic and geometry. These are, the *logic of propositions* and the *logic of terms*. The difference between the two consists in the fact that in the logic of propositions there appear, besides logical constants, only propositional variables, while in the logic of terms term-variables occur.

The simplest way of making this difference clear is to examine the Stoic and the Peripatetic versions of the law of identity. To avoid misunderstanding let me at once say that, so far as our sources indicate, the two laws of identity were only incidentally formulated by the ancients, and in no way belong to the basic principles of either logic. The *Stoic* law of identity reads 'if the first, then the first', and is to be found as a premiss in one of the inference-schemata cited by Sextus Empiricus.[1] The *Peripatetic* law of identity is '*a* belongs to all *a*', and is not mentioned by Aristotle, but can be inferred from a passage in Alexander's commentary on the *Prior Analytics*.[2] Using variable letters we can write the Stoic law of identity in the form 'if *p*

† This paper originally appeared under the title 'Z historii logiki zdań' in *Przegląd Filozoficzny* 37 (1934), pp. 417–37. It is reprinted in a collection of Łukasiewicz's papers entitled *Z zagadnień Logiki i Filozofii*, edited by J. Słupecki, Warsaw, 1961. A German translation by the author appeared as 'Zur Geschichte der Aussagenlogik' in *Erkenntnis* 5 (1935), pp. 111–31. Translated from the German version by S. McCall.

[1] Sextus, *Adv. Math.* viii. 292 (missing in Arnim): εἰ τὸ πρῶτον, τὸ πρῶτον. Good as H. von Arnim's collection is (*Stoicorum veterum fragmenta*, vol. ii, Leipzig 1903), it does not begin to serve as source material for Stoic dialectic.

[2] Alexander, *In anal. pr. comm.*, ed. Wallies, p. 34, l. 19: γίνεται . . . τὸ *A* τινὶ τῷ *A* μὴ ὑπάρχον, ὅπερ ἄτοπον.

then p'; the Peripatetic law can be recast in the form 'all a is a'. In the first law the expression 'if . . . then' is a logical constant, and 'p' a propositional variable; only *propositions* such as 'it is day' can be meaningfully substituted for 'p'. This substitution yields a special case of the Stoic law of identity: 'if it is day, it is day'. In the second law the expression 'all . . . is' is a logical constant, and 'a' a term-variable; 'a' can be meaningfully replaced only by a term, and, in accordance with a tacit assumption of Aristotelian logic, only by a general term at that, such as 'man'. Upon substitution we get a special case of the Peripatetic law of identity: 'all man is man'. The Stoic law of identity is a thesis of the *logic of propositions*, whereas the Peripatetic law is a thesis of the *logic of terms*.

This fundamental difference between the logic of propositions and the logic of terms was unknown to any of the older historians of logic. It explains why there has been, up to the present day, no history of the logic of propositions, and, consequently, no correct picture of the history of formal logic as a whole. Indispensable as Prantl's[1] work is, even today, as a collection of sources and material, it has scarcely any value as an historical presentation of logical problems and theories. The history of logic must be written anew, and by an historian who has fully mastered mathematical logic. I shall in this short paper touch upon only three main points in the history of propositional logic. Firstly I wish to show that the Stoic dialectic, in contrast to the Aristotelian syllogistic, is the *ancient* form of propositional logic; and, accordingly, that the hitherto wholly misunderstood and wrongly judged accomplishments of the Stoics should be restored their due honour. Secondly I shall try to show, by means of several examples, that the Stoic propositional logic lived on and was further developed in *medieval* times, particularly in the theory of 'consequences'. Thirdly I think it important to establish something that does not seem to be commonly known even in Germany, namely that the founder of *modern* propositional logic is Gottlob Frege.

[1] K. Prantl, *Geschichte der Logik im Abendlande*, vols. i–iv, Leipzig, 1855–70; vol. ii, 2nd edition, Leipzig, 1885.

The Stoic law of identity mentioned above, which belongs to propositional logic, bears witness that the Stoic dialectic is a logic of propositions. However, an isolated theorem proves nothing. We shall accordingly take into consideration the well-known inference-schema which the Stoics placed at the head of their dialectic as the first 'indemonstrable' syllogism:

> If the first, then the second;
> but the first;
> therefore the second.[1]

In this formula the words 'the first' and 'the second' are *variables*, for the Stoics denote variables not with letters, but with ordinal numbers.[2] It is clear that here too only *propositions* may be meaningfully substituted for these variables; e.g. 'it is day', and 'it is light'. When this substitution is made we get the inference which occurs again and again as a school example in Stoic texts: 'If it is day, then it is light; but it is day; therefore it is light.' That indeed *propositions* and not *terms* are to be substituted for the variables in the above formula is not only evident from its sense, but is clearly implied by the following example: 'If Plato lives, then Plato breathes; but the first; therefore the second.' Here 'the first' plainly refers to the *proposition* 'Plato lives', and 'the second' to the *proposition* 'Plato breathes'.[3]

The fundamental difference between Stoic and Aristotelian logic does not lie in the fact that hypothetical and disjunctive propositions occur in Stoic dialectic, while in Aristotelian syllogistic only categorical propositions appear. Strictly speaking, hypothetical propositions can be found in Aristotle's syllogistic also, for each proper Aristotelian syllogism is an *implication*, and hence a hypothetical proposition. For example, 'If a belongs to all b and c belongs to all a, then c belongs to all b'.[4]

---

[1] Sextus, *Adv. math.* viii. 227 (Arnim, ii. 242, p. 81, l. 22): εἰ τὸ πρῶτον, τὸ δεύτερον· τὸ δέ γε πρῶτον· τὸ ἄρα δεύτερον.

[2] Apuleius, *De interpr.* 279 (Arnim, ii, p. 81 note): 'Stoici porro pro litteris numeros usurpant, ut "si primum, secundum; atqui primum; secundum igitur".'

[3] Diogenes Laert. vii. 76 (quoted in Prantl, i, p. 471, note 177; missing in Arnim): εἰ ζῇ Πλάτων, ἀναπνεῖ Πλάτων· ἀλλὰ μὴν τὸ πρῶτον· τὸ ἄρα δεύτερον.

[4] Aristotle, *An. pr.* ii. 11. 61ᵇ34: εἰ γὰρ τὸ Α παντὶ τῷ Β καὶ τὸ Γ παντὶ τῷ Α, τὸ Γ παντὶ τῷ Β (sc. ὑπάρχει).

The main difference between the two ancient systems of logic lies rather in the fact that in the Stoic syllogisms the variables are *propositional variables*, while in Aristotle's they are *term variables*. This crucial difference is completely obliterated, however, if we translate the above-mentioned Stoic syllogism as Prantl does (i, p. 473):

> If the first is, the second is
> But the first is
> ─────────────────────
> Therefore the second is.

By adding to each variable the little word 'is', which occurs *nowhere* in the ancient texts, Prantl, without knowing or wishing it, falsely converts Stoic propositional logic into a logic of terms. For in Prantl's schema only terms, not propositions, can be meaningfully substituted for 'the first' and 'the second'. As far as we can judge from the fragmentary state of the Stoic dialectic that has come down to us, *all* Stoic inference-schema contain, besides logical constants, only propositional variables. Stoic logic is therefore a logic of propositions.[1]

There is yet a second important difference between the Aristotelian and the Stoic syllogisms. Aristotelian syllogisms are logical theses, and a *logical thesis* is a proposition which contains, besides logical constants, only propositional or term variables, and which is true for all values of its variables. Stoic syllogisms are inference-schemata, in the sense of rules of inference, and a *rule of inference* is a prescription empowering the reasoner to derive new propositions from ones already admitted. We should examine this difference somewhat more closely.

The Aristotelian syllogism quoted above, which can also be written 'if all *b* is *a* and all *a* is *c*, then all *b* is *c*', is an implication of the form 'if α and β, then γ', whose antecedent is a conjunction of the premisses α and β, and whose consequent is the

---

[1] I have defended this interpretation of the Stoic dialectic since 1923; see Łukasiewicz, 'Philosophische Bemerkungen zu mehrwertigen Systemen des Aussagenkalküls', *Comptes rendus des séances de la Société des Sciences et des Lettres de Varsovie* 23 (1930), Cl. iii, p. 77. [Paper 3 of this volume—Ed.] I rejoice in having found in H. Scholz, *Geschichte der Logik* (Berlin, 1931), p. 31, a supporter of this point of view.

conclusion γ. As an implication, this syllogism is a *proposition* which Aristotle recognizes as true; one that does indeed hold for all values of its variables 'a', 'b', and 'c'. If constant values are substituted for these variables, we get true *propositions*. Inasmuch as the syllogism in question contains, besides variables, only the logical constants 'if . . . then', 'and', and 'all . . . is', it is, like all other Aristotelian syllogisms, a logical thesis.

It is otherwise in Stoic logic. The Stoic syllogism given above, which with the help of letters can be written 'if p, then q; but p; therefore q', consists, as does the Aristotelian syllogism, of two premisses and a conclusion. But here the premisses are *not* bound up together with the conclusion in a single unified proposition. This is plain from the word 'therefore' which introduces the conclusion. The syllogism in question is consequently *not* a proposition. Since it is not a proposition, it can be neither true nor false; for it is acknowledged that truth and falsehood belong to propositions alone. Hence the Stoic syllogism is *not* a logical thesis: if constant values are substituted for its variables the result is not a proposition, but an *inference*. The syllogism is accordingly an inference-schema, having the force of a *rule of inference* which can be more accurately expressed in the following way: whoever accepts as true both the implication 'if p, then q' and its antecedent 'p', also has the right to accept as true the consequent 'q' of this implication—i.e. to detach 'q' from 'p'. This rule of inference, under the name of the 'rule of detachment', has become almost a classic in modern logic.

All Stoic syllogisms are formulated as *rules of inference*. In this way Stoic dialectic differs not only from Aristotelian syllogistic, but also from modern propositional logic, which is a system of *logical theses*.

However, the Stoics were acquainted with a clear and simple method of converting all their rules of inference into theses. This involves a distinction between binding and non-binding inferences. An inference with premisses α and β and conclusion γ they call *binding* [*bündig*], if the *implication*, whose antecedent is the conjunction of the two premisses α and β and whose consequent is the conclusion γ, is valid. For example, the following inference is binding: 'if it is day, then it is light; but it is day; therefore it is light', for the corresponding implication is correct: 'if it is day and if it is day then it is light, then it is light'.[1]

This just observation makes possible the conversion of inferences into propositions. When it is applied to the rule of inference 'if p then q; but p; therefore q' we obtain the implication 'if p and if p then q, then q', which is a thesis of propositional logic, since besides propositional variables only the logical constants 'if . . . then' and 'and' occur in it.

It is not possible for me to go into all the details of Stoic logic here. I wish only to comment upon the most important points. The Stoic logic of propositions is a *two-valued* logic. In it the basic principle holds, that every proposition is *either true or false*, or, as we say today, can take one of only *two* possible 'truth-values', 'the true' or 'the false'.[2] This principle is laid down in conscious opposition to the view that there are propositions which are *neither true nor false*, namely those which treat of future contingent events. This view, which was particularly widespread among the Epicureans, was also ascribed by the Stoics to Aristotle.[3]

In Stoic propositional logic the following functions occur: negation, implication, conjunction, and disjunction. The first three functions are defined, as is normally said nowadays, as 'truth-functions'. By a *truth-function* is meant a function whose

---

[1] Sextus, *Hyp. pyrrh.* ii. 137 (missing in Arnim, who nevertheless in ii. 239, p. 78, l. 15, quotes the parallel passage from *Adv. math.* viii. 415 (416)): ἐν τούτῳ τῷ ⟨λόγῳ⟩ "εἰ ἡμέρα ἔστι, φῶς ἔστιν· ἀλλὰ μὴν ἡμέρα ἔστιν· φῶς ἄρα ἔστιν" τὸ μὲν "φῶς ἄρα ἔστιν" συμπέρασμά ἐστι, τὰ δὲ λοιπὰ λήμματα. τῶν δὲ λόγων οἱ μέν εἰσι συνακτικοὶ οἱ δὲ ἀσύνακτοι, συνακτικοὶ μέν, ὅταν τὸ συνημμένον τὸ ἀρχόμενον μὲν ἀπὸ τοῦ διὰ τῶν τοῦ λόγου λημμάτων συμπεπλεγμένου, λῆγον δὲ εἰς τὴν ἐπιφορὰν αὐτοῦ, ὑγιὲς ᾖ, οἷον ὁ προειρημένος λόγος συνακτικός ἐστιν, ἐπεὶ τῇ διὰ τῶν λημμάτων αὐτοῦ συμπλοκῇ ταύτῃ "ἡμέρα ἔστι καὶ εἰ ἡμέρα ἔστι, φῶς ἔστιν" ἀκολουθεῖ τὸ "φῶς ἔστιν" ἐν τούτῳ τῷ συνημμένῳ "εἰ ἡμέρα ἔστι καὶ εἰ ἡμέρα ἔστι, φῶς ἔστι, φῶς ἔστιν".

[2] Cicero, *Acad. pr.* ii. 95 (Arnim, ii. 196, p. 63): 'Fundamentum dialecticae est, quidquid enuntietur, id autem appellant ἀξίωμα . . ., aut verum esse aut falsum.'

[3] Boethius, *Ad Arist. de interpr.* ed. secunda, Meiser, p. 208 (missing in Arnim): 'Putaverunt autem quidam, quorum Stoici quoque sunt, Aristotelem dicere in futuro contingentes nec veras esse nec falsas.' See on this matter my earlier paper cited above, pp. 75 ff. [This volume, pp. 63–65.—Ed.]

arguments are propositions, and whose truth-value depends only on the truth-value of its arguments.

According to the Stoics one obtains the *negation* or the contradictory of a proposition when the sign of negation is placed *in front of* the proposition.[1] This theoretically correct and practically valuable rule continues to be operative in the Middle Ages.[2] It is universally recognized in modern logic.

There are many disputes in ancient times over the meaning of the implication 'if *p* then *q*'.[3] The argument seems to have been started by Philo the Megarian, who was the first to define implication as a truth-function in much the same way as is done today. According to Philo an implication is true if and only if it does not begin with truth and end with falsehood. An implication is accordingly true in three cases: firstly, if its antecedent and its consequent are both true; secondly, if its antecedent and its consequent are both false; and thirdly, if the antecedent is false and the consequent true. Only in *one* case is the implication false, namely when the antecedent is true and the consequent false.[4] Another Megarian, Diodorus Cronus, maintained on the other hand that an implication is true if and only if it neither *was* nor *is possible* for it to begin with truth and end with falsehood.[5] This ancient dispute concerning the concept of implication, immortalized by Callimachus in an epigram ('Even the ravens on the roof tops are croaking about which conditionals are true'),[6] is reminiscent of the

---

[1] Apuleius, *De interpr.* 266 (Arnim, ii. 204a, p. 66): 'Solum autem abdicativum vocant, cui negativa particula praeponitur.' The word 'οὐχί' serves as the sign of propositional negation.

[2] See note 3, p. 81.

[3] Cicero, *Acad. pr.* ii. 143 (Arnim, ii. 285, p. 93): 'In hoc ipso, quod in elementis dialectici docent, quo modo iudicare oporteat, verum falsumne sit, siquid ita conexum est, ut hoc: "si dies est, lucet", quanta contentio est. Aliter Diodoro, aliter Philoni, Chrysippo aliter placet.'

[4] Sextus, *Adv. math.* viii. 113: ὁ μὲν Φίλων ἔλεγεν ἀληθὲς γίνεσθαι τὸ συνημμένον (= implication), ὅταν μὴ ἄρχηται ἀπ᾽ ἀληθοῦς καὶ λήγῃ ἐπὶ ψεῦδος, ὥστε τριχῶς μὲν γίνεσθαι κατ᾽ αὐτὸν ἀληθὲς συνημμένον, καθ᾽ ἕνα δὲ τρόπον ψεῦδος. There follows the enumeration of all four cases with examples.

[5] Sextus, *Adv. math.* viii. 115: Διόδωρος δὲ ἀληθὲς εἶναι φησι συνημμένον ὅπερ μήτε ἐνεδέχετο μήτε ἐνδέχεται ἀρχόμενον ἀπ᾽ ἀληθοῦς λήγειν ἐπὶ ψεῦδος.

[6] Sextus, *Adv. math.* i. 309: τὸ ὑπὸ τοῦ Καλλιμάχου εἰς Διόδωρον τὸν Κρόνον συγγραφέν (sc. ἐπιγραμμάτιον)· ἠνὶ δέ κου κόρακες τεγέων ἔπι κοῖα συνῆπται κρώζουσι. . . .

polemic waged by one of the modern followers of Diodorus, C. I. Lewis, against the other advocates of mathematical logic.[1] In the Stoic school, Philo's definition was accepted. At least, Sextus ascribes this concept directly to the Stoics.[2]

The *conjunction* '*p* and *q*' is defined by the Stoics as a truth-function. It is true if and only if both its members are true; otherwise it is false.[3]

An analogous definition of the *disjunction* '*p* or *q*' does not occur in the fragments of Stoic logic which have come down to us. We gather, from the rules of inference for disjunction laid down by Chrysippus, that he considered disjunction as an exclusive 'either-or' connective. Thus according to Chrysippus the two members of a true disjunction cannot both be true at the same time. This seems to have been changed later. The conviction arises that the expression '*p* or *q*' is synonymous with the implication 'if not *p* then *q*'.[4] In this case we would no longer be dealing with exclusive disjunction, but with non-exclusive *alternation*. In the Middle Ages, as we shall see later, the non-exclusive character of disjunction comes clearly to light.

All the above-mentioned logical functions are to be found in the inference-schemata of Stoic dialectic. Of these inference-schemata, some are considered to be 'indemonstrable', that is to say accepted *axiomatically* as correct, while the others are reduced to the indemonstrable ones. It is Chrysippus who is

---

[1] Being of the opinion that the concept of 'material implication', which comes from Philo, leads to paradoxes, such as 'a false proposition implies any proposition', and 'a true proposition is implied by any proposition' (compare the passage from Duns Scotus in note 1 of p. 83 below), Lewis wishes to replace 'material implication' by 'strict implication', the latter being defined in the following way. '*p* implies *q*' or '*p* strictly implies *q*' is to mean 'it is false that it is possible that *p* should be true and *q* false'. See C. I. Lewis and C. H. Langford, *Symbolic Logic*, New York and London, 1932, pp. 122 and 124.

[2] *Hyp. pyrrh.* ii. 104, and *Adv. math.* viii. 245 (Arnim, ii. 221, p. 72, l. 32). Cf. also Diogenes Laert. vii. 81 (Arnim, ii. 243, p. 81).

[3] Sextus, *Adv. math.* viii. 125 (Arnim, ii. 211, p. 69): (λέγουσιν) ὑγιὲς εἶναι συμπεπλεγμένον (= conjunction) τὸ πάντ' ἔχον ἐν αὑτῷ ἀληθῆ, οἷον τὸ "ἡμέρα ἐστι καί φῶς ἐστιν", ψεῦδος δὲ τὸ ἔχον ψεῦδος.

[4] Galen, *Institutio Logica*, ed. Kalbfleisch, p. 9, l. 13: τὸ τοιοῦτον εἶδος τῆς λέξεως "εἰ μὴ νύξ ἐστιν, ἡμέρα ἐστίν" διεζευγμένον ἐστὶν ἀξίωμα τῇ φύσει τῶν πραγμάτων αὐτῇ, συνημμένου δὲ ἰδέαν ἔχει τῇ λέξει. Exclusive disjunction is called "διεζευγμένον". For non-exclusive alternation Galen uses the expression "παραδιεζευγμένον" (p. 35. l. 6).

supposed to have laid down the indemonstrable inference-schemata or syllogisms. These consist of the following five (in which I denote the variables not by ordinal numerals, but by letters):

I. If p then q; but p; therefore q.

II. If p then q; but not-q; therefore not-p.

III. Not both p and q; but p; therefore not-q.

IV. Either p or q; but p; therefore not-q.

V. Either p or q; but not-q; therefore p.[1]

It is apparent from the fourth syllogism that disjunction is conceived of as an exclusive 'either-or' connective. For non-exclusive alternation this syllogism is not valid.[2]

The reduction of the derived inference-schemata to the indemonstrables is a masterpiece of logical acumen. The source of our information in this matter is Sextus, who thoroughly understands the dialectical technique of the Stoics and must be considered among the best sources of Stoic logic. With a clarity that leaves nothing to be desired he informs us, for example, how the Stoics reduced the inference-schema 'if p and q, then r; not-r, but p; therefore not-q' to the second and third indemonstrable syllogisms. From the premisses 'if p and q, then r' and 'not-r' we get, using the second syllogism, the conclusion 'not both p and q'. This conclusion and the remaining premiss 'p' yield, by the third syllogism, 'not-q'.[3]

---

[1] Galen, *Inst. log.*, ed. Kalbfleisch, p. 15 (Arnim, ii. 245, p. 82): ὃν ὁ Χρύσιππος ὀνομάζει πρῶτον ἀναπόδεικτον, ὁ τοιοῦτος τρόπος ἐστίν. "εἰ τὸ α΄, τὸ β΄· τὸ δὲ α΄· τὸ ἄρα β΄." ὃν ὁ Χρύσιππος δεύτερον ἀναπόδεικτον ὀνομάζει, τοιοῦτός ἐστιν· "εἰ τὸ α΄, τὸ β΄· οὐχὶ δὲ τὸ β΄· οὐκ ἄρα τὸ α΄".—Κἀπὶ τοῦ τρίτου κατὰ τοῦτον . . . τοιοῦτος ὁ τρόπος ἐστίν· "οὐχὶ τό τε α΄ καὶ τὸ β΄· ⟨τὸ δὲ α΄· οὐκ ἄρα τὸ β΄⟩".—κἀπὶ τοῦ τετάρτου κατὰ τὸν αὐτόν . . . τοιοῦτός τις ὁ τρόπος ἐστίν. "ἤτοι τὸ α΄ ἢ τὸ β΄· τὸ δὲ α΄· οὐκ ἄρα τὸ β΄".—κἀπὶ τοῦ πεμπτου . . . τοιοῦτός ἐστιν ὁ τρόπος. "ἤτοι τὸ α΄ ἢ τὸ β΄· ⟨οὐχὶ δὲ τὸ β΄· τὸ ἄρα α΄⟩". Cf. also Arnim, ii. 241 and 242, pp. 79–81.

[2] Prantl, who actually hates Stoic logic, writes on this matter as follows (i, p. 474): 'Here the enormous stupidity of the distinction between the moods IV and V does not have to be especially remarked upon.' It is disgraceful to encounter such an assertion in a learned work, particularly as it rests upon ignorance of logic. Prantl further supposes that Chrysippus took the five syllogisms from Theophrastus, and 'anyone who copies completely unfamiliar material thereby runs the risk of only displaying his own ignorance'. Herein lies another historical error. It cannot be shown from our sources that Theophrastus constructed or even knew of the above-mentioned syllogisms.

[3] Sextus, *Adv. math.* viii. 235, 236 (missing in Arnim). The inference-schema

Another example given by Sextus, in which the first syllogism is used twice, remained unintelligible to Prantl. The example, in general form, reads: 'if p, then if p then q; but p; therefore q'. The reduction proceeds as follows. From the premisses 'if p, then if p then q' and 'p' we get, by the first syllogism, the conclusion 'if p then q'. From this conclusion and the premiss 'p' we obtain, again by the first syllogism, 'q'.[1] The inference-schema dealt with here is most interesting: it corresponds to a thesis of the propositional calculus that was recently raised to the rank of an axiom by Hilbert and Bernays.[2]

The number of derived inference-schemata is supposed to have been very great.[3] Of those that have come down to us the following syllogism, quoted by Origen, merits our attention: 'If p then q; if p then not-q; therefore not-p.' The example of it, given in addition, is also very interesting: 'If you know that you are dead, then you are dead (for nothing false can be known); if you know that you are dead, then you are not dead (for the dead know nothing); therefore you do not know that you are dead.'[4] The above passage from Origen is also important in

reads:"εἰ τὸ πρῶτον καὶ τὸ δεύτερον, τὸ τρίτον· οὐχὶ δέ γε τὸ τρίτον, ἀλλὰ καὶ τὸ πρῶτον· οὐκ ἄρα τὸ δεύτερον." To the end of the reduction it runs: ὥστε δύο εἶναι ἀναποδείκτους, ἕνα μὲν τοιοῦτον· "εἰ τὸ πρῶτον καὶ τὸ δεύτερον, τὸ τρίτον· οὐχὶ δέ γε τὸ τρίτον· οὐκ ἄρα τὸ πρῶτον καὶ τὸ δεύτερον", ὅς ἐστι δεύτερος ἀναπόδεικτος, ἕτερον δὲ τρίτον τὸν οὕτως ἔχοντα· "οὐχὶ τὸ πρῶτον καὶ τὸ δεύτερον· ἀλλὰ μὴν τὸ πρῶτον· οὐκ ἄρα τὸ δεύτερον".

[1] Sextus, *Adv. math.* viii. 230–3 (missing in Arnim). The text is corrupt, although unambiguously clear. It was corrected first by E. Kochalsky in his dissertation *De Sexti Empirici adversus logicos quaestiones criticae*, Marburg, 1911, pp. 83–85. Nevertheless he finishes his corrections with an appeal to Zeller and Prantl in the following way: 'Nimirum huiusmodi argumentum non simplex indemonstrabile per se est absurdissimum, sed Stoicos in syllogismis inveniendis incredibilia paene gessisse inter omnes constat.' One sees from this how pernicious Prantl's influence was.

[2] Hilbert and Bernays, *Grundlagen der Mathematik* vol. i, Berlin, 1934, p. 66. The thesis in question is, in words, 'if [if p, then (if p then q)] then (if p then q)'.

[3] Cicero, *Topica* 14, 57 (quoted by Zeller, *Die Philosophie der Griechen*, iii. 1, 5th edition 1923, p. 114, note 1; missing in Arnim): 'ex iis modis conclusiones innumerabiles nascuntur.'

[4] Origen, *Contra Celsum*, vii. 15 (*Works*, vol. ii, ed. Koetschau, 1899, p. 166, missing in Arnim): ὅταν δὲ δύο συνημμένα λήγῃ εἰς τὰ ἀλλήλοις ἀντικείμενα τῷ καλουμένῳ "διὰ δύο τροπικῶν" θεωρήματι ἀναιρεῖται τὸ ἐν ἀμφοτέροις τοῖς συνημμένοις ἡγούμενον . . . καὶ ὑπάγεταί γε ὁ λόγος τρόπῳ τοιούτῳ· "εἰ τὸ πρῶτον, καὶ τὸ δεύτερον· εἰ τὸ πρῶτον, οὐ τὸ δεύτερον· οὐκ ἄρα τὸ πρῶτον". φέρουσι δὲ καὶ ἐπὶ ὕλης τὸν τρόπον τοῦτον οἱ ἀπὸ τῆς Στοᾶς λέγοντες τό· "εἰ ἐπίστασαι ὅτι τέθνηκας, ⟨τέθνηκας·

that it gives us information about the meaning of a hitherto erroneously interpreted expression of Stoic dialectic.[1]

In connexion with the Stoic logic of propositions I would like to touch upon one or two questions of a general nature. The Stoics are constantly criticized for the fact that in their logic the most trivial empiricism as well as the most empty formalism appears. Thus Prantl (i, p. 457) says, in citing the examples given by the Stoics for implication, that these are 'examples, from which it is sufficiently obvious both that the crudest empirical criterion is displayed, and that there is total lack of any understanding of the causal nexus between essences and inherences'. Prantl's unfavourable judgement is not justified. If empirical examples are given for logical formulae, the criterion of truth for these examples must also be in some way empirical. However, the examples do not belong to logic, and in Stoic logic itself we do not find the slightest trace of empiricism. When it is asserted that the Stoics lacked an understanding of the causal nexus, we may conclude only that Prantl fails to grasp the Philonian concept of implication accepted by the Stoa. In two-valued logic there *can* be no other concept of implication than the Philonian. This has nothing to do with either empiricism or the causal nexus, for the expression 'if p then q' does not mean the same as 'q follows from p'.

The accusation of formalism, which was often made even in ancient times,[2] is quite justified, only in our eyes it is not an accusation at all. Formalism, or better *formalization*, means the ideal of exactitude that each deductive system strives to attain. We say that a deductive, axiomatically constructed system is

εἰ ἐπίστασαι ὅτι τέθνηκας⟩, οὐ τέθνηκας". ἀκολουθεῖ τό· "οὐκ ἄρα ἐπίστασαι, ὅτι τέθνηκας".

[1] Neither Prantl nor Zeller knows the passage, although Fabricius had already referred to it (*Sexti Empirici Opera*, 2nd edition 1840, vol. i, p. 112). The expression in question is "διὰ δύο τροπικῶν" ("τροπικόν" is a non-simple premiss, e.g. an implication). It is wrongly interpreted by Prantl (i, p. 480) and Zeller (iii. 1, pp. 114–15 note 5) as meaning a syllogism in which two τροπικά, in this case two implications, occur as premisses.

[2] Galen, *Inst. log.*, ed. Kalbfleisch, p. 11, l. 6 (Arnim, ii. 208, p. 69, l. 4): ἀλλ' οἱ περὶ Χρύσιππον κἀνταῦθα τῇ λέξει μᾶλλον ἢ τοῖς πράγμασι προσέχοντες τὸν νοῦν. . . .

formalized when the correctness of the deductions in the system can be verified without having to refer back to the *meaning* of the expressions and symbols used in the deductions. They may be verified, that is, by anyone who understands the rules of inference of the system. In this sense the Stoics prepared the way for formalism, and they cannot be credited highly enough for that. They held strictly to *words* and not to their *meanings*, which is the principal requirement of formalization, and they even did so in conscious opposition to the Peripatetics. Alexander occasionally expresses the opinion that the essence of the syllogism lies not in words but in what the words mean.[1] The Stoics would undoubtedly maintain the opposite. For in spite of the fact that, for example, they took the expressions 'if *p* then *q*', and '*q* follows from *p*' to be synonymous (which is incorrect), they did *not* describe the inference-schema '*q* follows from *p*; but *p*; therefore *q*' as a syllogism, although the following schema, which in their opinion is synonymous with it, *is* a syllogism: 'if *p* then *q*; but *p*; therefore *q*'.[2]

In connexion with this controversy between the Stoic and the Peripatetic schools, we are ultimately confronted with the question, whether the Stoics understood anything about the meaning in principle of their propositional logic, and, in particular, whether they were aware of having created a system of logic different from Aristotle's. Scholz believes that we must answer the first part of this question in the negative.[3] For the second part of the question we have at our disposal two hitherto little-noticed accounts.

In his commentary on Aristotle's *Topics* Alexander enumerates, under the heading 'syncritical problems', certain controversial questions discussed in ancient times—as, for example,

---

[1] Alexander, *In anal. pr. comm.*, ed. Wallies, p. 372, l. 29: οὐκ ἐν ταῖς λέξεσιν ὁ συλλογισμὸς τὸ εἶναι ἔχει ἀλλ' ἐν τοῖς σημαινομένοις.

[2] Alexander, op. cit., p. 373, l. 29 (Arnim, ii. 253, p. 84): οἱ δὲ νεώτεροι ταῖς λέξεσιν ἐπακολουθοῦντες οὐκέτι δὲ σημαινομένοις, οὐ ταὐτόν φασι γίνεσθαι ἐν ταῖς εἰς τὰς ἰσοδυναμούσας λέξεις μεταλήψεσι τῶν ὅρων· ταὐτὸν γὰρ σημαίνοντος τοῦ "εἰ τὸ A, τὸ B" τῷ "ἀκολουθεῖ τῷ A τὸ B" συλλογιστικὸν μὲν λόγον φασὶν εἶναι τοιαύτης ληφθείσης τῆς λέξεως· "εἰ τὸ A, τὸ B· τὸ δὲ A· τὸ ἄρα B", οὐκέτι δὲ συλλογιστικὸν ἀλλὰ περαντικὸν τὸ "ἀκολουθεῖ τῷ A τὸ B· τὸ δὲ A· τὸ ἄρα B".

[3] *Geschichte der Logik*, p. 32.

whether the moon is bigger than the earth, or whether surgical treatment is to be preferred to medical. In so doing he also mentions the following comparable problems from logic: 'whether induction is more convincing than the syllogism; and which syllogism is the first, the categorical or the hypothetical; and which syllogistic figure is the first or the better'.[1]

It is the *second* question that interests us here: which syllogism is the first, the categorical or the hypothetical. Now the categorical syllogism is the Aristotelian, the hypothetical, the Stoic. Our controversy accordingly concerns the relation of Aristotelian to Stoic logic, and aims at establishing which of these systems is the *first*, i.e., as I understand it, the *logically prior*.

An answer to this question is to be found in the highly interesting introduction to logic written by Galen. Galen reports that Boethius, who according to Ammonius was the eleventh head of the Peripatetic school after Aristotle, and who was reckoned one of the most acute logicians of his time, himself considered, although he was a Peripatetic, the hypothetical and not the categorical syllogisms to be the first. Against this Galen raises the objection that the categorical premisses, as simple propositions, are logically prior to the hypothetical ones constructed out of them. However, he does not appear to attach any great importance either to this argument or to the whole controversy, for he thinks that there is not much to be gained or lost in the dispute. One should become as familiar with the one kind of syllogism as with the other, but in what order this should take place, or which of them should be referred to as primary, may be left to one's own discretion.[2]

---

[1] Alexander, *In top. comm.*, ed. Wallies, p. 218: ἔστιν ἔτι καὶ ἐν τῇ λογικῇ συγκριτικῶς τινα ζητούμενα, ὡς τὸ πότερον πειστικώτερον, ἐπαγωγὴ ἢ συλλογισμός, καὶ ποῖος πρῶτος συλλογισμός, ὁ κατηγορικὸς ἢ ὁ ὑποθετικός, καὶ ποῖον σχῆμα πρῶτον ἢ βέλτιον.

[2] Galen, *Inst. log.*, ed. Kalbfleisch, p. 17; καὶ μέντοι καὶ τῶν ἐκ τοῦ Περιπάτου τινὲς ὥσπερ καὶ Βοηθὸς οὐ μόνον ἀναποδείκτους ὀνομάζουσιν τοὺς ἐκ τῶν ἡγεμονικῶν λημμάτων συλλογισμούς, ἀλλὰ καὶ πρώτους· ὅσοι δὲ ἐκ κατηγορικῶν προτάσεων εἰσὶν ἀναπόδεικτοι συλλογισμοί, τούτους οὐκ ἔτι πρώτους ὀνομάζειν συγχωροῦσι· καίτοι καθ' ἕτερόν γε τρόπον οἱ τοιοῦτοι πρότεροι τῶν ὑποθετικῶν εἰσιν, εἴπερ γε καὶ αἱ προτάσεις αὐτῶν ἐξ ὧν σύγκεινται πρότεραι βεβαίως εἰσίν· οὐδεὶς γὰρ ἀμφισβητήσει τὸ μὴ οὐ πρότερον εἶναι τὸ ἁπλοῦν τοῦ συνθέτου. ἀλλὰ περὶ μὲν τῶν τοιούτων ἀμφισβητήσεων οὔτε εὑρεῖν οὔτε ἀγνοῆσαι μέγα· χρὴ γὰρ ἀμφότερα τὰ μέρη γιγνώσκειν τῶν συλλογισμῶν,

From these two fragments we may, in my opinion, conclude not only that the Stoics were aware of the difference between their own logical system and the Aristotelian, but also that they correctly judged the relation between the two systems. We know today that propositional logic is logically prior to the logic of terms. If we analyse the proofs that Aristotle uses in the *Analytics* to reduce syllogisms of the second and third figures to syllogisms of the first figure, we see clearly that theses of propositional logic must be employed throughout. The syllogism which later received the name 'Baroco' cannot be formally reduced to 'Barbara' without the propositional thesis 'if (if p and q, then r), then (if p and not-r, then not-q'). Now this thesis corresponds to an inference-schema which, as we saw above, was well known to the Stoics. It is highly probable that the application of this inference-schema to Aristotle's syllogisms did not escape the Stoics. We know also that the logic of propositions is of far greater importance than the meagre fragment of the logic of terms that is incorporated in Aristotle's syllogistic. The logic of propositions is the basis of all logical and mathematical systems. We must be thankful to the Stoics for having laid the foundations of this admirable theory.

A great deal about how Stoic influences continue to be at work in the Middle Ages may be found in Prantl. That, however, the *propositional logic* created by them undergoes a further development in that period seems to have been realized by no one up to now. Once again it is not possible for me to go into details here, especially since the sources for medieval logic are not easily accessible. I shall in what follows merely give a short account of what is to be found of propositional logic in the *Summulae logicales* of Petrus Hispanus, that classic manual of medieval logic, together with the commentary on it by Versorius; as well as what can be found in the writings of the subtle Duns Scotus. The Philonian criterion of a true implication, already disputed in ancient times, seems not to have been known to Petrus Hispanus. To make up for this there appears in his

καὶ τοῦτ᾽ ἔστι τὸ χρήσιμον, ὀνομάζειν δὲ τοὺς ἑτέρους ἢ διδάσκειν προτέρους ὡς ἑκάστῳ φίλον.

work, under the name of *disjunction* and replacing Chrysippus'
'either-or' connective, non-exclusive *alternation* as a truth-
function.[1] We learn that a disjunction, i.e. the joining together
of two propositions by means of the connective *'vel'*, is false if
and only if its two members are false. Otherwise it is true, even
when *both* its members are true—though this was admitted with
a certain reluctance.[2]

In the commentary the two following rules of inference are
laid down for disjunction. Firstly, from a disjunction and the
negation of one member the other member may be inferred; e.g.
'Man is an animal or a horse is a stone; but a horse is not a stone;
therefore man is an animal.' This is precisely the fifth indemonstrable syllogism of the Stoics; the fourth is, of course, missing,
since it is valid only for exclusive disjunction. Secondly, from
the truth of one member the truth of the disjunction may be
inferred; e.g. 'Man runs, therefore man is an ass or man runs'.[3]
The examples are grotesque, but none the less clear enough.
The second rule is new, not occurring in the Stoic texts. Moreover, it is correct only on condition that disjunction is taken as
non-exclusive alternation.

Conjunction, which here bears the name of *copulative* assertion, is defined by Petrus Hispanus as a truth-function, just as
it was by the Stoics. The only rule of inference which seems to

---

[1] Prantl (iii, p. 43) has nothing to report on this, for he is not aware of the
difference between disjunction and alternation.

[2] *Summulae*, tract. i, *De disiunctiva* (quoted only in abridged form in
Prantl iii, p. 43, note 158; I quote from a comparatively later edition, *Petri
Hispani Summulae Logicales cum Versorii Parisiensis clarissima expositione,
Venetiis* 1597 *apud Matthaeum Valentinum*, which differs from the text quoted
by Prantl in various places): 'Disiunctiva est illa, in qua coniunguntur duae
propositiones categoricae per hanc coniunctionem "vel" aut aliam sibi aequi-
valentem, ut "Socrates currit vel Plato disputat". Ad veritatem disiunctivae
sufficit, alteram partem esse veram, ut "homo est animal vel equus est asinus",
tamen permittitur, quod utraque pars eius sit vera, sed non ita proprie, ut
"homo est animal vel equus est hinnibilis". Ad falsitatem eius oportet,
utramque partem eius esse falsam, ut "homo est asinus vel equus est lapis".

[3] *Summulae*, loc. cit.: 'dupliciter arguitur a disiunctivis. Uno modo, a tota
disiunctiva cum destructione unius partis ad positionem alterius, ut "homo
est animal vel equus est lapis; sed equus non est lapis, igitur homo est animal".
Secundo modo, arguendo a veritate unius partis ad veritatem totius, et est bona
consequentia, unde bene sequitur, haec est vera "homo currit", igitur haec
est vera "homo est asinus vel homo currit".'

be new is one which is added in the commentary: from a conjunction each of its members may be inferred; e.g. 'Man is an
animal and God exists, therefore man is an animal.'[1]

In this connexion we find in the commentary on Petrus
Hispanus the following beautiful remark: a conjunction and
a disjunction with mutually contradictory members contradict
one another.[2] That is to say, the following propositions stand
in contradiction to one another: 'p and q' and 'not-p or not-q',
as well as 'p or q' and 'not-p and not-q'. In other words, 'p
and q' is equivalent to the negation of 'not-p or not-q', and 'p
or q' to the negation of 'not-p and not-q'. From which it follows
that the so-called De Morgan's laws were known long before
De Morgan.

Finally we read, at the same place in the commentary, that
the contradictory opposite of a proposition cannot be 'more
truly' formed than by *prefixing* the negation sign to the proposition.[3] Here the Stoic influence, which we mentioned above,
emerges particularly clearly. All the above rules of inference,
together with the last remark, are also found in Duns Scotus.
It seems therefore that they were generally recognized in the
Middle Ages.

The survival of Stoic propositional logic in the Middle Ages
is particularly evident in the theory of 'consequences'. By a
consequence the medieval logicians understand not only an implication, but also an inference-schema of the type 'p, therefore q',
in which 'p' and 'q' are propositions. As a rule, however, consequences are represented as *inference-schemata*.[4] Consequences

---

[1] *Summulae*, tract. i, *De copulativa*: 'arguendo a tota copulativa ad veritatem cuiuslibet partis eius seorsum, est bona consequentia. Ut bene sequitur
"homo est animal et Deus est, ergo homo est animal".

[2] *Summulae*, loc. cit.: 'copulativa et disiunctiva de partibus contradicentibus
contradicunt.' The same thought, which is here stated somewhat too concisely,
is expressed by Occam much more clearly: 'Opposita contradictoria disiunctivae
est una copulativa composita ex contradictoriis partium ipsius disiunctivae.'
(See Prantl, iii, p. 396, note 958.)

[3] *Summulae*, loc. cit.: 'non est verius dare contradictionem, quam toti
propositioni *praeponere* negationem.'

[4] Duns Scotus, *Quaestiones super anal. pr.* i, 10 (Prantl, iii, p. 139 note 614):
'Consequentia est propositio hypothetica composita ex antecedente et conse-
quente mediante coniunctione conditionali vel rationali.' As a *coniunctio
conditionalis* the word *si* is used; as a *coniunctio rationalis* either *igitur* or *ergo*.

are divided into material and formal. A consequence is *formal* if it holds for all terms in the same arrangement and form; otherwise it is *material*. Formal consequences are, as laws of logic, always correct. A material consequence is correct or 'good' (*bona*) only if it can be reduced to a formal consequence through the assumption of a true proposition as premiss. If the assumed proposition is necessarily true, the consequence is called *bona simpliciter*; if it is only contingently true, the consequence is called *bona ut nunc*. The latter distinction seems to me to be of no great significance.[1]

Later medieval handbooks of logic, in chapters entitled *De consequentiis*, introduced among various other formal consequences some which belong to propositional logic. Several of these consequences we have already become acquainted with above. It would be worth while for someone to take the trouble to assemble *all* of them, for then we would have a *complete* picture of the medieval logic of propositions.

The theory of consequences deserves the closest attention for another reason, however. From the concept of material consequence described above, the Philonian concept of implication, forgotten in the Middle Ages, can be derived in a logical but quite unexpected way. It is worth going into this derivation more closely.

The implication 'if *p* then *q*' corresponds to the inference-schema '*p*, therefore *q*'; both forms are even characterized in the same way as consequences. A true implication corresponds to a good consequence, and vice-versa. A material consequence is good if it can be transformed into a formal consequence by the assumption of a true premiss. It follows from this, first,

---

[1] Loc. cit. (Prantl, iii, p. 139 note 615, p. 140 notes 617, 619): 'Consequentia sic dividitur: quaedam est materialis, quaedam formalis. Consequentia formalis est illa, quae tenet in omnibus terminis stante consimili dispositione et forma terminorum.' (There follows a precise setting-forth of what belongs to the form of a consequence.) '—Consequentia materialis est illa, quae non tenet in omnibus terminis retenta consimili dispositione et forma. Et talis est duplex, quia quaedam est vera simpliciter, et alia est vera ut nunc. Consequentia vera simpliciter est illa, quae potest reduci ad formalem per assumptionem unius propositionis necessariae.—Consequentia materialis bona ut nunc est illa, quae potest reduci ad formalem per assumptionem alicuius propositionis contingentis verae.'

that every implication whose consequent is true must itself be true. Thus if 'q' is true, the material consequence 'p therefore q' is good for all p; for if the proposition 'q', true by assumption, be added as a premiss, we obtain the inference-schema 'p and q, therefore q', and this inference-schema is, as we have seen above, a *formal* consequence. Secondly it follows that every implication whose antecedent is false must also be true. Thus if 'p' is false, the material consequence 'p, therefore q' is good for all q; for if the true proposition 'not-p' (i.e. the contradictory of the proposition p, false by assumption) be added as a premiss, we get the rule of inference 'p and not-p, therefore q', and this rule of inference is a *formal* consequence, as we shall see below. In three cases therefore ('true-true', 'false-true', and 'false-false') is an implication true; in the fourth case ('true-false') it is of course false. Implication is therefore strictly defined as a truth-function, according to the Philonian model.

This conclusion seems to have escaped Duns Scotus. Still, he was clearly aware of all the assumptions that led up to it. He knows, that is, that from any false proposition any other proposition follows in a good material consequence, and that any true proposition results from any other proposition in a good material consequence. And finally he proves that, from a proposition which contains a formal contradiction, any proposition at all can be obtained in a *formal* consequence.[1] The proof is given by means of an example and goes as follows. The consequence 'Socrates runs and Socrates does not run, therefore you are in Rome' is formally correct. From the conjunction 'Socrates runs and Socrates does not run' the proposition 'Socrates runs', as well as the proposition 'Socrates does not run', follows in formal consequence. From the proposition 'Socrates runs' there follows further, in formal consequence, the disjunction 'Socrates runs or you are in Rome'. Finally, from this disjunction and

---

[1] Loc. cit. (Prantl, iii, p. 141 note 621): 'Ad quamlibet propositionem falsam sequitur quaelibet alia propositio in consequentia bona materiali ut nunc.— Omnis propositio vera sequitur ad quamcunque aliam propositionem in bona consequentia materiali ut nunc.—Ad quamlibet propositionem implicantem contradictionem de forma sequitur quaelibet alia propositio in consequentia *formali*.'

the negation of its first member we obtain, in formal consequence, the proposition 'you are in Rome'.[1] With the collapse
of medieval scholasticism all these fine investigations fell into
total oblivion.

The 'philosophical' logic of modern times is infected through
and through with psychology and epistemology. It has no
understanding of nor interest in questions of formal logic.
Aristotelian syllogistic is at best taken account of in its traditional distortion, and we find scarcely a trace of propositional
logic. In vain does one seek for problems that are new, precisely
formulated, and methodically solved. Everything dissolves in
vague philosophical speculations.

Modern logic is reborn out of the spirit of mathematics. With
'mathematical' logic or *logistic* a new logic arises and comes
into full bloom in the space of a few decades. With it the logic
of propositions again comes into its own. And here we encounter
all at once a phenomenon unique in the history of logic: sud-
denly, without any possible historical explanation, modern propositional logic springs with almost perfect completeness into
the gifted mind of Gottlob Frege, the greatest logician of our
time. In 1879 Frege published a small but weighty treatise
entitled *Begriffschrift, eine der arithmetischen nachgebildete For-
melsprache des reinen Denkens* ('Begriffschrift', a formalized
language of pure thought modelled upon arithmetic). In this
treatise the whole logic of propositions is for the first time laid
down as a deductive system in strict axiomatic form.[2] The
Fregean system of propositional logic is built upon two funda-

---

[1] Duns Scotus, *Quaestiones super anal. pr.* ii, 3 (not quoted by Prantl):
'"Socrates currit et Socrates non currit; igitur tu es Romae." Probatur, quia
ad dictam copulativam sequitur quaelibet eius pars gratia formae. Tunc re-
servata ista parte "Socrates non currit", arguatur ex alia sic: "Socrates currit,
igitur Socrates currit vel tu es Romae", quia quaelibet propositio infert seipsam
formaliter cum qualibet alia in una disiunctiva. Et ultra sequitur: "Socrates
currit vel tu es Romae; sed Socrates non currit (ut reservatum fuit); igitur
tu es Romae", quod fuit probatum per illam regulam: ex disiunctiva cum
contradictoria unius partis ad reliquam partem est bona consequentia.'

[2] See on this matter Łukasiewicz and Tarski, 'Untersuchungen über den
Aussagenkalkül', *Comptes rendus des séances de la Société des Sciences et des
Lettres de Varsovie* 23 (1930), Cl. iii, p. 35, note 9. [Translated in Tarski, *Logic,
Semantics, Metamathematics*, Oxford 1956—Ed.]

mental concepts, negation and implication. Implication is defined as a truth-function in just the same way as was done by Philo more than 2,000 years before. Other functions are not introduced, although the expression 'if not-p then q' can be read also as 'p or q', and the expression 'not-(if p then not-q)' as 'p and q'. With the help of the fundamental concepts six *Kernsatze* or axioms are laid down, from which all other theorems of propositional logic can be derived by means of rules of inference—the *rule of detachment*, which is explicitly formulated as a rule, and the *rule of substitution*, which is used without being formulated. Serving as the rule of detachment (the name does not originate with Frege) is the first indemonstrable syllogism of the Stoics: if the implication 'if α then β' together with the antecedent 'α' of this implication are admitted as theses of the system, then the consequent 'β' may also be admitted and detached from the implication as a new thesis. As for the rule of substitution, it allows meaningful expressions only to be substituted for the variables. *Meaningful* expressions (this concept does not appear in the *Begriffschrift*) include firstly variables, then negations of the type 'not-α', where α is a meaningful expression, and finally implications of the type 'if α then β', where α and β are meaningful expressions. The theses of the system, i.e. the axioms and theorems, are expressed in a symbolism consisting of vertical and horizontal lines that take up an excessive amount of space. This symbolism of Frege's does, however, have the advantage of avoiding all punctuation marks, such as brackets, dots, and so on. I have succeeded in devising a simpler bracket-free symbolism, requiring the least possible space. Brackets are eliminated by placing the functions 'if' and 'not' before their arguments. The expression 'if p then q' is represented in my symbolism by 'Cpq', and 'not-p' by 'Np'. Each 'C' has as arguments the two meaningful expressions immediately following it, and each 'N' has one such expression. Frege's axioms assume, in this symbolism, the following form:

I. $CpCqp$ IV. $CCpqCNqNp$

II. $CCpCqrCCpqCpr$ V. $CNNpp$

III. $CCpCqrCqCpr$ VI. $CpNNp$

This axiom system is complete: that is, all correct theses of propositional logic can be derived from it by means of the two rules of inference. It is deficient only in 'elegance': the system is not independent, for the third axiom can be deduced from the first two. The deduction, which is performed below, gives one an idea of how a modern formalized system of propositional logic appears. To explain the deductive technique used I add the following notes.[1] Before every thesis to be proved (each of which is provided with a consecutive number and can thereby be recognized as a thesis), there is an unnumbered line, which I shall call the 'derivational line'. Each derivational line consists of two parts, which are separated by the sign '×'. What stands before and after this separation designates the same formula, but in a different way. Before the separation sign is given the substitution which is to be performed on an already asserted thesis. For example, in the derivational line that belongs to thesis 1, the expression 'I $p/CCpCqrCCpqCpr$, q/Cqr' means that in I '$CCpCqrCCpqCpr$' is to be substituted for 'p' and 'Cqr' for 'q'. The thesis resulting from this substitution is omitted from the proof for the sake of brevity—it looks like this:

1′. $CCCpCqrCCpqCprCCqrCCpCqrCCpqCpr$.

The expression 'CII—1' after the separation sign indicates the construction of the same thesis 1′, and in such a way as to make it clear that the rule of detachment can be applied to 1′. We see, that is, that the thesis 1′ is of the type '$C\alpha\beta$', where 'α' denotes axiom II. Hence 'β', or 1, can be detached from it as a new thesis. Up to thesis 3 the deduction exactly follows Frege's train of thought.

I. $CpCqp$

II. $CCpCqrCCpqCpr$

       *

    I $p/CCpCqrCCpqCpr$, q/Cqr × CII—1

1. $CCqrCCpCqrCCpqCpr$

    II p/Cqr, $q/CpCqr$, $r/CCpqCpr$ × C1—2

---

[1] Cf. Łukasiewicz, 'Ein Vollständigkeitsbeweis des zweiwertigen Aussagenkalküls', *Comptes rendus des séances de la Société des Sciences et des Lettres de Varsovie* 24 (1931), Cl. iii, p. 157.

2. $CCCqrCpCqrCCqrCCpqCpr$
   $2 \times CI\,p/Cqr,\,q/p$—3
3. $CCqrCCpqCpr$
   $\Pi\,p/Cqr,\,q/Cpq,\,r/Cpr \times C3$—4
4. $CCCqrCpqCCqrCpr$
   $I\,p/CpCqp,\,q/r \times CI$—5
5. $CrCpCqp$
   $4\,q/Cpq,\,p/q \times C5\,r/CCpqr,\,p/q,\,q/p$—6
6. $CCCpqrCqr$
   $3\,q/CCpqr,\,r/Cqr,\,p/s \times C6$—7
7. $CCsCCpqrCsCqr$
   $7\,s/CpCqr,\,r/Cpr \times C\Pi$—8
8. $CCpCqrCqCpr.$ \hspace{2em} (III)

The two-valued logic of propositions, founded by the Stoics, carried on by the Scholastics, and axiomatized by Frege, stands now as a completed system before us. Scholarly research, however, knows no limits. With 'many-valued' systems of propositional logic a new domain of investigation has, in recent years, come into being; a domain which opens up surprising and unsuspected vistas. History, however, need only report about this new logic in the future.

5

THE EQUIVALENTIAL CALCULUS†

JAN ŁUKASIEWICZ

CONTENTS

§ 1. Equivalence and the equivalential system

B y an *equivalence* I mean an expression of the type 'α if and only if β', in symbols '*Eαβ*', where α and β are propositions or propositional functions. The equivalence is true if α and β have the same truth-value, i.e. either both are true or both are false;

---

† This paper was intended to appear, under the title 'Der Äquivalenzen-kalkül', in vol. 1 of the Polish periodical *Collectanea Logica* (Warsaw, 1939), pp. 145–69. The following short history of this periodical is taken from the introduction to B. Sobociński's 'An investigation of protothetic', published as no. 5 of the *Cahiers de l'Institut d'Études polonaises en Belgique*, Brussels, 1949.

'In 1937, at the suggestion of Mr. Jan Łukasiewicz, we founded in Poland a periodical devoted to Logic, its history and its applications, under the title *Collectanea Logica*. It was to be issued as one large volume each year, and would be international in character, containing different papers in Polish, English, French, German, Italian, and Latin. The editor of *Collectanea Logica* was Łukasiewicz, and its managing editor myself. . . . On the first of September 1939 the first part of the volume, which would have had 500 pages, was printed, the second part already collected and in proof. Moreover, the first five papers from the prepared part were already published as offprints. At the siege of Warsaw in September 1939 the printing-house of the periodical was completely burned, with all the prepared type, blocks, and offprints. The final proofs of the first volume, most of the prepared offprints, and the archives of the publication escaped in my flat, but all this was destroyed in August 1944 during the Warsaw Insurrection.' Sobociński follows this by giving a brief description of the contents of the first volume.

Only one copy of 'Der Äquivalenzenkalkül', sent as a review copy to Scholz in Münster, is known to have survived the war, and is now in Poland. Translated by P. Woodruff.

otherwise it is false. If we denote the True by '1' and the False by '2', the following equations hold:

$$E11 = 1 \qquad E12 = 2$$
$$E21 = 2 \qquad E22 = 1.$$

These equations are indicated in the following matrix

E	1	2
1	1	2
2	2	1

which I shall call the *normal* matrix for equivalence. The first argument is written to the left of the vertical stroke, the second above the horizontal line.

An equivalence $E\alpha\beta$, in which besides propositional variables only functors of the propositional calculus appear, is called an equivalence of *propositional logic*. For example $EKpqKqp$, in words 'p and q if and only if q and p', is an equivalence of propositional logic. If in an equivalence of propositional logic no functor of the propositional calculus other than E appears, I shall call it a *pure* equivalence of propositional logic.

By the ordinary or two-valued equivalential system I mean the set of all pure equivalences of propositional logic which satisfy the normal matrix for equivalence. The matrix is said to be satisfied by a given equivalence, if all replacements of the propositional variables of the equivalence with the values 1 or 2 yield expressions which after reduction according to the matrix assume the value 1. For example, the equivalence Epp satisfies the matrix, for we get

$$\text{for } p/1 \qquad E11 = 1$$
$$\text{for } p/2 \qquad E22 = 1.$$

Likewise $EEpqEqp$ satisfies the matrix, since the following equations hold:

$$\text{for } p/1, q/1 \qquad EE11E11 = E11 = 1$$
$$\text{for } p/1, q/2 \qquad EE12E21 = E22 = 1$$
$$\text{for } p/2, q/1 \qquad EE21E12 = E22 = 1$$
$$\text{for } p/2, q/2 \qquad EE22E22 = E11 = 1.$$

On the other hand, the matrix is not satisfied by $EpEqp$, for we have:

$$\text{for } p/2,\ q/2 \qquad E2E22 = E21 = 2.$$

The two-valued equivalential system is one of the simplest sub-systems of the propositional calculus. Numerous methodological questions can be formulated with particular clarity and simplicity in this system, and can be easily solved. For this reason it is worth while to subject the system to a detailed examination, for we obtain thereby an easily grasped introduction to the problems and methodology of the propositional calculus.

§ 2. On the history of the equivalential calculus

Two pure equivalences are to be found in *Principia Mathematica*,[1] namely

$$p \equiv p \quad \text{and} \quad p \equiv q \, . \equiv . \, q \equiv p,$$

to which correspond the following theses in my bracket-free symbolism:

$$Epp \quad \text{and} \quad EEpqEqp.$$

The first thesis says that equivalence is *reflexive*, the second, that it is *commutative*. Now, I long ago noted that equivalence is also *associative*, and accordingly I established the following thesis in the symbolism of *Principia*:

$$p \equiv . q \equiv r \colon \equiv \colon p \equiv q \, . \equiv r.$$

This thesis, which in my symbolism can be expressed by

$$EEpEqrEEpqr,$$

is cited by Tarski[2] in his doctoral thesis of 1923.

Leśniewski[3] was in 1929 the first to recognize that the two-valued equivalential system can be axiomatized. In particular,

---

[1] A. N. Whitehead and B. Russell, *Principia Mathematica*, vol. i (Cambridge, 1910), p. 121, theorems *4.2 and *4.21.

[2] A. Tajtelbaum-Tarski, *O wyrazie pierwotnym logistyki* (On the primitive term of logistic), Doctoral thesis, *Przegląd Filozoficzny* 26 (Warsaw, 1923), p. 72 n. See also A. Tajtelbaum, 'Sur le terme primitif de la logistique', *Fundamenta Mathematicae* 4 (Warsaw, 1923), p. 199 n.

[3] S. Leśniewski 'Grundzüge eines neuen Systems der Grundlagen der Mathematik', *Fundamenta Mathematicae* 14 (Warsaw, 1929), § 3, pp. 15–30.

this can be done with the help of two rules of inference; the rule of substitution and the rule of detachment. The rule of detachment for equivalence is analogous to that for implication. Leśniewski characterizes this rule approximately thus: If an equivalence A belongs to the system whose right side is equiform with S, and if a theorem belongs to the system which is equiform with the left side of the equivalence A, then a theorem equiform with S may be added to the system. In other words, if the expressions '$E\alpha\beta$' and 'α' belong to the system, then 'β' may also be added to it. Leśniewski shows that with the help of this rule of detachment and the rule of substitution all pure equivalences provable in the ordinary propositional calculus can be deduced from the following two axioms:

A1. $p \equiv r \, . \equiv . \, q \equiv p : \equiv . \, r \equiv q$

A2. $p \equiv . \, q \equiv r : \equiv : p \equiv q \, . \equiv r.$

The second axiom is the law of associativity for equivalence, discovered by myself. From these axioms, which in my symbolism read

$$EEEprEqpErq \quad \text{and} \quad EEpEqrEEpqr,$$

Leśniewski first derives seventy-nine theses in symbolic form, and then proves with the help of reasoning conducted in ordinary language that the above axiomatic system is complete. Proofs of consistency and independence are not found in Leśniewski.

After Leśniewski, Wajsberg[1] in 1932 published simpler axiom-systems for the equivalential system, at first without completeness-proofs. Two of these systems consist of two axioms each, two others of one axiom apiece. The four systems of Wajsberg, given by the writer in my symbolism, are

(a) $EEEpqrEpEqr$ and $EEpqEqp$,
(b) $EEpEqrErEqp$ and $EEEpppp$,
(c) $EEEpEqrEErssEpq$,
(d) $EEEEpqrsEsEpEqr$.

Wajsberg gives completeness-proofs for these four axiom-

[1] M. Wajsberg, 'Ein neues Axiom des Aussagenkalküls in der Symbolik von Sheffer', *Monatshefte für Mathematik and Physik* 39 (Leipzig, 1932), p. 262.

systems in a later work[1] by reducing them in part directly and in part indirectly to that of Leśniewski.

To Wajsberg belongs the credit of having first shown that the equivalential calculus can be based on a *single* axiom. The two single axioms of Wajsberg consist of fifteen letters each. It soon turned out that there are other equivalences, each of fifteen letters, which can be postulated as single axioms of the system. In a paper of 1932 Sobociński[2] gives the following six axioms, which were discovered by various authors:

(e) $EEpEqrEEqEsrEsp$ discovered by Bryman

(f) $EEpEqrEEqErsEsp$,,   ,,   Łukasiewicz

(g) $EEsEpEqrEEpqErs$,,   ,,   ,,

(h) $EEpEqrEEpErsEsq$,,   ,,   Sobociński

(i) $EEpEqrEEpEsrEsq$,,   ,,   ,,

(j) $EEpEqrEEpErsEqs$,,   ,,   ,,

Axiom (g) is obtained as a first detachment from Wajsberg's axiom (d). Completeness-proofs are not given by Sobociński.

In 1937 the Rumanian mathematical logician E. Gh. Mihailescu published a paper devoted specifically to the equivalential calculus.[3] Mihailescu bases his work on the above-mentioned work of Leśniewski, and uses the bracket-free notation which I introduced. In metalogical investigations he makes use of Tarski's terminology. Wajsberg's work is apparently unknown to him. In his essay, the equivalential calculus is based on the two axioms $EEpqEqp$ and $EEEpqrEpEqr$, discovered by Wajsberg. For this axiom-system he gives a new completeness-proof

---

[1] M. Wajsberg, 'Metalogische Beiträge', *Wiadomości Matematyczme* 43 (Warsaw, 1936), pp. 132–3 and 163–6. [See pp. 286 and 314–16 of paper 14 of this volume—Ed.] Instead of axiom-system (a) the writer considers here the following axiom-system:

(a') $EEpEqrEEpqr$ and $EEpqEqp$,

which is obviously deductively equivalent to (a).

[2] B. Sobociński, 'Z badań nad teorią dedukcji' (Investigations into the theory of deduction), *Przegląd Filozoficzny* 35 (Warsaw, 1932), pp. 186–7 and 192–3, nn. 35–37.

[3] E. Gh. Mihailescu, 'Recherches sur un sous-système du calcul des propositions' *Annales scientifiques de l'Université de Jassy* 23 (Jassy, 1937), pp. 106–24.

by reducing all meaningful expression to certain formal forms. For this purpose ninety-three theses are deduced from the axioms by substitution and detachment. The consistency and independence of the axiom-system is shown by the matrix method.

The present work contains my results from the year 1933. Among the most important results I wish to present are the discovery of the shortest axiom of the equivalential system as well as a new completeness-proof, which seems to me to be simpler than those of Leśniewski and Mihailescu.

§ 3. Meaningful expressions and the rule of substitution

All expressions of our system are formed by juxtaposition of capital E's, the sign of equivalence, and small latin letters, the propositional variables. Not every expression thus formed is, however, *meaningful*; i.e. not every expression represents a proposition, or, more precisely, a propositional function. For example, 'pq', 'E', 'EE', 'pEq', '$Epqr$' are not meaningful expressions, for they do not represent propositional functions. On the other hand, propositional variables such as 'p', 'q', 'r', etc., as well as equivalences both of whose members are meaningful expressions, such as 'Epq', '$EEpqr$', '$EpEqr$', etc., are evidently meaningful. In the following I give a purely *structural* definition of 'meaningful expression' for the equivalential system, by slightly modifying a definition found for the implication-negation system by Jaśkowski:[1]

An expression made up of the letter 'E' and small Latin letters is *meaningful* if, and only if, it fulfils the following two conditions:

1. The number of 'E's' occurring in the expression must be *one less* than the number of small letters.

2. In every segment, which begins at an arbitrary point in the expression and reaches to the end of the expression, the number of 'E's' must be *less* then the number of small letters.

---

[1] See J. Łukasiewicz, 'Ein Vollständigkeitsbeweis des zweiwertigen Aussagenkalküls', *Comptes rendus des séances de la Societé des Sciences et des Lettres de Varsovie*, Cl. iii, 24 (1931), p. 156, n. 5.

The two conditions are independent, as examples readily show. Thus, '*EpEqr*' fulfils both conditions and is therefore meaningful. The expressions '*EpqEr*' fulfils the first condition but not the second, since in the segment beginning with the second '*E*' the number of '*E*'s' is not less than the number of small letters. The expression '*pEEqrs*' satisfies the second condition but not the first, since the number of '*E*'s' in the expression is not one but two less than the number of small letters. Finally, it is clear that the expression '*pqEErs*' fulfils neither the first nor the second condition. The last three expressions are meaningless.

These conditions yield, among others, the following consequences:

(*a*) Propositional variables are meaningful expressions, for they satisfy both conditions.

(*b*) All composite meaningful expressions must begin with an '*E*'. For if the second condition is to be fulfilled, then in the segment beginning with the second letter the number of '*E*'s' must be at least one less than the number of small letters. If a small letter is now added at the beginning, then the number of '*E*'s' in the whole expression must be less than that of the small letters by at least two letters, which contradicts the first condition.

In connexion with this definition there is a simple practical rule which enables us to decide at once if a given expression, composed of the letter '*E*' and small letters, is meaningful or not.[1] One first assigns each '*E*' the number −1 and each small letter the number +1. Then one adds these numbers sequentially, starting with the number assigned to the last letter on the right of the expression and proceeding by steps to the left, to the beginning of the expression. The following example illustrates this process:

$$EEEpqErsEtu$$
$$1\ 2\ 343\ 2\ 32\ 121$$

---

[1] The idea behind this rule is not mine, but rather—as far as I know—that of a student of L. Chwistek.

The 'u' is assigned $+1$, also the 't'; 1 plus 1 is 2, 'E' is -1, $2-1=1$, etc. If the expression is meaningful, then the first condition says that the *sum*, which corresponds to the whole expression and stands at the very beginning, must be equal to 1; the second condition says that all *partial sums*, which correspond to single segments, must be positive, i.e. greater than 0. A glance at the number-series which belongs to the expression in the above example suffices to determine that this expression is meaningful. If such a number-series does not begin with 1, if a 0 or even a negative number appears, then the expression is meaningless; e.g.

$$pEEqrs \qquad EpqEr$$

$$2\,1\;2\,3\,2\,1 \qquad 1\,2\,1\,0\,1$$

In the first example the sum which corresponds to the whole expression equals 2; which is incompatible with the first condition. In the second example the partial sum which corresponds to the segment 'Er' equals 0, contrary to the second condition.

Not all meaningful expressions belong to the system. Those which do I call theses. In our equivalential system the theses are distinguished by the fact that they satisfy the normal matrix for equivalence.

Since we now have the concept of a meaningful expression at our disposal, we can formulate the *rule of substitution* precisely. On the basis of this rule one obtains a new thesis from a given thesis by replacing one or more of the propositional variables of the given thesis by *meaningful* expressions, where all equiform variables must be replaced by equiform expressions. For example, if in the thesis with which we are already familiar,

$$EEpqEqp,$$

the meaningful expression 'Eqr' is substituted for 'q' (which transformation I denote by 'q/Eqr') we obtain a new thesis:

$$EEpEqrEEqrp.$$

The other rule which is used to derive theses is the previously characterized *rule of detachment*: If $E\alpha\beta$ and α are theses, β is also a thesis and hence can be detached from $E\alpha\beta$. For example, let the following two theses be given:

1. $EEpqEqp$.
2. $EEpEqrEEpqr$.

If in 1 the substitution '$p/EpEqr, q/EEpqr$' is made, one obtains

1'. $E\quad EEpEqrEEpqr\quad EEEpqrEpEqr$.
$\qquad\qquad\quad 2 \qquad\qquad\qquad 3$

Thesis 1' begins with an 'E' followed by thesis 2 as its first member; consequently its second member, in accordance with the rule of detachment, can be detached as a new thesis:

3. $EEEpqrEpEqr$.

This derivation I indicate briefly as follows:

$\qquad 1\,p/EpEqr,\ q/EEpqr \times E2-3$

3. $EEEpqrEpEqr$.

In the *derivational line* which precedes thesis 3, the series of expressions both before and after the *separation sign* '$\times$' designates the thesis 1', which is omitted for the sake of brevity.

§ 4. The shortest axiom

The shortest axiom for the equivalential system, which I discovered, consists of eleven letters and reads: $EEpqEErqEpr$. From this axiom I will first derive Leśniewski's two axioms, as well as those theses which are required for the completeness-proof to be given later. All these theses are marked by an asterisk. I will then prove that no shorter axiom possesses the property of being a single axiom of the system. The following deductions, which are constructed with the sole use of the rules of substitution and detachment mentioned above, should be clear enough after what I have said above.[1]

*1. $EEpqEErqEpr$
$\qquad 1\,p/Epq,\ q/EErqEpr,\ r/s \times E1-2$
2. $EEsEErqEprEEpqs$
$\qquad 2\,s/Epq \times E1-3$

---

[1] For the derivational technique see p. 157 of my article cited in note 1, p. 93, above, as well as my essay 'Zur Geschichte der Aussagenlogik', *Erkenntnis* 5 (Leipzig, 1935), p. 126. [P. 86 of paper 4 of this volume—Ed.]

3. $EEpqEpq$
   $1\,p/Epq,\ q/Epq \times E3{-}4$
4. $EErEpqEEpqr$
   $4\,r/Epq,\ p/Erq,\ q/Epr \times E1{-}5$
5. $EEErqEprEpq$
   $5\,r/p,\ q/p \times E3\,q/p{-}6$
*6. Epp
   $1\,p/q,\ r/p \times E6\,p/q{-}7$
*7. $EEpqEqp$
   $1\,p/Epq,\ q/Eqp \times E7{-}8$
8. $EErEqpEEpqr$
   $7\,p/ErEqp,\ q/EEpqr \times E8{-}9$
9. $EEEpqrErEqp$
   $2\,s/EEprEpq,\ r/p,\ p/r \times E9\,q/r,\ r/Epq{-}10$
10. $EErqEEprEpq$
   $5\,r/Epq,\ p/EpEpq \times E10\,r/Epq{-}11$
*11. $EEpEpqq$
   $7\,p/EpEpq \times E11{-}12$
*12. $EqEpEpq$
   $1\,p/EpEpq \times E11{-}13$
13. $EErqEEpEpqr$
   $2\,s/EEpqr,\ r/q,\ q/Eqr \times E13\,r/Epq,\ q/r,\ p/q{-}14$
*14. $EEpEqrEEpqr$
   $7\,p/EpEqr,\ q/EEpqr \times E14{-}15$
*15. $EEEpqrEpEqr$
   $9\,p/Erq,\ q/p,\ r/EpEqr \times E9\,p/r,\ r/p{-}16$
*16. $EEpEqrEpErq$
   $16\,p/EEprEqp \times E5\,r/p,\ q/r,\ p/q{-}17$
*17. $EEEprEqpErq$
   $16\,p/EEpqr,\ q/r,\ r/Eqp \times E9{-}18$
18. $EEEpqrEEqpr$
   $10\,r/EEqrs,\ q/EErqs \times E18\,p/q,\ q/r,\ r/s{-}19$
*19. $EEpEEqrsEpEErqs$
   $10\,r/EEqrs,\ q/EqErs \times E15\,p/q,\ q/r,\ r/s{-}20$
*20. $EEpEEqrsEpEqErs$
   $7\,p/EpEEqrs,\ q/EpEqErs \times E20{-}21$
*21. $EEpEqErsEpEEqrs.$

H

Of the derived theses, *14 and *17 are Leśniewski's axioms. Herewith the proof is given, indirectly, that our axiom comprehends all theses of the system. It is, moreover, not the only 'shortest' axiom of the equivalential system; I have found two other theses of eleven letters which can be likewise postulated as single axioms of the system. These are $EEpqEEprErq$, and $EEpqEErpEqr$. From each of these theses *1 can be derived in the following manner:

A

1. $EEpqEEprErq$
    $1\,p/Epq,\ q/EEprErq,\ r/s \times E1-2$
2. $EEEpqsEsEEprErq$
    $2\,p/Epq,\ q/s,\ s/EsEEprErq \times E2-3$
3. $EEsEEprErqEEEpqrErs$
    $3\,s/Epq \times E1-4$
4. $EEEpqrErEpq$
    $4\,r/EEprErq \times E1-5$
5. $EEEprErqEpq$
    $5\,r/Epr,\ q/ErEpr \times E1\,q/Epr-6$
6. $EpErEpr$
    $1\,q/ErEpr,\ r/q \times E6-7$
7. $EEpqEqErEpr$
    $2\,s/EqErEpr \times E7-8$
8. $EEqErEprEEprErq$
    $8\,q/p \times E6-9$
9. $EEprErp$
    $2\,s/Eqp \times E9\,r/q-10$
10. $EEqpEEprErq$
    $3\,s/Eqp \times E10-11$
11. $EEEpqrErEqp$
    $11\,p/Epr,\ q/Erq,\ r/Epq \times E5-12$
12. $EEpqEErqEpr.$

B

1. $EEpqEErpEqr$
    $1\,p/Epq,\ q/EErpEqr,\ r/s \times E1-2$
2. $EEEsEpqEEErpEqrs$
    $2\,s/Epq,\ p/Erp,\ q/Eqr,\ r/s \times E1-3$

3. $EEEsErpEEqrsEpq$
    $3\,r/Eqr,\,p/q,\,q/ErEqr \times E2\,p/Eqr-4$
4. $EqErEqr$
    $2\,s/q,\,p/r,\,q/Eqr,\,r/s \times E4-5$
5. $EEEsrEEqrsq$
    $5\,r/Eqq,\,q/Eqq \times E2\,p/q,\,r/q-6$
6. Eqq
    $1\,p/q \times E6-7$
7. $EErqEqr$
    $2\,s/Epq,\,p/q,\,q/p \times E7r/p-8$
8. $EEErqEprEpq$
    $7\,r/EErqEpr,\,q/Epq \times E8-9$
9. $EEpqEErqEpr.$

§ 5. The completeness-proof

The new completeness-proof, which I intend to give here, rests on a concept of completeness which is essentially due to the American mathematical logician Post.[1] I intend to prove the following:

Every meaningful expression in the equivalential system has either the property that it can be derived by the rules of inference from axiom *1 or the property that, when it is added to axiom *1, every meaningful expression is derivable.

The first property I call ξ_1, the second ξ_2.

The 'either-or' in this case is non-exclusive. However, it will later turn out that the equivalential system constructed on the basis of our axiom is *consistent*, i.e. does not include all meaningful expressions of the system. Thus the properties ξ_1 and ξ_2 do in fact exclude each other.

The proof is based essentially on the previously stressed fact that every meaningful expression either is a propositional variable or begins with an 'E', thus is of the type '$E\alpha\beta$', where 'α' and 'β' are understood to be meaningful expressions. In

---

[1] See in this connexion p. 161, n. 10 of my article cited in note 1, p. 93 above, as well as the essay of H. Hermes and H. Scholz, 'Ein neuer Vollständigkeitsbeweis für das reduzierte Fregesche Axiomensystem des Aussagenkalküls', *Forschungen zur Logik und zur Grundlegung der exakten Wisssenschaften*, New Series, vol. 1 (Leipzig, 1937), p. 6, n. 5.

general, I will designate arbitrary meaningful expressions by the first few letters of the Greek alphabet, while designating propositional variables by 'π'.

The proof divides into eight sections, (a) to (h). These exhaust all possible cases which can occur when any meaningful expression is given.

(a) The given expression is a propositional variable. Then it has the property ξ_2, for from a variable all meaningful expressions can be derived through substitution.

(b) The given expression begins with more than one 'E'. Then on the basis of the theses:

*15. $EEEpqrEpEqr$

*14. $EEpEqrEEpqr$

which are derivable from axiom *1, it can be transformed into a deductively equivalent and not longer expression which begins with only one 'E'.

Proof. Two expressions are called *deductively equivalent*[1] with respect to axiom *1, if on the basis of this axiom either expression can be derived from the other by means of the established rules of inference. Expressions which begin with more than one 'E', i.e. have the form '$EE\alpha\beta\gamma$', are deductively equivalent to expressions of the form $E\alpha E\beta\gamma$, for in view of *15 and *14 we have:

I. $EE\alpha\beta\gamma$
    *15 $p/\alpha, q/\beta, r/\gamma \times EI-II$
II. $E\alpha E\beta\gamma$

II. $E\alpha E\beta\gamma$
    *14 $p/\alpha, q/\beta, r/\gamma \times EII-I$
I. $EE\alpha\beta\gamma$.

'$E\alpha E\beta\gamma$' is no longer than '$EE\alpha\beta\gamma$' and has one 'E' less at the beginning. If 'α' again begins with an 'E', the same transformation can be made and repeated until one obtains an expression which begins with one 'E', and is hence of the form '$E\pi\delta$'.

(c) The given expression begins with one 'E' followed by a propositional variable, i.e. is of the type '$E\pi\delta$', where in 'δ' no variable equiform with 'π' occurs. Then the expression has the

---

[1] The term 'deductively equivalent' I owe to the above-mentioned paper of Hermes and Scholz.

property ξ_2, i.e. if it is taken together with the axiom, all meaningful expressions are derivable from it.

Proof. If '$E\pi\delta$' is conjoined to the axiom, one obtains on the basis of thesis

*7. $EEpqEqp$

the expression '$E\delta\pi$'. On the other hand, we can derive the expression 'δ' from '$E\pi\delta$' by the substitution $\pi/E\pi\delta$, since 'δ' contains no variable of the same shape as 'π' and hence is not changed by the substitution. From '$E\delta\pi$' and 'δ' we get the variable 'π' by detachment, and from 'π' by substitution all meaningful expressions. The formal derivation has the form:

   I. $E\pi\delta$
       *7 $p/\pi,\ q/\delta \times E$I−II
  II. $E\delta\pi$
       I $\pi/E\pi\delta \times E$I−III
 III. δ
       II $\times E$III−IV
 IV. π.

In the following it is assumed that in expressions of the type '$E\pi\delta$' the expression 'δ' always contains a variable equiform with 'π'. Furthermore, 'δ' is either a variable or an expression of the form '$E\alpha\beta$'. We examine first the latter case.

(*d*) The given expression has the form '$E\pi E\alpha\beta$', where the equivalence beginning with the second 'E', i.e. '$E\alpha\beta$', contains a variable equiform with 'π'. If this variable is in the second but not in the first member of '$E\alpha\beta$', then the expression '$E\pi E\alpha\beta$' can be transformed on the basis of the thesis

*16. $EEpEqrEpErq$

into a deductively equivalent and not longer expression, namely '$E\pi E\beta\alpha$', in which the variable equiform with 'π' appears in the first member of the equivalence beginning with the second 'E'.

Proof. '$E\pi E\alpha\beta$' and '$E\pi E\beta\alpha$' are by *16 deductively equivalent, for we have

I. $E\pi E\alpha\beta$	II. $E\pi E\beta\alpha$
*16 $p/\pi, q/\alpha, r/\beta \times E$I−II	*16 $p/\pi, q/\beta, r/\alpha \times E$II−I
II. $E\pi E\beta\alpha$	I. $E\pi E\alpha\beta$.

If 'π' does not occur in 'α', it must be contained in 'β', since *ex hypothesi* it occurs in '$E\alpha\beta$'.

On the basis of this section we may assume subsequently that in expressions of the form '$E\pi E\alpha\beta$' the variable equiform with 'π' occurs in the first member of the equivalence '$E\alpha\beta$', i.e. in 'α'. Now 'α' is either a variable or an expression of the form '$E\alpha\beta$'. We consider first the latter case.

(e) The given expression has the form '$E\pi EE\alpha\beta\gamma$' where the equivalence beginning with the third 'E', i.e. '$E\alpha\beta$', contains a variable equiform with 'π'. If this variable occurs not in the first but in the second member of the equivalence beginning with the third 'E', then the expression '$E\pi EE\alpha\beta\gamma$' can be transformed, in virtue of thesis

*19. $EEpEEqrsEpEErqs$,

into a deductively equivalent not longer expression, namely '$E\pi EE\beta\alpha\gamma$', in which the variable equiform with 'π' appears in the first member of the equivalence beginning with the third 'E'.

Proof. '$E\pi EE\alpha\beta\gamma$' and '$E\pi EE\beta\alpha\gamma$' are deductively equivalent by *19, for we have:

I. $E\pi EE\alpha\beta\gamma$	II. $E\pi EE\beta\alpha\gamma$
*19 $p/\pi, q/\alpha, r/\beta, s/\gamma \times EI-$II	*19 $p/\pi, q/\beta, r/\alpha, s/\gamma \times EII-$I
II. $E\pi EE\beta\alpha\gamma$	I. $E\pi EE\alpha\beta\gamma$.

If 'π' does not appear in 'α', it must in 'β', since by assumption the equivalence '$E\alpha\beta$' contains a variable equiform with 'π'.

By reason of this section we may assume in what follows that, in expressions of the form '$E\pi EE\alpha\beta\gamma$', the variable equiform with 'π' is contained in 'α'.

(f) The given expression has the form '$E\pi EE\alpha\beta\gamma$', where 'α' contains a variable equiform with 'π'. Then by the theses

*20. $EEpEEqrsEpEqErs$

*21. $EEpEqErsEpEEqrs$

the expression '$E\pi EE\alpha\beta\gamma$' can be transformed into the deductively equivalent not longer expression '$E\pi E\alpha E\beta\gamma$'.

Proof. '$E\pi EE\alpha\beta\gamma$' and '$E\pi E\alpha E\beta\gamma$' are deductively equivalent, since by *20 and *21 we obtain:

I. $E\pi EE\alpha\beta\gamma$
 *20 $p/\pi,q/\alpha,r/\beta,s/\gamma \times EI-$II
II. $E\pi E\alpha E\beta\gamma$

II. $E\pi E\alpha E\beta\gamma$
 *21 $p/\pi,q/\alpha,r/\beta,s/\gamma \times EII-$I
I. $E\pi EE\alpha\beta\gamma$.

In the expression '$E\pi E\alpha E\beta\gamma$', '$\alpha$' contains a variable of the same form as 'π'. If 'α' is an equivalence, the transformations described under (e) and (f) can again be carried out, and repeated until a propositional variable is obtained in place of 'α'. Herewith we come back to the unresolved case mentioned at the end of section (d): the given expression has the form '$E\pi E\alpha\beta$' where 'α' is a variable and also contains a variable equiform with 'π'. 'α' must then, of course, be equiform with 'π', and we have the case:

(g) The given expression has the form $E\pi E\pi\alpha$. Then on the basis of the theses

*11. $EEpEpqq$

*12. $EqEpEpq$

it can be transformed into the deductively equivalent and *shorter* expression 'α'.

Proof. '$E\pi E\pi\alpha$' and 'α' are deductively equivalent by reason of *11 and *12:

I. $E\pi E\pi\alpha$
 *11 $p/\pi, q/\alpha \times E$I$-$II
II. α

II. α
 *12 $q/\alpha, p/\pi \times E$II$-$I
I. $E\pi E\pi\alpha$.

To this shorter expression we can again apply the transformation rules mentioned in (a) to (g). If sections (a) or (c) are applicable, then the investigation is finished, for it is clear that the given expression has the property ξ_2. If this does not happen, one obtains progressively shorter expressions, until one reaches the shortest expression which possesses property ξ_1. This is, of course, the expression '$E\pi\pi$'. And therewith the final outstanding case is resolved which was mentioned at the end of section (c): the given expression has the form '$E\pi\delta$', where 'δ' is a variable and contains a variable equiform with 'π'. 'δ' must then be equiform with 'π', and we obtain the case:

(*h*) The given expression has the form '$E\pi\pi$'. Then it has property ξ_1 as a substitution-instance of the thesis

　　*6. *Epp.*

With this the completeness-proof is finished. The proof rests on ten theses, all of which are deducible from our axiom *1: *6, *7, *11, *12, *14, *15, *16, *19, *20, *21. Thus the proof is *effective*; i.e. with the aid of the enumerated theses it can always be decided whether a given expression has property ξ_1 or property ξ_2, and in each case the procedure which one must follow to exhibit one or the other property is exactly specified. This will be clarified subsequently by two examples, which are furthermore intended to make the completeness-proof here presented more intelligible.

§ 6. Examples for the completeness-proof

As examples I choose two expressions, of which one exhibits property ξ_1, the other property ξ_2. The examples are so chosen that all the transformation rules enumerated in sections (*a*) to (*h*) are used in one or the other.

The first expression reads '$EEEpEqpEqrr$', and thus begins with more than one 'E'. Hence it falls under section (*b*). On the grounds of the schema:

$$EE\alpha\beta\gamma \sim E\alpha E\beta\gamma$$

where 'α' is '$EpEqp$', 'β' is 'Eqr', 'γ' is 'r', and the sign '$\sim$' denotes deductive equivalence, we have:

$EEEpEqpEqrr \sim EEpEqpEEqrr$ [section (*b*),

　　　　　　　　　　　　　　　　　theses *15 and *14].

The expression on the right obtained by the transformation still begins with more than one 'E'. So we apply rule (*b*) a second time:

$EEpEqpEEqrr \sim EpEEqpEEqrr$ [section (*b*),

　　　　　　　　　　　　　　　　　theses *15 and *14].

Now we have obtained an expression of the form '$E\pi EE\alpha\beta\gamma$' where the variable equiform with 'π' is contained in 'β'. This comes under section (e), therefore under the schema:

$$E\pi EE\alpha\beta\gamma \sim E\pi EE\beta\alpha\gamma.$$

This schema yields the deductive equivalence:

$EpEEqpEEqrr \sim EpEEpqEEqrr$ [section (e), thesis *19].

The new expression is of the form '$E\pi EE\alpha\beta\gamma$', the variable equiform with 'π' being contained in 'α'. Hence we must now apply section (f), i.e. the schema:

$$E\pi EE\alpha\beta\gamma \sim E\pi E\alpha E\beta\gamma,$$

from which we get:

$EpEEpqEEqrr \sim EpEpEqEEqrr$ [section (f),
$\qquad\qquad\qquad\qquad\qquad\qquad$ theses *20 and *21].

Now it is the turn of rule (g), since the new expression is of the type '$E\pi E\pi\alpha$' and according to the schema

$$E\pi E\pi\alpha \sim \alpha$$

can be transformed into the shorter expression 'α'. Thus we have:

$EpEpEqEEqrr \sim EqEEqrr$ [section (g), theses *11
$\qquad\qquad\qquad\qquad\qquad\qquad\qquad\qquad$ and *12].

This shorter expression yields after two transformations:

$EqEEqrr \sim EqEqErr$ [section (f), theses *20 and *21],

$EqEqErr \sim Err$ [section (g), theses *11 and *12],

the shortest expression with the property ξ_1, namely

Err [section (h), thesis *6].

The analysis is ended. Now comes the synthesis, namely the derivation of the given expression, which has property ξ_1, from the theses adduced in the analysis, and hence indirectly from our axiom *1. We begin with the last expression to which the analysis led and climb back up step by step. In this process we shall not, however, use all the theses mentioned, but in every case when two theses are given in connexion with a deductive equivalence, we shall use only the second. Thus the derivation is based on theses *6, *12, *14, *19, and *21, which appear in this order: *12, *6, *21, *12, *21, *19, *14, and *14.

$$*12 \, q/Err, \, p/q \times E*6 \, p/r - \mathrm{I}$$

I. $EqEqErr$

$$*21 \, p/q, \, s/r \times E\mathrm{I} - \mathrm{II}$$

II. $EqEEqrr$

$$*12 \, q/EqEEqrr \times E\mathrm{II} - \mathrm{III}$$

III. $EpEpEqEEqrr$

$$*21 \, q/p, \, r/q, \, s/EEqrr \times E\mathrm{III} - \mathrm{IV}$$

IV. $EpEEpqEEqrr$

$$*19 \, q/p, \, r/q, \, s/EEqrr \times E\mathrm{IV} - \mathrm{V}$$

V. $EpEEqpEEqrr$

$$*14 \, q/Eqp, \, r/EEqrr \times E\mathrm{V} - \mathrm{VI}$$

VI. $EEpEqpEEqrr$

$$*14 \, p/EpEqp, \, q/Eqr \times E\mathrm{VI} - \mathrm{VII}$$

VII. $EEEpEqpEqrr.$

This completes the proof that the given expression

$$`EEEpEqpEqrr\text{'}$$

possesses the property ξ_1.

As a second example I choose the expression '$EEpEqrEps$' which, as it will turn out, has the property ξ_2. That is, added to the axioms, it entails the derivability of all meaningful expressions. After what I have said above, the following should be clear without further ado:

$EEpEqrEps \sim EpEEqrEps$ [section (b), theses *15 and *14],
$EpEEqrEps \sim EpEEpsEqr$ [,, (d), ,, *16],
$EpEEpsEEqr \sim EpEpEsEqr$ [,, (f), ,, *20 and *21],
$EpEpEsEqr \sim EsEqr$ [,, (g), ,, *11 and *12],
$EsEqr \sim EEqrs$ [,, (c), ,, *7].

'$EEqrs$' yields in conjunction with 'Eqr', which follows from '$EsEqr$', the variable 's', and hence by (a) all meaningful expressions.

The synthetic construction begins with the given expression '$EEpEqrEps$' and descends to the variable 's'. If, in the process, two theses are mentioned in a deductive equivalence, we use only the first. The deduction is thus based on theses *15, *16, *20, *11, and *7:

I. $EEpEqrEps$
$*15\,q/Eqr,\ r/Eps \times EI-II$

II. $EpEEqrEps$
$*16\,q/Eqr,\ r/Eps \times EII-III$

III. $EpEEpsEqr$
$*20\,q/p,\ r/s,\ s/Eqr \times EIII-IV$

IV. $EpEpEsEqr$
$*11\,q/EsEqr \times EIV-V$

V. $EsEqr$
$*7\,p/s,\ q/Eqr \times EV-VI$

VI. $EEqrs$
$V\,s/EsEqr \times EV-VII$

VII. Eqr
$VI \times EVII-VIII$

VIII. s.

This constitutes the proof that the given expression

$$\text{`}EEpEqrEps\text{'}$$

possesses the property ξ_2.

§ 7. Consistency of the equivalential system

As noted above, the equivalential system is *consistent*, i.e. not all meaningful expressions belong to the system, nor are they all, accordingly, derivable from our axiom. A simple proof of consistency is provided by the normal matrix for equivalence. However, I will here give in addition a purely *structural* proof of the consistency of the system, using an idea of Leśniewski's.

Leśniewski was the first to note that in all theses of the equivalential system the number of equiform variables of each shape, e.g. the number of 'p's, the number of 'q's, etc., is *even*.[1] Let us designate this property by 'G'. It can now easily be shown, as Leśniewski and Tarski long since realized, that the property G is *hereditary* with respect to the rules of substitution and detachment. This means that all expressions which are derived

---

[1] Cf. op. cit., p. 26, point 6 and p. 29, point 11.

from given G-expressions by means of these rules of inference also have the property G. This is evident in the case of the rule of substitution. For if an arbitrary variable 'π' appears an even number of times in an expression, and for 'π' any meaningful expression 'α' is substituted, the number of equiform variables of any shape is changed by an even number. In the case of the rule of detachment, the assertion can be proved as follows. If the number of variables of each shape in '$E\alpha\beta$' and 'α' is even, two cases can be distinguished. First, 'β' contains a variable 'π' such that no variable in 'α' is equiform with 'π'. Then 'π' must appear an even number of times in 'β', since the total number of equiform variables in '$E\alpha\beta$' is even. Second, 'β' contains a variable 'π' equiform with some variables in 'α'. Since in '$E\alpha\beta$' as well as 'α' all equiform variables appear an even number of times, the number of variables 'π' in 'β' must also be even; for evens subtracted from evens yield evens. Hence if the expressions '$E\alpha\beta$' and 'α' have property G, then 'β', which follows from them by detachment, also has property G.

We now determine that in our axiom *1 all equiform variables of any shape, i.e. all 'p's, all 'q's, and all 'r's appear exactly twice, hence an even number of times. Therefore axiom *1 has the property G. Since this property is hereditary with respect to the rules of inference assumed in the system, it must belong to all consequences of the axioms, i.e. all theses of the equivalential system. From this it follows that not all meaningful expressions of the system are derivable from our axiom. For expressions such as 'p', 'Epq', '$EpEqp$' and, in general, expressions in which at least one variable occurs an odd number of times, cannot be derived from the axiom. With this the consistency of our system is proved.

§ 8. Proof that axiom *1 is the shortest

The proof that $EEpqEErqEpr$ is the shortest axiom of the equivalential system is divided into two parts. First I set down all theses which are shorter than axiom *1, i.e. number less than eleven letters; then I show that none of these theses can be the axiom.

To obtain all theses which number less than eleven letters, we must first remind ourselves of the following two points. Firstly we ascertained in § 3 that in all meaningful expressions of our system, hence in all theses thereof, the number of 'E's is one less than the number of small letters. This yields the conclusion that every thesis of our system consists of an *odd* number of letters, hence must number 1, 3, 5, 7, or 9 letters if it is to be shorter than our axiom. Secondly we know from the previous paragraph that in all theses of the equivalential system the number of variables is *even*. It follows that no thesis of our system may consist of 1, 5, or 9 letters, for in all such expressions the number of variables is odd, being respectively 1, 3, or 5 letters. Thus we see that theses which number less than 11 letters must consist of either 3 or 7 letters.

There is only one thesis of three letters, namely

$$Epp.$$

Theses of seven letters divide according to the order of the functors into the following five groups:

I. *EEExxxx*
II. *EExExxx*
III. *EExxExx*
IV. *ExEExxx*
V. *ExExExx.*

In each group the variables 'p' and 'q' (more than two different variables cannot occur) can be ordered in three ways: *ppqq*, *pqpq*, and *pqqp*. The remaining three orderings, *qppq*, *qpqp*, and *qqpp*, result from the first three by a change of variables. We thus obtain the following fifteen theses:

I₁. *EEEppqq*   II₁. *EEpEpqq*   III₁. *EEppEqq*
I₂. *EEEpqpq*   II₂. *EEpEqpq*   III₂. *EEpqEpq*
I₃. *EEEpqqp*   II₃. *EEpEqqp*   III₃. *EEpqEqp*

IV₁. *EpEEpqq*   V₁. *EpEpEqq*
IV₂. *EpEEqpq*   V₂. *EpEqEpq*
IV₃. *EpEEqqp*   V₃. *EpEqEqp.*

These are *all* the theses which come under consideration, for theses which result from the above by identification of variables are weaker and hence may be disregarded.

Now we must show that none of these theses can be the axiom of our system. We do this in the following manner. For each thesis we give a matrix, preserving the property of deducibility, which is so constituted as to be satisfied by the given thesis but *not* by our axiom. This suffices to prove that from such a thesis our axiom cannot be derived. But if even *one* thesis of our system is not deducible from a given thesis, the latter can certainly not be the axiom of the system.

In all the matrices below, the first argument is written at the left, the second at the top. All contain only one designated value, denoted by '1'. A matrix M is *satisfied* by a given thesis, if this thesis, for all assignments of values to its variables, yields an expression which, after reduction according to matrix M, yields the value 1. The designated value 1 appears only once in the one-line of the matrix, and always in the first position, so that only $E11 = 1$ while, for all β other than 1, $E1\beta \neq 1$. This suffices to ensure that satisfaction of the matrix is preserved by deductions using the rule of detachment. For if the expressions '$E\alpha\beta$' and 'α' equal 1, so must 'β'. Satisfaction of all matrices is preserved in deductions using the rule of substitution. Thus if such a matrix is satisfied by a given thesis, it must be satisfied by all consequences of this thesis. Our axiom cannot be among these consequences if it does not satisfy the matrix.

To begin with, it is clear that our axiom

$$EEpqEErqEpr$$

does not satisfy the two-valued matrix M_1 below—the normal matrix for implication.

E	1	2
1	1	2
2	1	1

$$M_1$$

For $p/1$, $q/1$, $r/2$ we get:

$$EE11EE21E12 = E1E12 = E12 = 2.$$

On the other hand this matrix is satisfied by Epp as well as the theses: I_1. $EEppqq$, III_1. $EEppEqq$, III_2. $EEpqEpq$, IV_1. $EpEEpqq$, IV_3. $EpEEqqp$, V_1. $EpEpEqq$, V_2. $EpEqEpq$, V_3. $EpEqEqp$. For one sees at once that all these theses retain their validity when E is interpreted as the sign of implication. From this it follows that none of these theses can be the axiom.

The four-valued matrix M_2 is not satisfied by our axiom:

E	1	2	3	4
1	1	2	3	4
2	4	1	1	3
3	2	4	1	1
4	3	1	2	1

$$M_2$$

For $p/1$, $q/3$, $r/2$ we get:

$$EE13EE23E12 = E3E12 = E32 = 4.$$

On the other hand it is satisfied by theses I_2. $EEEpqpq$ and II_2. $EEpEqpq$, as may be seen from the following tables:

p	1	1	1	1	2	2	2	2	3	3	3	3	4	4	4	4
q	1	2	3	4	1	2	3	4	1	2	3	4	1	2	3	4
Epq	1	2	3	4	4	1	1	3	2	4	1	1	3	1	2	1
$EEpqp$	1	4	2	3	1	2	2	4	1	2	3	3	1	4	3	4
$EEEpqpq$	1	1	1	1	1	1	1	1	1	1	1	1	1	1	1	1

p	1	1	1	1	2	2	2	2	3	3	3	3	4	4	4	4
q	1	2	3	4	1	2	3	4	1	2	3	4	1	2	3	4
Eqp	1	4	2	3	2	1	4	1	3	1	1	2	4	3	1	1
$EpEqp$	1	4	2	3	1	4	3	4	1	2	2	4	1	2	3	3
$EEpEqpq$	1	1	1	1	1	1	1	1	1	1	1	1	1	1	1	1

The three-valued matrix M_3 is not satisfied by the axiom for $p/1$, $q/3$, $r/2$; for we have:

$$EE13EE23E12 = E3E22 = E31 = 3,$$

but it is satisfied by thesis I_3. $EEEpqqp$.

E	1	2	3
1	1	2	3
2	2	1	2
3	3	3	1

$$M_3$$

p	1	1	1	2	2	2	3	3	3
q	1	2	3	1	2	3	1	2	3
Epq	1	2	3	2	1	2	3	3	1
$EEpqq$	1	1	1	2	2	2	3	3	3
$EEEpqqp$	1	1	1	1	1	1	1	1	1

Likewise the axiom does not satisfy the three-valued matrix M_4, since we have for $p/1$, $q/3$, $r/2$:

$$EE13EE23E12 = E3E32 = E32 = 2,$$

while the matrix is satisfied by theses II_1. $EEpEpqq$ and II_3. $EEpEqqp$.

E	1	2	3
1	1	2	3
2	2	1	3
3	3	2	1
M_4			

p	1 1 1 2 2 2 3 3 3
q	1 2 3 1 2 3 1 2 3
Epq	1 2 3 2 1 3 3 2 1
$EpEpq$	1 2 3 1 2 3 1 2 3
$EEpEpqq$	1 1 1 1 1 1 1 1 1

p	1	2	3
Eqq	1	1	1
$EpEqq$	1	2	3
$EEpEqqp$	1	1	1

Furthermore, the three-valued matrix M_5 is not satisfied by the axiom; for $p/1$, $q/3$, $r/2$ yields:

$$EE13EE23E12 = E3E22 = E31 = 3.$$

However, it is satisfied by thesis III_3. $EEpqEqp$.

E	1	2	3
1	1	2	3
2	2	1	2
3	3	2	1
M_5			

p	1 1 1 2 2 2 3 3 3
q	1 2 3 1 2 3 1 2 3
Epq	1 2 3 2 1 2 3 2 1
Eqp	1 2 3 2 1 2 3 2 1
$EEpqEqp$	1 1 1 1 1 1 1 1 1

Finally, the four-valued matrix M_6 is not satisfied by the axiom for $p/1$, $q/3$, $r/2$; for we have:

$$EE13EE23E12 = E2E44 = E21 = 3,$$

but is satisfied by thesis IV_2. $EpEEqpq$.

E	1	2	3	4
1	1	4	2	4
2	3	1	4	1
3	3	1	1	2
4	4	3	3	1
M_6				

p	1 1 1 1 2 2 2 2 3 3 3 3 4 4 4 4
q	1 2 3 4 1 2 3 4 1 2 3 4 1 2 3 4
Eqp	1 3 3 4 4 1 1 3 2 4 1 3 4 1 2 1
$EEqpq$	1 1 1 1 4 4 2 2 3 3 2 2 4 4 4 4
$EpEEqpq$	1 1 1 1 1 1 1 1 1 1 1 1 1 1 1 1

From all this it follows that none of the sixteen shorter theses can be the axiom. Hence our axiom *1 is the shortest axiom for the equivalential system.

§ 9. 'Creative' definitions

In closing I would like to touch on an important methodological question which stands out with particular clarity in the
context of the equivalential system.

In mathematical logic definitions are normally introduced by
means of a special sign of definition. So, for example, one could
introduce into the equivalential calculus the expression 'Vp',
read 'verum of p', in accordance with the usage of *Principia
Mathematica*[1] as follows:

I. $Vp = Epp$ Df.

Here the identity-sign together with the following letters 'Df.'
indicate that the *definiendum* 'Vp' means the same as the 'Epp'.
Thus one may always replace 'Epp' by 'Vp' and vice versa,
and every substitution-instance of the one expression may be
replaced by a corresponding substitution-instance of the other.

There are, however, mathematical logicians who, in order to
avoid a special definition-sign, introduce definitions as equivalences. This can happen in systems in which equivalence occurs
as a primitive concept. In such systems the above definition
of 'Vp' can be written in the following manner:

II. $EVpEpp$.

Now, this definition is methodologically different from the
first. For cases may occur in which the second definition yields
more than the first, in that it can have—I can find no better
term for it—'creative' effects. The following example makes
clear what is to be understood by this.

The expression

III. $EEsEppEEsEppEEpqEErqEpr$

is easily verified to be a thesis of the equivalential system. This
thesis has the peculiarity that its consequences can be obtained
only by substitution, not by detachment. It is 'undetachable',
as can be shown by a method deriving from Tarski. If the above
thesis is to yield a new one by detachment, it must be possible
to obtain two substitution-instances of thesis III, of which one

[1] Op. cit., p. 11.

is of the type '$E\alpha\beta$' and the other of the type 'α'. These conditions may be expressed as follows:

(a) $E\alpha\beta \cong EE\gamma E\delta\delta EE\gamma E\delta\delta EE\delta\epsilon EE\zeta\epsilon E\delta\zeta,$

(b) $\alpha \cong EE\rho E\sigma\sigma EE\rho E\sigma\sigma EE\sigma\tau EE\upsilon\tau E\sigma\upsilon.$

The sign of congruence '$\cong$' means here that the left expression is equiform with that on the right. Now in the first congruence the expression '$E\gamma E\delta\delta$' corresponds to the letter 'α'. Hence the following congruences must also hold:

(c) $\alpha \cong E\gamma E\delta\delta \cong EE\rho E\sigma\sigma EE\rho E\sigma\sigma EE\sigma\tau EE\upsilon\tau E\sigma\upsilon.$

This yields the further result, that the following expressions must be equiform:

(d) $\gamma \cong E\rho E\sigma\sigma$ (e) $\delta \cong E\rho E\sigma\sigma$ (f) $\delta \cong EE\sigma\tau EE\upsilon\tau E\sigma\upsilon.$

From (e) and (f) finally we get the following congruences:

(g) $\rho \cong E\sigma\tau$ (h) $\sigma \cong E\upsilon\tau$ (i) $\sigma \cong E\sigma\upsilon.$

The last congruence yields an absurdity; for it is impossible for σ to be equiform with an expression which contains σ as a proper part. From this we conclude that it is not impossible to find two substitution-instances of thesis III of the forms '$E\alpha\beta$' and 'α'. Hence thesis III is undetachable. From this it follows immediately that no *shorter* thesis can be derived from thesis III, in particular not our axiom $EEpqEErqEpr$.

If, however, definitions of the second sort are now introduced in the equivalential system, e.g. definition II, then it is easily shown that thesis III can be postulated as the axiom of the system. For we have:

III. $EEsEppEEsEppEEpqEErqEpr$

II. $EVpEpp$

III $s/Vp \times E$II$-E$II$-$IV

IV. $EEpqEErqEpr.$

Thesis IV is the axiom of our system. From III by itself it cannot be derived; it can be inferred only with the help of definition II. The new term 'V', however, which was introduced by the definition, does not appear in this thesis. Hence definition II leads to theses which cannot be derived from the thesis assumed as an axiom, although in these theses only primitive concepts of the system occur. Such definitions I call 'creative'. It is clear that one cannot get thesis IV from III if one introduces 'Vp' by a definition of the first sort.

In deductive systems the role of definitions would seem to consist mainly in allowing us to replace longer and more complicated expressions by shorter and simpler ones. Moreover, some definitions can bring with them new, intuitively valuable insights. Under no circumstances, however, do definitions seem to be intended to give new properties to the undefined primitive concepts of the system. Primitive concepts should be characterized solely by *axioms*. If one takes this position, one should avoid the use of creative definitions whenever possible.

6

INTRODUCTORY REMARKS TO THE CONTINUATION OF MY ARTICLE: 'GRUNDZÜGE EINES NEUEN SYSTEMS DER GRUNDLAGEN DER MATHEMATIK'†

STANISŁAW LEŚNIEWSKI

In 1929 I published in *Fundamenta Mathematicae* the beginning of the article referred to in the title.[1] The continuation of this article has not yet appeared in print. This fact derives from circumstances about which I wrote in 1930: 'The succeeding part of the above-mentioned article in German which I had already submitted in 1929, and which had been accepted by the editors of *Fundamenta Mathematicae*, I withdrew for personal reasons. In the circumstances it is difficult for me to foresee whether, where, and when, I can find place for its publication.'[2] The withdrawn manuscript remained for more than seven years in my desk, where it awaited a more auspicious occasion for its publication.

The editors of *Collectanea Logica* have obligingly offered me space for the continuation of my article. I am naturally taking speedy advantage of this kind offer, submitting the above-

† This paper was to have appeared, under the title 'Einleitende Bemerkungen zur Fortsetzung meiner Mitteilung u.d.T. "Grundzüge eines neuen Systems der Grundlagen der Mathematik"', in vol. 1 of the periodical *Collectanea Logica* (Warsaw, 1939), pp. 1–60. (For the fate of this journal see the bibliographical footnote to paper 5 of this volume.) An offprint copy of Leśniewski's paper survives in the Harvard College Library, together with the continuation (§12) of his original article, which was also to have appeared in *Collectanea Logica*. Translated by W. Teichmann and S. McCall.

[1] Stanisław Leśniewski, 'Grundzüge eines neuen Systems der Grundlagen der Mathematik', *Fundamenta Mathematicae* 14 (1929), pp. 1–81.

[2] Stanisław Leśniewski, 'Über die Grundlagen der Ontologie', *Comptes rendus des Séances de la Société des Sciences et des Lettres de Varsovie*, Cl. iii, 23 (1930), p. 112. Paper presented by J. Łukasiewicz at the meeting of 22 May 1930.

mentioned manuscript for the first volume of the new journal with only slight symbolic improvements and bibliographical additions. Because of the long delay in the printing of my article, and the change of place of publication, I have, for the convenience of the reader of this further part of the intended whole, and in accordance with the wishes of the editors, decided to preface it with a résumé of its beginning, published in *Fundamenta Mathematicae*. Since I do not wish to ruin the architectonic structure of my article by including this résumé, I have decided to bring out these 'Introductory Remarks' separately. In them will be found, besides the résumé, various minor observations of an explicative, informative, and polemical nature connected with my article.

The already published part of my *Grundzüge eines neuen Systems der Grundlagen der Mathematik* consists of an introduction and the first eleven paragraphs of section I, entitled 'The Foundations of Protothetic'. The more important items contained in this section are summarized in what follows.

Introduction to my article

The object of the paper is a succinct presentation of my system of the foundations of mathematics. This system consists of three deductive theories, whose union forms one of the possible bases of the whole structure of mathematics. The theories in question are the following: (1) What I call *Protothetic*, which is the result of a certain peculiar enlargement of the well-known theory which goes by the name of the 'propositional calculus', or 'theory of deduction'. (2) What I call *Ontology*, which forms a type of modernized 'traditional logic' and which most closely resembles in its content and power Schröder's 'logic of classes', regarded as including the theory of 'individuals'. (3) What I call *Mereology*, whose first outline was published by me in a work of 1916 entitled *Die Grundlagen der allgemeinen Mengenlehre. I*.

§ 1 of the article

In 1912 Henry Maurice Sheffer showed that in the theory of deduction of Whitehead and Russell there could be defined two functions of two propositional variables, in terms of either of which as sole primitive the two primitive functions of Whitehead and Russell, namely alternation and negation, could be defined. One of these functions of Sheffer's is equivalent for all values of its variables to the function '$\sim (p \lor q)$'; the other to the function '$\sim p . \lor . \sim q$'. In 1916 J. G. P. Nicod built up the theory of deduction from a single axiom, which apart from variables contained only the sign for the second of Sheffer's functions. For this sign Nicod used the vertical stroke '$|$'.

In the definition of non-primitive functions in the theory of deduction, both Sheffer and Nicod make use of a special definitional sign of identity, which is not itself defined in terms of the primitive functions of the system. This fact makes it difficult to say that Nicod's theory of deduction is really based upon the sole primitive sign '$|$'. In 1921 I remarked that if one wishes really to base a system of the theory of deduction which contains definitions upon a single primitive term, one must write definitions using this primitive term without resorting to a special definitional sign of identity. In particular, if one were to make such a reform in Nicod's system, the definitions occurring in the system could be written, for example, in the form of an expression of the type

$$\Big(\big(p \mid \langle q \mid q \rangle\big) \mid \big\{q \mid \langle p \mid p \rangle\big\}\big) . \big\{\big(p \mid \langle q \mid q \rangle\big) \mid \{ q \mid \langle p \mid p, \rangle\big\}\Big)$$

which, as it is easy to verify, is equivalent to the corresponding equivalence '$p \equiv q$'.

In 1922 Alfred Tarski established that, by employing functional variables and quantifiers, all the familiar functions of the theory of deduction could be defined using the equivalence function as the sole primitive function. The central point in Tarski's arguments consists in the proof of a theorem stating that

$$[p,q] :: p . q . \equiv . \therefore [f] . \therefore p \equiv : [r] . p \equiv f(r) . \equiv . [r] . q \equiv f(r),$$

and in the demonstration that in systems of logistic in which the following thesis holds:

$$[p,q,f] : p \equiv q . f(p) . \supset . f(q)$$

the following thesis must also hold:

$$[p,q] \therefore p \cdot q \cdot \equiv \; : [f] : p \equiv \cdot f(p) \equiv f(q).$$

§ 2 of the article

In 1922 I sketched my conception of 'semantic categories' and constructed for the fundamental mathematical theories, especially for 'Protothetic' and 'Ontology', directives for definition and inference adapted to this conception. In my axiomatic investigations concerning the directives of protothetic I concentrated upon the task of axiomatizing as simply as possible a system based upon the sign of equivalence as the only primitive term. Tarski's above-mentioned work had made such a system possible, but it had not yet been realized in fact.

§ 3 of the article

Terminological note. I say of an expression X that it is an 'equivalence proposition' when it satisfies the following conditions: (*a*) X is a propositional variable or an equivalence; (*b*) if any Y is an equivalence forming a proper or improper part of the expression X, then each of the arguments of the equivalence Y is either a propositional variable or an equivalence.

I began the construction of an axiomatic system of protothetic, based on the sign of equivalence as the sole primitive term, with the construction of a weaker system. This system was to consist of all the equivalence propositions that can be proved in the ordinary theory of deduction. As the axioms of this weaker system I took the following two propositions:

A1. $p \equiv r \cdot \equiv \cdot q \equiv p : \equiv \cdot r \equiv q$

A2. $p \equiv \cdot q \equiv r : \equiv \; : p \equiv q \cdot \equiv r.$

(The thesis which here appears as axiom *A2* had already been proved before the year 1922 by Jan Łukasiewicz.) As for the directives, I made use in this system (i) of the directive for substitution of equivalence propositions for variables in propositions that already belong to the system, and (ii) of the directive for detachment, permitting the addition to the system of a proposition S when the system already has both an equivalence A, whose second argument is equiform with S, and a proposition equiform with the first argument of the equivalence A. This system I call the system $\mathfrak{S}$.

§ 4 of the article

I obtained a further stage of development of protothetic by considering the following question: with which axioms and directives would the system $\mathfrak{S}$ have to be strengthened, in order to obtain from it a system of the ordinary propositional calculus, completed by the addition of the thesis

$$[p,q,f] \therefore p \equiv q . \supset : f(p) . \equiv . f(q)?$$

The axioms of an enriched propositional calculus named $\mathfrak{S}_1$, constructed by me in answering this question, may be written in Peano–Whitehead–Russell style as follows:

Ax. I. $[p,q,r] \therefore p \equiv r . \equiv . q \equiv p : \equiv . r \equiv q$

Ax. II. $[p,q,r] \therefore p \equiv . q \equiv r : \equiv : p \equiv q . \equiv r$

Ax. III. $[g,p] : \because [f] :: g(p,p) . \equiv \therefore [r] : f(r,r) . \equiv . g(p,p) :$
$\equiv : [r] : f(r,r) . \equiv . g(p \equiv . [q] . q,p) :: \equiv . [q] . g(q,p).$

In the authentic symbolism of protothetic these axioms have the following form (expressions of the type '$\phi(pq)$' here replace the corresponding expressions of the type '$p \equiv q$' in *Ax. I*– *Ax. III*):

A1. $\llcorner pqr \lrcorner \ulcorner \phi\big(\phi(\phi(pr)\phi(qp))\phi(rq)\big) \urcorner$

A2. $\llcorner pqr \lrcorner \ulcorner \phi\big(\phi(p\phi(qr))\phi(\phi(pq)r)\big) \urcorner$

A3. $\llcorner gp \lrcorner \ \phi\Big(\llcorner f \lrcorner \ulcorner \ \phi\Big(g(pp)\phi\big(\llcorner r \lrcorner \ulcorner \phi(f(rr)g(pp))\urcorner \llcorner r \lrcorner \ulcorner \phi\big(f(rr)$
$g(\phi(p\llcorner q \lrcorner \ulcorner q \urcorner)p))\urcorner\big)\Big)\urcorner\Big) \llcorner q \lrcorner \ulcorner g(qp) \urcorner \Big)$

I obtained new propositions from propositions already belonging to the system by the use of six directives, which may be informally characterized as follows:

(α) The directive for detachment, as in the system $\mathfrak{S}$.

(β) The directive for substitution.

(γ) The directive for the 'distribution of the quantifier', which, if the system already contains a thesis T consisting of a universal quantifier Q and an equivalence A within the scope of the quantifier, permits the addition of a new thesis formed from the thesis T through the distribution—in a definite and in practice unambiguous way—of all or some variables occurring in the quantifier Q into quantifiers standing before the two arguments of the equivalence A.

(δ) The directive for the writing of definitions in the form of equivalences, the *definiendum* occurring as the first argument.

(ϵ) The directive for the writing of definitions consisting of a universal generalization of an equivalence, containing the *definiendum* as the first argument.

(ζ) The directive concerning universal quantifiers, which, in conjunction with the other rules, permits in practice the carrying out of all the familiar operations with these quantifiers.

§ 5 of the article

The axiom $A3$ made it possible for me to employ, from a certain point in the system $\mathfrak{S}_1$, a method of proving or disproving propositions beginning with universal quantifiers containing propositional variables. This method makes appeal to certain corresponding propositions, already proved or disproved in the system, which are made up of propositions to be proved or disproved by the substitution for the propositional variables occurring in them of the expressions

$$\text{`}\llcorner q \lrcorner \ulcorner q \urcorner\text{'} \quad \text{and} \quad \text{`}\phi(\llcorner q \lrcorner \ulcorner q \urcorner \llcorner q \lrcorner \ulcorner q \urcorner)\text{'},$$

which correspond in $\mathfrak{S}_1$ to the 'zero' and 'one' of the traditional propositional calculus. The axiom $A3$ was thus a sort of axiomatic correlate for the well-known method of verification of formulae of the propositional calculus, i.e. by substitution of 'zero' and 'one' for the variables contained in them. Regarded genetically, this axiom corresponded to the verification rules of Łukasiewicz's 1921 paper 'Two-valued Logic'.

I did not know how to prove, in the system $\mathfrak{S}_1$, a certain sequence of meaningful propositions which I consider as valid as the familiar theses of the ordinary propositional calculus. Because I wanted to construct a system in which this would be possible, which would contain the system $\mathfrak{S}_1$, and at the same time contain no meaningful proposition that I did not know either how to prove or how to disprove, in 1922 I completed the system $\mathfrak{S}_1$ by a new directive η. This directive was formed on the pattern of one of Łukasiewicz's rules, except that, instead of propositional variables, it concerned all and only those variables occurring in $\mathfrak{S}_1$ which were *not* propositional variables. The directive η permitted me to join to the system a new thesis T, beginning with a universal quantifier governing variable function-signs of any 'semantic category', when the system already included all theses which could be obtained from T by substituting for its above-mentioned variable function-signs certain constant signs, the method of definition of the latter having been completely laid down for all semantic categories in advance. The completed system obtained from the system $\mathfrak{S}_1$ by the addition of directive η I call the system $\mathfrak{S}_2$. This is one of the many possible mutually equivalent systems of the theory that I name prototthetic.

To formulate the directive η precisely I needed a complicated apparatus of numerous terminological explanations. This circumstance induced me to look for some other directive that would accomplish the same theoretical effect as the directive η, and at the same time could be precisely formulated in a simpler way. I concentrated mainly on the question, whether the directive η could not be replaced by a directive which would permit a direct determination of the 'extensionality' of the different categories of functions occurring in prototthetic.

§ 6 of the article

When I formulated the directive ζ of the system $\mathfrak{S}_1$ I was aiming above all at the simplest method of guaranteeing the demonstrability in the system of propositions of the type:

$$[f,p] \mathbin{.\,.} p \supset . [q] . f(q) : \equiv : [q] : p \supset . f(q),$$
$$[f,p] \mathbin{.\,.} p \supset . [q,r] . f(q,r) : \equiv : [q,r] : p \supset . f(q,r),$$
$$[f,p] \mathbin{.\,.} p \supset . [q,r,s] . f(q,r,s) : \equiv : [q,r,s] : p \supset . f(q,r,s),$$

etc.,

the signs here appearing in the expressions '$f(q)$', '$f(q,r)$', '$f(q,r,s)$', etc., being of any number or any semantic category. These propositions assured me of the possibility of carrying out all the familiar operations with universal quantifiers. Directive ζ was somewhat simplified by Tarski, and his result was further simplified by me. However, the whole question was finally liquidated in 1922 by Tarski, who showed that no special directive at all was necessary, each of the above-mentioned propositions being provable in the system $\mathfrak{S}_1$ with the help of the directives already in that system. Tarski's result was based, among other things, upon an earlier result obtained by me; namely the possibility of proving in the system $\mathfrak{S}_1$, without the help of directive ζ,

(a) all theses corresponding to the theses of $\mathfrak{S}$;

(b) theses corresponding to the traditional theorems

$$'[p] . 0 \supset p' \quad \text{and} \quad '[p] : 1 \supset p . \equiv p';$$

(c) the availability in practice of the method of proving theses beginning with universal quantifiers containing propositional variables by appealing to theses, already proved in the system, which may be constructed out of the theses to be proved by substituting 'zero' and 'one' for the propositional variables of those theses.

In showing that the directive ζ is dispensable in the system $\mathfrak{S}_1$, Tarski also noted that analogous directives in the systems based on $\mathfrak{S}_1$ may be dispensed with. In particular this was true of my 'Ontology'. Tarski extended the method of proving the propositions mentioned at the beginning of this section without the help of directive ζ to all analogous meaningful propositions in which, in place of the expressions '$f(q)$', '$f(q,r)$', etc., expressions of any construction whatever occur.

§ 7 of the article

On the basis of the system $\mathfrak{S}_2$ I could obtain, besides all theses of $\mathfrak{S}_1$ (including all theses of the usual theory of deduction) theses of the following two kinds:

(a) Theses which determined, in a universal form, the 'extensionality' of all functions occurring in the system, independent of the semantic category of expressions occurring in these functions. The following proposition will serve as an example:

$$[f, g] \therefore [p, q] : f(p, q) . \equiv . g(p, q) : \equiv : [\phi] : \phi\{f\} . \equiv . \phi\{g\}.$$

(b) Theses which established, in a universal form, that every propositional function occurring in the system of the type '$\phi\{f\}$', '$\phi\{f, g\}$', '$\phi\{f, g, h\}$', etc., in which at least one argument is not a proposition, is satisfied for all values of its arguments whenever the corresponding 'logical product' of certain propositions is satisfied. This logical product is the product of those propositions which are the values of the function in question for certain values of its arguments, such values for each semantic category being finite in number and specifiable in advance. (As values of arguments of different semantic categories there occur besides the 'zero' and 'one' of the traditional calculus those constant function-signs referred to above in the résumé of § 5.) An example of this kind of thesis is:

$$[\phi] : [f] . \phi\{f\} . \equiv . \phi\{vr\} . \phi\{as\} . \phi\{\sim\} . \phi\{fl\},$$

in which the expressions 'vr', 'as', '$\sim$' and 'fl' are constant function-signs of propositional functions of one propositional argument.

The problem I mentioned in the résumé of § 5, namely whether the directive η of the system $\mathfrak{S}_2$ could not be replaced by an 'extensionality directive', was for me in effect equivalent to the problem, whether the addition to the system $\mathfrak{S}_1$ of all theses of kind (a) made it possible to obtain in this system, without the use of directive η, all propositions of kind (b). This question was answered in the affirmative by Tarski, who in 1922 sketched a general method of proving in $\mathfrak{S}_1$ individual propositions of kind (b), given corresponding propositions of kind (a). For reasons already given I decided to replace the directive η by a new 'extensionality directive' η^*. The system of prototothetic based on the axioms A1–3 and the directives α, β, γ, δ, ϵ, and η^* I call the system $\mathfrak{S}_3$.

§ 8 of the article

From the beginning I took pains to formulate the directives of protothetic in such a way that they could be easily adapted to systems constructed on the same pattern but on the basis of different primitive terms. In this connexion I took account of the fact that the system $\mathfrak{S}_2$ could be transformed almost automatically into a system of protothetic in which the implication sign occurs as the sole primitive term. The axioms and directives of this system may be given in outline as follows:

A. *Directives*

(α_1) The directive for detachment, permitting the addition to the system of a proposition S when the system already contains both a conditional K, whose consequent is equiform with S, and a proposition equiform with the antecedent of K.

(β_1) The directive for substitution.

(γ_1) The directive for the distribution of the quantifier, analogous to rule γ of the system $\mathfrak{S}_2$ except in that it concerns conditional propositions rather than equivalences.

(δ_1) and (ϵ_1) The directives which permit the writing of definitions analogous to those allowed for by the directives δ and ϵ of the system $\mathfrak{S}_2$. Unlike the definitions of directives δ and ϵ, however, these definitions are expressed in the form not of an equivalence, nor of an equivalence preceded by a universal quantifier, but in the form of any other function stipulated in advance for all definitions, this function being expressed in terms of the implication sign and being equivalent either to the equivalence in question or to the equivalence preceded by a universal quantifier.

(ζ_1) The directive analogous to and serving the same purpose as the directive ζ of $\mathfrak{S}_2$, except that it is framed in terms of conditional propositions instead of equivalences.

(η_1) The directive exactly analogous to η of $\mathfrak{S}_2$.

B. *Axioms*

(*I*) Any axiom set consisting of theses of the classical theory of deduction which contain no other constants besides the implication sign, and from which, using the directives of the system, the whole of the ordinary theory of deduction can be derived.

(II) An axiom analogous to the axiom $A3$ of $\mathfrak{S}_2$. This axiom makes possible in the system a method of proving or of disproving propositions beginning with universal quantifiers containing propositional variables. The method makes appeal to corresponding propositions, already proved or disproved in the system, which can be constructed out of the propositions to be proved or disproved by substituting for the propositional variables they contain the expressions '$\llcorner q \lrcorner \ulcorner q \urcorner$' and '$\diamondsuit(\llcorner q \lrcorner \ulcorner q \urcorner \llcorner q \lrcorner \ulcorner q \urcorner)$', corresponding to the 'zero' and 'one' of the traditional propositional calculus. (The sign '$\diamondsuit$' appearing in the second expression plays the role of the implication sign in my symbolism.)

The axiom in question could, for example, take the following form:

$$\llcorner gpq \lrcorner \ulcorner \diamondsuit\Big(g(pp)\diamondsuit\big(g(\diamondsuit(p\llcorner q \lrcorner \ulcorner q \urcorner)p)g(qp)\big)\Big)\urcorner.$$

With the discovery of the system $\mathfrak{S}_3$ it became clear that a system of protothetic based on the implication sign could have simpler directives if it were modelled on $\mathfrak{S}_3$ rather than $\mathfrak{S}_2$. This would involve discarding the directive ζ_1, and replacing η_1 by η_1^*, the latter 'extensionality directive' being analogous to the directive η^*. A certain system of this kind, which I constructed on the basis of axioms of the types I and II above, and directives of types α_1, β_1, γ_1, δ_1, ϵ_1, and η_1^*, I named the system $\mathfrak{S}_4$.

In 1922 Tarski established that, however many axioms a given set A sufficient for $\mathfrak{S}_4$ may have, it may be replaced by a set of only two axioms without altering the directives of the system. These axioms are (a) the proposition '$\llcorner pq \lrcorner \ulcorner \diamondsuit(p\diamondsuit(qp))\urcorner$', and ($b$) the 'logical product' of all propositions belonging to A which are distinct from (a), the 'logical product' of the propositions 'P' and 'Q' being taken to be a proposition of the type

$$\llcorner r \lrcorner \ulcorner \diamondsuit(\diamondsuit(P\diamondsuit(Qr))r)\urcorner.$$

In conversation with Tarski I then conjectured that his result could be improved upon, and that for protothetic two axioms could suffice, one of which would have approximately the same form as the proposition given as an example under *II* above, while the other would be a simple thesis of the ordinary theory of deduction. In the same year (1922) Tarski established that:

(*A*) For the construction of a system of protothetic with the directives of the system $\mathfrak{S}_4$ the following two axioms suffice:

(1) $\llcorner pq \lrcorner ^{\ulcorner} \diamondsuit (p \diamondsuit (qp)) ^{\urcorner}$

(2) $\llcorner pqrf \lrcorner ^{\ulcorner} \diamondsuit \Big(f(rp) \diamondsuit \big(f(r \diamondsuit (p \llcorner s \lrcorner ^{\ulcorner} s^{\urcorner})) f(rq) \big) \Big) ^{\urcorner}.$

(*B*) A system of protothetic could also be based on a single axiom if one took, besides the directives $\alpha_1, \beta_1, \gamma_1, \eta_1^*$ of $\mathfrak{S}_4$, new directives δ_1^* and ϵ_1^* in place of δ_1 and ϵ_1. These new directives would permit the introduction into the system of definitions in the form of two converse conditionals corresponding to one equivalence (directive δ_1^*), or two such conditionals preceded by universal quantifiers (directive ϵ_1^*). The sole axiom of the system might, for example, be axiom 2 of *A* above.

§ 9 of the article

In 1923 Tarski noted that however many axioms a given set *A* sufficient for $\mathfrak{S}_3$ may have, it may be replaced by a set of only two axioms without altering the directives of the system. Of these one is the thesis

$$\llcorner pq \lrcorner ^{\ulcorner} \diamondsuit \big(\diamondsuit (pq) \diamondsuit (qp) \big) ^{\urcorner},$$

while the other is the 'logical product', expressed in terms of the function '$\diamondsuit(pq)$', of all propositions belonging to *A* which are distinct from the above thesis. Tarski also established at the same time that such a 'logical product' of the axioms *A1–A3* introduced in the résumé of § 4 could take the form of the following expression:

$$\llcorner hs \lrcorner ^{\ulcorner} \diamondsuit \Big(h(Ps) \diamondsuit \big(h \big(\llcorner kt \lrcorner ^{\ulcorner} \diamondsuit (k(Qt) \diamondsuit (k(Rt)Q)) ^{\urcorner} s \big) P \big) \Big) ^{\urcorner}$$

where the letters '*P*', '*Q*', and '*R*' are to be replaced by the axioms *A1*, *A2*, and *A3*.

§ 10 of the article

In the same year 1923 I observed in addition that if I replaced the directives δ and ϵ of $\mathfrak{S}_3$ by the corresponding directives δ^* and ϵ^*, prescribing that the *definiendum* of a definition should occur as the second rather than as the first argument of the stipulated equivalence, I would obtain a system of protothetic equivalent to the system $\mathfrak{S}_3$. The system based upon axioms *A1–A3* and directives α, β, γ, δ^*, ϵ^*, and η^* I call $\mathfrak{S}_5$. I showed that with the help of its directives a system equivalent to it could be based upon a single axiom. This axiom consists of the 'logical product', after the pattern of the expression

$$\llcorner fp\lrcorner\ulcorner \phi\big(f(Pp)\phi\big(f(Qp)P\big)\big)\urcorner,$$

of two propositions; one of these—corresponding to 'P' in the above expression—being the proposition

$$\llcorner pq\lrcorner\ulcorner \phi(\phi(pq)\phi(qp))\urcorner,$$

while the other —corresponding to 'Q'—is again one of those 'logical products' which occur as the second axiom of the axiom sets constructed by the method of Tarski referred to in the résumé of § 9. I noted at the same time that it would be simple to produce such an axiom, sufficient as a basis for protothetic, if one retained the directives β, γ, δ, ϵ, and η^* but replaced α by α^*. This new directive would permit the addition to the system of a proposition S, when there already belonged to the system both an equivalence A, whose first argument was equiform with S, and a proposition equiform with the second argument of A. I know of no proposition which would serve as the sole axiom of a system of protothetic based either on the directives α, β, γ, δ, ϵ, and η^*, or the directives α^*, β, γ, δ^*, ϵ^*, and η^*.

In 1925 Tarski gave a method of reducing to a single axiom the axiomatic basis of any system of protothetic with the directives of $\mathfrak{S}_4$ and the implication sign as its primitive term. In connexion with this he also showed how systems of the classical theory of deduction, when they contain the implication sign among their primitive terms, can be based upon a single axiom.

§ 11 of the article

The single axiom of the system of protothetic with the directives of $\mathfrak{S}_5$ was, in the years 1923–6, successively simplified by me and Mordchaj Wajsberg. The following axiom of such a protothetic, devised by me in 1926, has not yet, as far as I know, been further shortened by anybody:†

$$\llcorner fpqrst \lrcorner \; \phi \bigg(\phi(pq) \llcorner g \lrcorner \; \phi \Big(f\big(pf(p \llcorner u \lrcorner \ulcorner u \urcorner)\big)\phi \big(\llcorner u \lrcorner \ulcorner f(qu) \urcorner \phi$$
$$\big(g(\phi(\phi(rs)t)q)g(\phi(\phi(st)r)p)\big)\Big)\bigg)^{\urcorner} \bigg)^{\urcorner} .$$

The central position in § 11 of my article is occupied by the directives of the system $\mathfrak{S}_5$, which were formulated with all the precision of which I was capable. In the statement of these directives there occurs a sequence of terms whose meaning I first establish with the help of forty-nine 'terminological explanations'. As I see no possibility of summarizing these explanations, I am compelled at this point to refer the reader to my original article (pp. 59–76). In my résumé of the contents of § 11 I shall confine myself to a few general remarks.

I first expounded the directives of the system $\mathfrak{S}_5$ in my lectures on 'Logistic' in the academic years 1924–5 and 1925–6. These directives were given in considerably simpler form in my lectures on the 'Foundations of Ontology' in the academic year 1926–7. For some of the simplifications in the exposition of the directives of protothetic that I introduced at that time I am indebted to Adolf Lindenbaum.

In the final statement of the directives of the system $\mathfrak{S}_5$, I unified in suitably 'organic' fashion directives δ^* and ϵ^* in a single directive for writing definitions. After this simplification the whole system of directives could be assembled in the following brief schema:

(1) the directive for writing definitions (union of δ^* and ϵ^*);

† [*Ed. note.* Sobociński in 1945 reduced protothetic to a single axiom containing only fifty-four signs. See E. C. Luschei, *The Logical Systems of Leśniewski*, Amsterdam 1962, §6.2.2, and the references therein.]

(2) the directive for distribution of the quantifier (γ);

(3) the directive for detachment (α);

(4) the directive for substitution (β);

(5) the extensionality directive (η^*).

The directives of the system $\mathfrak{S}_5$ presuppose no special shape for the constant terms of the system as opposed to the variables. All signs, with the exception of parentheses and the signs '$\llcorner$', '$\lrcorner$', '$\ulcorner$', and '$\urcorner$', can occur in the system either as constants or variables, the character they have in such and such a formula depending on the variety and position of the quantifiers contained in the formula in question.

The directives of the system presuppose no special shape for the constants or variable signs of one semantic category as contrasted with the signs of another.

Function signs may be placed, according to the directives of the system, only in front of the parentheses enclosing their arguments.

The different arguments need not, according to the directives, be separated from each other by commas.

The directives allow no possibility of introducing into the system any kind of quantifier other than the previously mentioned universal quantifier governing any number of variables.

The directives do not allow us to obtain in the system any thesis containing a universal proposition of the type

$$\llcorner ab \ldots \lrcorner \ulcorner \llcorner kl \ldots \lrcorner \ulcorner f(ab \ldots kl \ldots) \urcorner \urcorner$$

In the cases in which we would normally have to deal with propositions of this type we are confronted in my system only with corresponding expressions of the type

$$\llcorner ab \ldots kl \ldots \lrcorner \ulcorner f(ab \ldots kl \ldots) \urcorner,$$

in which, in view of a sufficiently 'liberal' formulation of the directives of the system, the variables occurring in the quantifier may easily be permuted.

The system's directive for substitution, while it permits different substitutions for variables, does not allow anything to be substituted for a whole expression of the type '$f(ab \ldots)$'.

In the last long footnote to § 11 of my article I state that, among those works which take as their task the construction of the foundations of mathematics, I do not know of a single one which establishes, in a way that raises no doubt as to interpretation, a combination of rules which is sufficient for the derivation of all theses effectively recognized in the system, and which does not at the same time lead to a contradiction in some way or other not foreseen by the author. I construct explicitly two such contradictions, unforeseen by their authors, in the systems of Leon Chwistek (*The Theory of Constructive Types*) and of J. von Neumann (*Zur Hilbertschen Beweistheorie*).

I pass now to the minor observations, mentioned at the beginning of this paper, with which I would like to supplement the résumé I have just given of the part of my article published in *Fundamenta Mathematicae*.

Supplementary remark I

In the course of lectures, entitled 'Introduction to Mathematical Logic', which I delivered in the University of Warsaw in the academic year 1933–4, I remarked, as it is in fact easy to verify, that

$$p \equiv q \, . \, \equiv \, \therefore \, p \mid q \, . \mid : p \mid p \, . \mid . \, q \mid q,$$

so that in a system of the theory of deduction based upon the primitive term '|' definitions can be written by means of corresponding formulae of the type '$p \mid q \, . \mid : p \mid p \, . \mid . \, q \mid q$'. Such formulae are considerably simpler than those mentioned in § 1 of my article of the type

$$p \mid . \, q \mid q : \mid : q \mid . \, p \mid p \, \therefore \mid \, \therefore \, p \mid . \, q \mid q : \mid : q \mid . \, p \mid p.$$

Supplementary remark II

The sketch of the argument of § 3, which aims at proving that the system $\mathfrak{S}$ consists of all equivalence propositions provable in the usual theory of deduction, begins with a direct derivation in the system $\mathfrak{S}$ of the theses *T1–T79*. Of these only *T7*, *T19–T21*, *T69*, *T70*, and *T79* are necessary for the continuation of my argument. All the others are only auxiliary premises for them.

In 1929 Łukasiewicz considerably shortened my deductions. He deduced in the system $\mathfrak{S}$ the theses $T7$, $T19$–21, $T69$, $T70$ and a correlate of my thesis $T79$ with the help of forty-eight successive applications of the system's directives instead of my seventy-nine applications. The thesis which I call here a correlate of the thesis $T79$ differs from this thesis in variables only, and, for my proof, this difference is insignificant. I give here Łukasiewicz's derivation copied from the author's original. Łukasiewicz's theses are indicated by the signs '$Ł1$', '$Ł2$', etc., to '$Ł48$'. The thesis $Ł48$ is the correlate of my thesis $T79$, mentioned above. $A1$ and $A2$ are, of course, the axioms of the system $\mathfrak{S}$, introduced above.

$A1.\ p \equiv r\ .\ \equiv\ .\ q \equiv p : \equiv\ .\ r \equiv q$

$A2.\ p \equiv\ .\ q \equiv r : \equiv\ : p \equiv q\ .\ \equiv r$

$Ł1$ (my thesis $T1$). $q \equiv r\ .\ \equiv\ .\ r \equiv q : \equiv\ .\ r \equiv r \quad A1\left[\dfrac{q}{p},\dfrac{r}{q}\right]$

$Ł2$ (my $T2$). $r \equiv\ .\ q \equiv r : \equiv\ : r \equiv q\ .\ \equiv r \therefore \equiv\ : q \equiv r\ .$

$\equiv\ .\ r \equiv q \qquad\qquad \left[A1\,\dfrac{r}{p},\dfrac{r \equiv q}{q},\dfrac{q \equiv r}{r}\right]$

$Ł3$ ($T4$). $r \equiv\ .\ q \equiv r : \equiv\ : r \equiv q\ .\ \equiv r \qquad\qquad \left[A2\,\dfrac{r}{p}\right]$

$Ł4$ ($T7$). $q \equiv r\ .\ \equiv\ .\ r \equiv q \qquad\qquad\qquad [Ł2, Ł3]$

$Ł5$ ($T12$). $p \equiv r\ .\ \equiv\ .\ q \equiv p : \equiv\ .\ r \equiv q \therefore \equiv \therefore r \equiv q\ .$

$\equiv\ : p \equiv r\ .\ \equiv\ .\ q \equiv p \qquad \left[Ł4\,\dfrac{p \equiv r\ .\ \equiv\ .\ q \equiv p}{q},\dfrac{r \equiv q}{r}\right]$

$Ł6$ ($T19$). $r \equiv r \qquad\qquad\qquad\qquad\qquad [Ł1, Ł4]$

$Ł7$ ($T20$). $q \equiv r\ .\ \equiv\ .\ q \equiv r \qquad\qquad \left[Ł6\,\dfrac{q \equiv r}{r}\right]$

$Ł8$ ($T21$). $r \equiv r\ .\ \equiv\ .\ r \equiv r \qquad\qquad \left[Ł7\,\dfrac{r}{q}\right]$

$Ł9$ ($T22$). $r \equiv q\ .\ \equiv\ : p \equiv r\ .\ \equiv\ .\ q \equiv p \qquad [Ł5, A1]$

$Ł10.\ q \equiv r\ .\ \equiv\ .\ r \equiv q : \equiv \therefore p \equiv\ .\ q \equiv r : \equiv\ : r \equiv q\ .\ \equiv p$

$\left[Ł9\,\dfrac{r \equiv q}{q},\dfrac{q \equiv r}{r}\right]$

Ł11. $p \equiv . q \equiv r : \equiv : r \equiv q . \equiv p$ [Ł10, Ł4]

Ł12. $p \equiv . q \equiv r : \equiv : p \equiv q . \equiv r \therefore \equiv \therefore r \equiv . p \equiv q :$

     $\equiv : p \equiv . q \equiv r$ $\left[\text{Ł11} \dfrac{p \equiv . q \equiv r}{p}, \dfrac{p \equiv q}{q} \right]$

Ł13. $p \equiv r . \equiv . q \equiv p : \equiv . r \equiv q \therefore \equiv \therefore q \equiv r . \equiv : p$

     $\equiv r . \equiv . q \equiv p$ $\left[\text{Ł11} \dfrac{p \equiv r . \equiv . q \equiv p}{p}, \dfrac{r}{q}, \dfrac{q}{r} \right]$

Ł14. $r \equiv . p \equiv q : \equiv : p \equiv . q \equiv r$ [Ł12, A2]

Ł15. $q \equiv r . \equiv : p \equiv r . \equiv . q \equiv p \therefore \equiv \therefore p \equiv r . \equiv : q$

     $\equiv p . \equiv . q \equiv r$ $\left[\text{Ł14} \dfrac{p \equiv r}{p}, \dfrac{q \equiv p}{q}, \dfrac{q \equiv r}{r} \right]$

Ł16. $p \equiv r . \equiv : q \equiv p . \equiv . q \equiv r \therefore \equiv \therefore q \equiv p . \equiv : q$

     $\equiv r . \equiv . p \equiv r$ $\left[\text{Ł14} \dfrac{q \equiv p}{p}, \dfrac{q \equiv r}{q}, \dfrac{p \equiv r}{r} \right]$

Ł17. $s \equiv : p \equiv q . \equiv r \therefore \equiv : p \equiv q . \equiv . r \equiv s$

             $\left[\text{Ł14} \dfrac{p \equiv q}{p}, \dfrac{r}{q}, \dfrac{s}{r} \right]$

Ł18. $q \equiv r . \equiv : p \equiv r . \equiv . q \equiv p$ [Ł13, A1]

Ł19. $p \equiv . q \equiv r : \equiv : p \equiv q . \equiv r \therefore \equiv :: s \equiv : p \equiv q . \equiv r$

$\therefore \equiv \therefore p \equiv . q \equiv r : \equiv s$

           $\left[\text{Ł18} \dfrac{s}{p}, \dfrac{p \equiv . q \equiv r}{q}, \dfrac{p \equiv q . \equiv r}{r} \right]$

Ł20 (T28). $p \equiv r . \equiv : q \equiv p . \equiv . q \equiv r$ [Ł15, Ł18]

Ł21. $s \equiv . p \equiv r : \equiv \therefore q \equiv s . \equiv : q \equiv . p \equiv r$

               $\left[\text{Ł20} \dfrac{s}{p}, \dfrac{p \equiv r}{r} \right]$

Ł22. $q \equiv r . \equiv s : \equiv : r \equiv q . \equiv s \therefore \equiv :: p \equiv : q \equiv r . \equiv s$

$\therefore \equiv \therefore p \equiv : r \equiv q . \equiv s$

            $\left[\text{Ł20} \dfrac{q \equiv r . \equiv s}{p}, \dfrac{p}{q}, \dfrac{r \equiv q . \equiv s}{r} \right]$

Ł23. $p \equiv . q \equiv r : \equiv : q \equiv p . \equiv r \therefore \equiv :: s \equiv : p \equiv . q$

$\equiv r \therefore \equiv \therefore s \equiv : q \equiv p . \equiv r$

            $\left[\text{Ł20} \dfrac{p \equiv . q \equiv r}{p}, \dfrac{s}{q}, \dfrac{q \equiv p . \equiv r}{r} \right]$

Ł24. $q \equiv r . \equiv : p \equiv r . \equiv . q \equiv p \therefore \equiv :: s \equiv . q \equiv r :$
$\equiv \therefore s \equiv : p \equiv r . \equiv . q \equiv p$

$$\left[Ł20\, \frac{q \equiv r}{p}, \frac{s}{q}, \frac{p \equiv r . \equiv . q \equiv p}{r} \right]$$

Ł25. $s \equiv : p \equiv . q \equiv r \therefore \equiv \therefore s \equiv : q \equiv p . \equiv r :: \equiv ::: t$
$\equiv \therefore s \equiv : p \equiv . q \equiv r :: \equiv :: t \equiv \therefore s \equiv : q \equiv p . \equiv r$

$$\left[Ł20\, \frac{s \equiv : p \equiv . q \equiv r}{p}, \frac{t}{q}, \frac{s \equiv : q \equiv p . \equiv r}{r} \right]$$

Ł26. $p \equiv . q \equiv r : \equiv s \therefore \equiv : p \equiv q . \equiv . r \equiv s :: \equiv ::: t$
$\equiv \therefore p \equiv . q \equiv r : \equiv s :: \equiv \therefore t \equiv : p \equiv q . \equiv . r \equiv s$

$$\left[Ł20\, \frac{p \equiv . q \equiv r : \equiv s}{p}, \frac{t}{q}, \frac{p \equiv q . \equiv . r \equiv s}{r} \right]$$

Ł27 (T44). $q \equiv p . \equiv : q \equiv r . \equiv . p \equiv r$ [*Ł16, Ł20*]

Ł28. $q \equiv r . \equiv . r \equiv q : \equiv \therefore q \equiv r . \equiv s : \equiv : r \equiv q . \equiv s$

$$\left[Ł27\, \frac{r \equiv q}{p}, \frac{q \equiv r}{q}, \frac{s}{r} \right]$$

Ł29. $s \equiv : p \equiv q . \equiv r \therefore \equiv \therefore p \equiv . q \equiv r : \equiv s :: \equiv ::: s$
$\equiv : p \equiv q . \equiv r \therefore \equiv : p \equiv q . \equiv . r \equiv s :: \equiv :: p \equiv . q$
$\equiv r : \equiv s \therefore \equiv : p \equiv q . \equiv . r \equiv s$

$$\left[Ł27\, \frac{p \equiv . q \equiv r : \equiv s}{p}, \frac{s \equiv : p \equiv q . \equiv r}{q}, \frac{p \equiv q . \equiv . r \equiv s}{r} \right]$$

Ł30. $q \equiv r . \equiv s : \equiv : r \equiv q . \equiv s$ [*Ł28, Ł4*]

Ł31. $p \equiv : q \equiv r . \equiv s \therefore \equiv \therefore p \equiv : r \equiv q . \equiv s$
            [*Ł22, Ł30*]

Ł32 (T55). $p \equiv . q \equiv r : \equiv : p \equiv q . \equiv r \therefore \equiv \therefore p \equiv . q$
$\equiv r : \equiv : q \equiv p . \equiv r$ $\left[Ł31\, \dfrac{p \equiv . q \equiv r}{p}, \dfrac{p}{q}, \dfrac{q}{r}, \dfrac{r}{s} \right]$

Ł33 (T56). $s \equiv . p \equiv r : \equiv \therefore q \equiv s . \equiv : p \equiv q . \equiv r ::$
$\equiv :: s \equiv . p \equiv r : \equiv \therefore s \equiv q . \equiv : p \equiv q . \equiv r$

$$\left[Ł31\, \frac{s \equiv . p \equiv r}{p}, \frac{s}{r}, \frac{p \equiv q . \equiv r}{s} \right]$$

Ł34 (T59). $p \equiv . q \equiv r : \equiv : q \equiv p . \equiv r$ [*Ł32, A2*]

Ł35. $s \equiv : p \equiv . q \equiv r \therefore \equiv \therefore s \equiv : q \equiv p . \equiv r$
            [*Ł23, Ł34*]

Ł36. $t \equiv \therefore s \equiv : p \equiv . q \equiv r :: \equiv :: t \equiv \therefore s \equiv : q \equiv p . \equiv r$

[Ł25, Ł35]

Ł37. $s \equiv . p \equiv r : \equiv \therefore q \equiv s . \equiv : q \equiv . p \equiv r :: \equiv :: s$

$\equiv . p \equiv r : \equiv \therefore q \equiv s . \equiv : p \equiv q . \equiv r$

$$\left[Ł36 \frac{q}{p}, \frac{p}{q}, \frac{q \equiv s}{s}, \frac{s \equiv . p \equiv r}{t} \right]$$

Ł38 (T69). $s \equiv . p \equiv r : \equiv \therefore q \equiv s . \equiv : p \equiv q . \equiv r$

[Ł37, Ł21]

Ł39 (T70). $s \equiv . p \equiv r : \equiv \therefore s \equiv q . \equiv : p \equiv q . \equiv r$

[Ł33, Ł38]

Ł40. $s \equiv : p \equiv q . \equiv r \therefore \equiv \therefore p \equiv . q \equiv r : \equiv s$ [Ł19, A2]

Ł41. $s \equiv : p \equiv q . \equiv r \therefore \equiv : p \equiv q . \equiv . r \equiv s :: \equiv :: p$

$\equiv . q \equiv r : \equiv s \therefore \equiv : p \equiv q . \equiv . r \equiv s$ [Ł29, Ł40]

Ł42. $p \equiv . q \equiv r : \equiv s \therefore \equiv : p \equiv q . \equiv . r \equiv s$ [Ł41, Ł17]

Ł43. $t \equiv \therefore p \equiv . q \equiv r : \equiv s :: \equiv \therefore t \equiv : p \equiv q . \equiv . r \equiv s$

[Ł26, Ł42]

Ł44. $s \equiv . q \equiv r : \equiv :: t \equiv : p \equiv r . \equiv . q \equiv p \therefore \equiv . s \equiv t$

$:: \equiv :: s \equiv . q \equiv r : \equiv \therefore t \equiv . p \equiv r : \equiv : q \equiv p . \equiv . s \equiv t$

$$\left[Ł43 \frac{t}{p}, \frac{p \equiv r}{q}, \frac{q \equiv p}{r}, \frac{s \equiv t}{s}, \frac{s \equiv . q \equiv r}{t} \right]$$

Ł45. $s \equiv . q \equiv r : \equiv \therefore s \equiv : p \equiv r . \equiv . q \equiv p$ [Ł24, Ł18]

Ł46. $s \equiv . q \equiv r : \equiv \therefore s \equiv : p \equiv r . \equiv . q \equiv p :: \equiv :: s$

$\equiv . q \equiv r : \equiv :: t \equiv : p \equiv r . \equiv . q \equiv p \therefore \equiv . s \equiv t$

$$\left[Ł45 \frac{t}{p}, \frac{s}{q}, \frac{p \equiv r . \equiv . q \equiv p}{r}, \frac{s \equiv . q \equiv r}{s} \right]$$

Ł47. $s \equiv . q \equiv r : \equiv :: t \equiv : p \equiv r . \equiv . q \equiv p \therefore \equiv . s \equiv t$

[Ł46, Ł45]

Ł48. $s \equiv . q \equiv r : \equiv \therefore t \equiv . p \equiv r : \equiv : q \equiv p . \equiv . s \equiv t$

[Ł44, Ł47].

Supplementary remark III

Here I would like to comment, without any pretence to exactitude, on the form I used for the constant function signs of propositional functions of the type '$f(p)$' and '$f(pq)$' with one and with two propositional arguments. What leads me to make these comments is the fact that I do not set down these function signs at random, but rather construct them according to a general scheme. To avoid possible misunderstandings, I remark that the scheme in question possesses an entirely 'unofficial' character and is in no way a consequence of the directives of any of my systems.

Each one of my constant signs for propositional functions of the type '$f(p)$' or '$f(pq)$' with one or with two propositional arguments is a sign which consists of a 'basic outline', and possibly also of an indicator of the type '1', '2', '3', etc., placed under the basic outline. The basic outline for a function of one argument has always one of the following four forms: '⊢⊣', '⊣', '⊢', '–'; the basic outline for a function of two arguments, one of the following sixteen forms: '-o̩̍-', '-o̩-', '-o̩̍', 'o̩̍-', '-o̍-', 'o̩-', '-o̩', 'o̩̍', '-o-', '-o̍', 'o̍-', 'o̩', '-o', 'o-', 'o̍', 'o'. In the basic outline for a function of one argument, the perpendicular stroke on the left occurs if and only if the given function with a false argument becomes a true proposition; the perpendicular stroke on the right occurs if and only if the function with a true argument becomes a true proposition. In the basic outline for a function of two arguments, the left-hand horizontal bar occurs if and only if the given function with a true first and false second argument becomes a true proposition. The upper vertical stroke occurs if and only if the function with two false arguments becomes a true proposition. The right-hand horizontal bar occurs if and only if the function with a false first and true second argument becomes a true proposition. The lower vertical stroke occurs if and only if the function with two true arguments becomes a true proposition.

By considering the principles for the construction of function signs of propositional functions of the type '$f(p)$' and '$f(pq)$' with one and with two propositional arguments, we can quite easily construct from the basic outlines the familiar two-valued truth tables of the corresponding functions, and vice versa. We can establish further elementary correlations between the basic outlines and the logical characteristics of the corresponding functions of the following kind:

(1) Two function signs have the same basic outline, as for example the signs '⊢' and '⊢', or the signs -◇- and -◇-, if and
only if the functions corresponding to these signs are equivalent to one another for the same values of corresponding arguments.

(2) The basic outline of the sign of any function F is 'contained' in the basic outline of the sign of a function G, as for example the basic outline of the sign '–' in the basic outline of '⊢', or the basic outline of 'φ' in the basic outline of '-◇-', if
and only if G holds for any given values of its arguments provided F holds for the same values of corresponding arguments.

(3) The basic outline of the sign of the function F 'complements' the basic outline of the sign of any other function G, as for example the outline '⊢' complements the outline '⊣', or the
outline 'φ' complements the outline '-◇-', if and only if the
function F is equivalent to the negation of the function G for the same values of corresponding arguments.

(4) The basic outline of the sign of any function F is the 'sum' of the outlines of the signs of two functions G and H, as for example the outline '⊢⊣' is the sum of the outlines '⊢' and '⊣',
or the outline '-◇-' of the outlines 'φ' and '-φ-', if and only if
the function F is equivalent to the logical sum of the functions G and H for the same values of corresponding arguments.

(5) The basic outline of the sign of any function F is the 'intersection' of the outlines of the signs of the functions G and H, as for example the outline '⊢' is the intersection of the outlines '⊢⊣' and '⊢', or the outline 'o' of the outlines 'φ' and '-o-', if and only if the function F is equivalent to the logical product of the functions G and H for the same values of corresponding arguments.

Supplementary remark IV

The single axiom of protothetic constructed according to the directives of the system $\mathfrak{S}_5$, which I gave in the résumé of § 11 of my article, has withstood eleven years of research by me and others without having been shortened by so much as a single word. However, even to this day ideas for the solution of this problem continue to be brought to my attention. In the interests of the reader, who may wish to do independent research on this matter, I shall give a survey of the most important theoretical considerations which contribute to the evolution of a single axiom of protothetic and which illustrate different auxiliary devices. I am impelled to do this without waiting to give a proper discussion, in the continuation of my article, of problems connected with the origin of this axiom and with its further minor alterations.

(1) Reflecting on Tarski's thesis, mentioned above in the résumé of § 1 of my article, which states that

$$[p,q] \therefore p \cdot q \equiv : [f] : p \equiv . f(p) \equiv f(q),$$

I remarked in 1923 that in the propositional calculus, as completed by the thesis

$$[p,q,f] : p \equiv q \cdot f(p) \cdot \supset . f(q),$$

and consequently also in the system $\mathfrak{S}_1$, the following related thesis holds true:

$$[p,q,r] \therefore p \cdot q \equiv r \cdot \equiv : [f] : p \equiv . f(q) \equiv f(r).$$

From this I concluded that in the theories in question the following thesis also holds:

$$[p,q] \therefore p \cdot q \cdot \equiv : [f] : p \equiv . f(q) \equiv f(1).$$

In this thesis any true proposition can replace the sign '1'.

(2) In the year 1923 I established that from the proposition

$$A1^*. \quad \llcorner pqr \lrcorner \ulcorner \phi\big(\phi(pq)\phi(\phi(rq)\phi(pr))\big) \urcorner$$

together with the directives of the system $\mathfrak{S}_5$ and in particular with the help of the auxiliary definition

$$\llcorner p \lrcorner \ulcorner \phi(p \dashv (p)) \urcorner,$$

the propositions which state that

$$\llcorner p \lrcorner \ulcorner \phi(pp) \urcorner \qquad \text{(Law of Identity)}$$

and $\qquad \llcorner pq \lrcorner \ulcorner \phi(\phi(pq)\phi(qp)) \urcorner$

(Law of Commutativity)

and the axiom $A1$ can be successively derived. From this I

concluded that the axiom system *A1–A3* could be replaced by the axiom system *A1**, *A2*, *A3*.

(3) After finding the method of construction, stated in the résumé of § 10 of my article, for single axioms of protothetic based on the directives of the system $\mathfrak{S}_5$, I based my deductions for some time, in practice, on the following axiom composed of 290 signs:

$$(A_a)\ \llcorner fp \lrcorner \phi \left(f \left(\llcorner pq \lrcorner \ulcorner \phi(\phi(pq)\phi(qp))\urcorner p \right) \phi \left(f \left(\llcorner hs \lrcorner\ \phi \left(h \left(\llcorner pqr \lrcorner \right. \right. \right. \right. \right.$$

$$\ulcorner \phi(\phi(\phi(pr)\phi(qp))\phi(rq))\urcorner s \right) \phi \left(h \left(\llcorner kt \lrcorner\ \phi \left(k \left(\llcorner pqr \lrcorner \ulcorner \phi(\phi(p\phi(qr)) \right. \right. \right. \right.$$

$$\phi(\phi(pq)r))\urcorner t \right) \phi \left(k \left(\llcorner gp \lrcorner\ \phi \left(\llcorner f \lrcorner\ \ulcorner \phi \left(g(pp)\phi \left(\llcorner r \lrcorner \ulcorner \phi(f(rr)g(pp))\urcorner \llcorner r \lrcorner \right. \right. \right. \right. \right. \right.$$

$$\ulcorner \phi(f(rr)g(\phi(p \llcorner q \lrcorner \ulcorner q\urcorner)p))\urcorner \right) \right) \urcorner \llcorner q \lrcorner \ulcorner g(qp)\urcorner \right) \urcorner t \right) \llcorner pqr \lrcorner \ulcorner \phi(\phi(p\phi(qr))$$

$$\phi(\phi(pq)r))\urcorner \right) \right) \urcorner s \right) \llcorner pqr \lrcorner \ulcorner \phi(\phi(\phi(pr)\phi(qp))\phi(rq))\urcorner \right) \right) p \right) \llcorner pq \lrcorner$$

$$\ulcorner \phi(\phi(pq)\phi(qp))\urcorner \right) \right) .$$

The observations referred to in (2) enabled me to replace this axiom by a shorter one, consisting of 232 signs, which is the logical product of the proposition $A1^*$ and the axioms $A2$ and $A3$:

$$(A_b) \quad \llcorner hs \lrcorner \quad \phi \left(h \left(\llcorner pqr \lrcorner \ulcorner \phi(\phi(pq)\phi(\phi(rq)\phi(pr))) \urcorner s \right) \phi \left(h \left(\llcorner kt \lrcorner \right. \right. \right.$$

$$\phi \left(k \left(\llcorner pqr \lrcorner \ulcorner \phi(\phi(p\phi(qr))\phi(\phi(pq)r)) \urcorner t \right) \phi \left(k \left(\llcorner gp \lrcorner \phi \left(\llcorner f \lrcorner \phi \left(g(pp) \right. \right. \right. \right. \right.$$

$$\phi \left(\llcorner r \lrcorner \ulcorner \phi(f(rr)g(pp)) \urcorner \llcorner r \lrcorner \ulcorner \phi(f(rr)g(\phi(p \llcorner q \lrcorner \ulcorner q \urcorner)p)) \urcorner \right) \right) \llcorner q \lrcorner \ulcorner g(qp) \urcorner \right) t \right)$$

$$\llcorner pqr \lrcorner \ulcorner \phi(\phi(p\phi(qr))\phi(\phi(pq)r)) \urcorner \right) s \right) \llcorner pqr \lrcorner \ulcorner \phi(\phi(pq)$$

$$\phi(\phi(rq)\phi(pr))) \urcorner \quad .$$

From the axiom A_b, which dates from the year 1923, one can deduce the proposition $A1^*$ in the way that 'P' was deduced from the formula a, as sketched in § 10 of my article. The Law of Commutativity and the axiom $A1$ can be obtained from $A1^*$, as I established in (2). Once the Law of Commutativity is available, the 'logical factors' $A2$ and $A3$, contained in A_b, can be easily obtained by the methods given in § 10 of my article.

(4) In 1923, drawing on all the results summarized under (1), as well as other theoretical constructions conceived *ad hoc* which revealed definite possibilities for simplifying the axiom of proto-

thetic, I discovered an axiom A_c composed of 156 signs and thus shorter than axiom A_b. As the subsequent simplifications of the axiom of protothetic pertained to the axiom A_b and, in this way, to a large extent deprived the axiom A_c of theoretical importance, I cite this axiom without further comment:

$$(A_c) \; \ulcorner fpqr \lrcorner \; \phi \left(f(\phi(pp)q)\phi \left(f \left(\ulcorner \lrcorner g \lrcorner \; \phi \left(g(pp)\phi \left(g(\phi(r\phi(pr))p) \right. \right. \right. \right. \right.$$

$$\ulcorner \lrcorner h \lrcorner \; \phi \left(\ulcorner \lrcorner k \lrcorner \; \phi \left(\ulcorner \lrcorner s \lrcorner \; \ulcorner \phi(k(ss)h(pp)) \urcorner \phi \left(h(pp) \ulcorner \lrcorner s \lrcorner \; \ulcorner \phi(k(ss) \right. \right. \right.$$

$$\left. \left. \left. h(\phi(p \ulcorner \lrcorner t \lrcorner \; \ulcorner t \urcorner)p)) \urcorner \right) \urcorner \; \ulcorner \lrcorner t \lrcorner \; \ulcorner h(tp) \urcorner \right) \urcorner \right) \right) \; q \right) \phi(\phi(pq)\phi(\phi(rq)\phi(pr))) \right) \right) .$$

(5) Once Tarski learned of the results referred to in (1), he proved in 1924, and on the same basis on which he had earlier constructed the proof of his theorem mentioned in the résumé of § 1 of my article, which states:

$$[p,q] :: p \cdot q \cdot \equiv \therefore [f] \therefore p \equiv : [r] \cdot p \equiv f(r) \cdot \equiv \cdot [r] \cdot q \equiv f(r),$$

the proposition analogous to this theorem, namely:

$$[p,q] :: p \cdot q \cdot \equiv \therefore [f] \therefore p \equiv : [r] \cdot f(r) \cdot \equiv \cdot [r] \cdot q \equiv f(r).$$

(6) In the year 1926 Wajsberg informed me of his proof that the axiom system of the system $\mathfrak{S}$, given by me in the résumé of § 3 of my article, could be replaced by either one of the two following propositions:

$$(W) \quad p \equiv q \cdot \equiv r : \equiv s \therefore \equiv \therefore s \equiv : p \equiv \cdot q \equiv r$$
$$(W^*) \quad p \equiv \cdot q \equiv r : \equiv : r \equiv s \cdot \equiv s \therefore \equiv \cdot p \equiv q^1$$

---

[1] See M. Wajsberg, *Metalogische Beiträge* (Warsaw, 1936), pp. 34–36.
[Translated as paper 14 of this volume–Ed.]

(7) Wajsberg's result, mentioned in (6), enabled me in 1926 to replace A_c by a shorter axiom of 124 signs:

$$(A_d) \quad {}_\llcorner fhpqrx_\lrcorner \; \phi\left(f\left({}_\llcorner k_\lrcorner{}^\ulcorner \phi\left({}_\llcorner s_\lrcorner{}^\ulcorner \phi(k(ss)h(pp))^\urcorner \phi\left(h(pp){}_\llcorner s_\lrcorner\right.\right.\right.\right.$$

$$\left.\left.{}^\ulcorner \phi\left(k(ss)h(\phi(p{}_\llcorner t_\lrcorner{}^\ulcorner t^\urcorner)p))^\urcorner\right)\right)^\urcorner q\right)\phi\left(f({}_\llcorner t_\lrcorner{}^\ulcorner h(tp)^\urcorner q)\phi\left(\phi(\phi(p\phi(qr))\right.\right.$$

$$\left.\left.\left.\phi(\phi(rx)x))\phi(pq)\right)\right)^\urcorner\right).$$

The axiom A_d is a 'logical product' of my axiom $A3$ and of Wajsberg's axiom W^*. This logical product is related to those formulae which, according to one of the propositions introduced in (1), are equivalent to the corresponding formulae of the type '$p \cdot q \equiv r$'. If we wish to derive the axioms $A1$–$A3$ from A_d, we can do so in the following stages:

(a) We define, as in § 10 of my article, a propositional function of the type '$\Phi(pq)$' which is satisfied for all values of its arguments.

(b) We substitute the sign of the function mentioned in (a) for the variable 'f' in the axiom A_d. Then, applying several times the directives for distribution of the quantifier, substitution and detachment, we easily obtain Wajsberg's axiom W^*, which forms one of the 'logical factors' of A_d. This axiom is obtained according to a pattern which approximates to the one already familiar to us.

(c) We derive the axioms $A1$ and $A2$ from the axiom W^*, using Wajsberg's method.

(d) Since, as a result of (c), the whole system $\mathfrak{S}$ stands at our disposal, we transform the axiom A_d in such a way that by detachment we obtain the proposition:

$$ {}_\llcorner fhpq_\lrcorner \; \phi\left(f\left({}_\llcorner k_\lrcorner{}^\ulcorner \phi\left({}_\llcorner s_\lrcorner{}^\ulcorner \phi(k(ss)h(pp))^\urcorner \phi\left(h(pp){}_\llcorner s_\lrcorner{}^\ulcorner \phi(k(ss)\right.\right.\right.\right.$$

$$\left.\left.h(\phi(p{}_\llcorner t_\lrcorner{}^\ulcorner t^\urcorner)p))^\urcorner\right)\right)^\urcorner q\right)f({}_\llcorner t_\lrcorner{}^\ulcorner h(tp)^\urcorner q)\right).$$

(e) We substitute the sign 'ϕ' for the variable 'f' in the proposition just obtained. Similarly, using the theorems of the system and the directives α–γ, we quite easily obtain a proposition which differs from axiom $A3$ in variables but in no important way, and which can be proved equivalent to $A3$ with the help of $A1$ and $A2$.

(8) In one of my 1926 university lectures, attended by Wajsberg, I remarked that if the correlate of Wajsberg's proposition $W*$ in the axiom A_d were replaced by some other formula, from which followed not only all theorems of the system $\mathfrak{S}$, but also some correspondingly formulated 'extensionality proposition' analogous to the proposition

$$\llcorner pq \lrcorner \ulcorner \phi \big(\phi(pq) \llcorner f \lrcorner \ulcorner \phi(f(p)f(q)) \urcorner \big) \urcorner,$$

one could likewise make a change in A_d, without altering the deductive power of this axiom. This change would replace the universal proposition which begins with the quantifier $\llcorner k \lrcorner$, and agrees more or less with the first of Tarski's formulae for the 'logical product' (see the résumé of § 1 of my article), by a corresponding logical product constructed like the second, and shorter, of these formulae. At the time I mentioned that I saw no reason why A_d could not be thus shortened.

(9) Considering my remarks reviewed in (8), Wajsberg established in 1926 that all theorems of the system $\mathfrak{S}$ could be derived from the proposition

$$\llcorner pqrst \lrcorner \ulcorner \phi \Big(\phi(pq) \llcorner g \lrcorner \ulcorner \phi \big(g(\phi(\phi(\phi(rs)t)q)g(\phi(\phi(st)r)p)) \big) \urcorner \Big) \urcorner.$$

This proposition, although no 'extensionality proposition' in the strict sense, nevertheless makes it possible—as Wajsberg proved —to replace the longer by the shorter form of the logical product in the axiom A_d. As a result of Wajsberg's changes in A_d, he obtained the following shorter axiom consisting of 120 signs:

$$(A_e) \ \llcorner fhpqrst \lrcorner \ \ulcorner \phi \bigg(f \Big(\llcorner g \lrcorner \ulcorner \phi \big(h(pp) \phi \big(g(qh(pp)) g(qh(\phi(\phi(p \llcorner t \lrcorner$$

$$\ulcorner t\urcorner)p)))\big)^{\urcorner} r\Big)\phi\Big(f(\llcorner t\lrcorner\ulcorner h(tp)\urcorner r)\phi\Big(\phi(pq)\llcorner g\lrcorner^{\ulcorner}\phi\Big(g(\phi(\phi(\phi(rs)t)q)$$

$$g(\phi(\phi(st)r)p)\big)^{\urcorner}\Big)\Big)\Big).$$

(10) Starting again directly with the axiom A_d and using Tarski's proposition introduced in (5), I constructed, in 1926, an axiom of 116 signs, shorter than Wajsberg's axiom A_e. Having introduced in the axiom A_d, as we saw, a correlate of the proposition W^* given in (6), I was able to use in a similar way the proposition W in the new axiom. However, I could proceed vice versa with equal success. The new axiom runs thus:

$$(A_f)\ \llcorner fhpqrs\lrcorner\ \phi\Bigg(f(\llcorner t\lrcorner\ulcorner h(tp)\urcorner q)\phi\Bigg(f\Big(\llcorner k\lrcorner^{\ulcorner}\phi\Big(h(pp)\phi\big(\llcorner s\lrcorner^{\ulcorner}k(ss)\urcorner$$

$$\llcorner s\lrcorner^{\ulcorner}\phi\big(k(ss)h(\phi(\phi(p\llcorner t\lrcorner\ulcorner t\urcorner)p))^{\urcorner}\big)\Big)^{\urcorner}q\Big)\phi\Big(\phi(\phi(\phi(pq)r)s)\phi(s\phi(p\phi(qr)))\Big)\Bigg)\Bigg)$$

(11) During the same year, 1926, Wajsberg reduced the 116 signs of the axiom A_f to 106. The shortened axiom is:

$$(A_g)\ \llcorner fp\lrcorner\ \phi\Bigg(\llcorner s\lrcorner^{\ulcorner}f(sp)\urcorner\llcorner g\lrcorner\ \phi\Bigg(f(pp)\phi\Big(\llcorner t\lrcorner\ulcorner g(tt)\urcorner\llcorner qrt\lrcorner^{\ulcorner}\phi\Big(g(\phi$$

$$(\phi(tt)t)t)\phi\Big(f(\phi(\phi(p\llcorner s\lrcorner^{\ulcorner}s\urcorner)p)\phi(\phi(p\phi(qr))\phi(r\phi(qp))))\Big)^{\urcorner}\Big)\Bigg)^{\urcorner}\Bigg)^{\urcorner}.$$

(12) Making essential use of Wajsberg's axiom A_e and at the same time appealing to Tarski's theorem, according to which every truth function f is such that

$$[p]\cdot f(p)\cdot\equiv\cdot f(f(Fl))^1,$$

¹ See (i) Alfred Tajtelbaum-Tarski, 'O wyrazie pierwotnym logistyki', (On the primitive term of logistic), Doctoral thesis. Offprint from *Przegląd*

I constructed the 82-sign axiom introduced in the résumé of § 11 of my article. To preserve consistency with the names of the other axioms considered here, I call this axiom A_h. It happened that in one and the same conversation Wajsberg and I informed each other of the two axioms A_g and A_h, which had been discovered independently of one another, yet at approximately the same time.

As I have already mentioned, the axiom A_h possesses the peculiarity of being the shortest anyone has yet thought of. However, it has in practice a disadvantage, quite similar to that of Wajsberg's axiom A_e, its genetic predecessor: namely that the method of deriving any familiar basis of the system $\mathfrak{S}$ from this axiom is rather too complicated. The result is that it is not as quickly understood how this axiom provides an adequate basis for protothetic. In this respect the axiom A_i has the advantage, as will be discussed later. The remarks concerning this axiom should throw some light on the deductions which can be made from the axioms A_e, A_g, and A_h.

(13) Łukasiewicz proved in 1933 that Wajsberg's propositions W and W^* (see (6)), each of which can be the sole axiom of the system $\mathfrak{S}$, may be further shortened, and that it is possible to derive this system from any one of the following three axioms:

Ł. $p \equiv q \,.\, \equiv \,:\, r \equiv q \,.\, \equiv \,.\, p \equiv r$

(This axiom corresponds to my proposition $A1^*$, given in (2).)

Ł*. $p \equiv q \,.\, \equiv \,:\, p \equiv r \,.\, \equiv \,.\, r \equiv q$

Ł**. $p \equiv q \,.\, \equiv \,:\, r \equiv p \,.\, \equiv \,.\, q \equiv r$

From this result of Łukasiewicz's it became clear to me that:

(a) to carry out the derivations from the proposition $A1^*$ mentioned in (2), I needed no auxiliary definition;

(b) the axioms $A1^*$ and $A3$ suffice for the construction of a system of protothetic;

(c) the axioms A_b and A_c, which constitute different forms of 'logical product', contain superfluous factors.

Filozoficzny (1923), pp. 19 and 20. (ii) Alfred Tajtelbaum-Tarski, 'Sur les truth-functions au sens de MM. Russell et Whitehead', *Fundamenta Mathematicae* 5 (1924), p. 69.

(14) In the same year, 1933, making use of Łukasiewicz's discovery referred to in (13), I constructed the axiom A_i mentioned in (12). This axiom, like the axiom A_h, consists of eighty-two signs and reads thus:

$$(A_i) \quad {}_{\llcorner}fpqrst_{\lrcorner} \; \phi\Big(\phi(pq){}_{\llcorner}g_{\lrcorner} \; \phi\Big(f(pf(p{}_{\llcorner}u_{\lrcorner}{}^{\ulcorner}u^{\urcorner}))\phi\big({}_{\llcorner}u_{\lrcorner}{}^{\ulcorner}f(qu)^{\urcorner}\phi\big($$

$$g\big(\phi(\phi(rt)\phi(sr))q)g(\phi(\phi(st)p)\big)\Big)\Big)^{\urcorner}\Big).$$

The basic initial derivations from this axiom can be roughly outlined in the following way:

(a) We introduce, as in (2), an auxiliary definition which states that ${}_{\llcorner}p_{\lrcorner}{}^{\ulcorner}\phi(p \dashv (p))^{\urcorner}$.

(b) We define, as in the derivations from the axiom A_d, a propositional function '$\Phi(pq)$' which is satisfied for all values of its arguments.

(c) Having substituted in the axiom A_i the expression '$\dashv(p)$' for the variable 'q', the sign 'Φ' for the variable 'f', and, in the proposition obtained by the first detachment, the expression '$\phi(\phi(rt)\phi(sr))$' for the variable 'p', and the sign 'ϕ' for the variable 'g', we get the proposition

$${}_{\llcorner}rst_{\lrcorner}{}^{\ulcorner}\phi\big(\phi(st)\phi(\phi(rt)\phi(sr))\big)^{\urcorner},$$

namely $A1^*$ with different variables. This gives us the entire theory $\mathfrak{S}$, by Łukasiewicz's result of (13). In the derivation of the above proposition we exploit the idea used also by Wajsberg in proving, from the following proposition referred to in (9):

$${}_{\llcorner}pqrst_{\lrcorner}{}^{\ulcorner}\phi\Big(\phi(pq){}_{\llcorner}g_{\lrcorner}{}^{\ulcorner}\phi\big(g(\phi(\phi(rs)t)q)g(\phi(\phi(st)r)p)\big)^{\urcorner}\Big)^{\urcorner}$$

the proposition

$${}_{\llcorner}rst_{\lrcorner}{}^{\ulcorner}\phi\big(\phi(\phi(st)r)\phi(\phi(rs)t)\big)^{\urcorner}.$$

(d) Substituting 'p' for 'q' and 'Φ' for 'f' in the axiom A_i we obtain, just as easily, the proposition

$${}_{\llcorner}gprst_{\lrcorner}{}^{\ulcorner}\phi\Big(g\big(\phi(\phi(rt)\phi(sr))p\big)g(\phi(st)p)\Big)^{\urcorner}.$$

(e) Substituting 'p' for 'q' in A_i we manipulate according to the system $\mathfrak{S}$ the proposition obtained by detachment, in such a way that we can further detach from it the proposition

$$\llcorner fp \lrcorner \phi\Big(f\big(pf(p\llcorner u\lrcorner \ulcorner u\urcorner)\big)\llcorner u\lrcorner \ulcorner f(pu)\urcorner\Big)\urcorner.$$

(f) We introduce the following three definitions:

$$(D_a)\quad \llcorner pq \lrcorner \;\phi\left(\llcorner gr \lrcorner \;\ulcorner\phi\Big(p\phi\big(g(\phi(\phi(rr)\phi(rr))q)g(\phi(rr)p)\big)\Big)\;\urcorner\varphi(pq)\right)^{\urcorner}$$

$$(D_b)\quad \llcorner pq \lrcorner \ulcorner \phi\big(\phi(p\phi(qq))\text{-}\varphi(pq)\big)\urcorner$$

$$(D_c)\quad \llcorner pq \lrcorner \ulcorner \phi\big(\phi(\Phi(pq)\llcorner u\lrcorner \ulcorner u\urcorner)\text{o}(pq)\big)\urcorner.$$

The first of these three is one of the possible definitions of a 'logical product'. The second is the definition of a function equivalent to its own first argument. And the third, as is seen in connexion with (b), is the definition of a function which yields a false proposition for all values of its arguments.

(g) Substituting in the proposition

$$\llcorner pq \lrcorner \ulcorner \phi\big(q\phi(p\phi(pq))\big)\urcorner,$$

the expression

$$\phi\Big(g\big(\phi(\phi(rr)\phi(rr))p\big)g(\phi(rr)p)\Big)$$

for the variable 'q', and making use of the proposition introduced in (d), we obtain by detachment, distribution of the quantifier, and application of the definition D_a, the proposition

$$\llcorner p \lrcorner \ulcorner \phi\big(p\varphi(pp)\big)\urcorner.$$

(h) Substituting in A_i the expression '$\llcorner u\lrcorner \ulcorner u\urcorner$' for '$p$', the expression '$\phi(\llcorner u\lrcorner \ulcorner u\urcorner \llcorner u\lrcorner \ulcorner u\urcorner)$' for '$q$', the sign '-$\varphi$' for '$f$', and the sign '$r$' for '$s$' and '$t$', we obtain, with the help of theorems of the system $\mathfrak{S}$ and the definitions D_b and D_a, the theorem

$$\phi\Big(\llcorner u\lrcorner \ulcorner u\urcorner \varphi\big(\llcorner u\lrcorner \ulcorner u\urcorner \phi(\llcorner u\lrcorner \ulcorner u\urcorner \llcorner u\lrcorner \ulcorner u\urcorner)\big)\Big).$$

(i) Similarly, substituting in A_i the expression

$$\text{'}\phi(\llcorner u\lrcorner \ulcorner u\urcorner \llcorner u\lrcorner \ulcorner u\urcorner)\text{'}$$

for 'p', the expression '$\llcorner u\lrcorner \ulcorner u\urcorner$' for '$q$', the sign 'o' for '$f$', and the sign '*r*' for '*s*' and '*t*' we obtain, with the help of the definitions D_c and D_a, the analogous theorem

$$\phi\big(\llcorner u\lrcorner\ulcorner u\urcorner\varphi(\phi(\llcorner u\lrcorner\ulcorner u\urcorner\llcorner u\lrcorner\ulcorner u\urcorner)\llcorner u\lrcorner\ulcorner u\urcorner)\big).$$

If an arbitrary propositional function '$F(u)$' of one propositional argument possesses the characteristic that its values '$F(\llcorner u\lrcorner\ulcorner u\urcorner)$' and '$F(\phi(\llcorner u\lrcorner\ulcorner u\urcorner\llcorner u\lrcorner\ulcorner u\urcorner))$' have already been proved in the system, then it is possible to obtain in the system the universal proposition '$\llcorner u\lrcorner\ulcorner F(u)\urcorner$'. The procedure to be adopted may be sketched as follows. In accordance with our hypothesis we have that

α. $F(\llcorner u\lrcorner\ulcorner u\urcorner)$

and

β. $F(\phi(\llcorner u\lrcorner\ulcorner u\urcorner\llcorner u\lrcorner\ulcorner u\urcorner))$.

We define a new functional sign, let us say 'G', by means of an auxiliary definition of the type

γ. $\llcorner pq\lrcorner\ulcorner\phi\big(\phi(F(q)\phi(pp))G(pq)\big)\urcorner$.

From γ we deduce, in accordance with the system $\mathfrak{S}$,

δ. $\llcorner pq\lrcorner\ulcorner\phi\big(F(q)G(pq)\big)\urcorner$.

From δ and α we infer that

ε. $G(\phi(\llcorner u\lrcorner\ulcorner u\urcorner\llcorner u\lrcorner\ulcorner u\urcorner)\llcorner u\lrcorner\ulcorner u\urcorner)$;

from ε, that

ζ. $\phi\big(G(\phi(\llcorner u\lrcorner\ulcorner u\urcorner\llcorner u\lrcorner\ulcorner u\urcorner)\llcorner u\lrcorner\ulcorner u\urcorner)\phi(\llcorner u\lrcorner\ulcorner u\urcorner\llcorner u\lrcorner\ulcorner u\urcorner)\big)$;

and from δ and β, that

η. $G\big(\phi(\phi(\llcorner u\lrcorner\ulcorner u\urcorner\llcorner u\lrcorner\ulcorner u\urcorner)\phi(\llcorner u\lrcorner\ulcorner u\urcorner\llcorner u\lrcorner\ulcorner u\urcorner))\phi(\llcorner u\lrcorner\ulcorner u\urcorner\llcorner u\lrcorner\ulcorner u\urcorner)\big)$.

We substitute in A_i the expression '$G(\phi(\llcorner u\lrcorner\ulcorner u\urcorner\llcorner u\lrcorner\ulcorner u\urcorner)\llcorner u\lrcorner\ulcorner u\urcorner)$' for '*p*', the expression '$\phi(\llcorner u\lrcorner\ulcorner u\urcorner\llcorner u\lrcorner\ulcorner u\urcorner)$' for '*q*', the sign '$\Phi$' for '*f*', and the expression '$\llcorner u\lrcorner\ulcorner u\urcorner$' for '*r*', '*s*', and '*t*'. Then in the proposition resulting from a single detachment using ζ, we substitute the sign 'G' for '*g*'. We can then assert, in virtue of the definition mentioned in (*b*) and the proposition η, that

θ. $G\big(\phi(\llcorner u\lrcorner\ulcorner u\urcorner\llcorner u\lrcorner\ulcorner u\urcorner)G(\phi(\llcorner u\lrcorner\ulcorner u\urcorner\llcorner u\lrcorner\ulcorner u\urcorner)\llcorner u\lrcorner\ulcorner u\urcorner)\big)$.

Finally, from the proposition introduced in (e) and the proposition θ, we conclude that

$$\llcorner u \lrcorner \ulcorner G\big(\varphi(\llcorner u \lrcorner \ulcorner u \urcorner \llcorner u \lrcorner \ulcorner u \urcorner)u\big) \urcorner,$$

from which, on the basis of the proposition δ, it follows that

$$\llcorner u \lrcorner \ulcorner F(u) \urcorner.$$

The results of the deductions sketched here being at our disposal, we can see without difficulty that the axiom A_i suffices for the construction of the given system of protothetic.

Supplementary remark V

In the course of my seminar on the foundations of mathematics in the academic year 1924–5 at the University of Warsaw, I and the other participants analysed Łukasiewicz's *Two-valued Logic*, mentioned above in the résumé of § 5 of my article. In this connexion I constructed in 1924 a system of protothetic based on principles quite different from those according to which the systems $\mathfrak{S}_2$–$\mathfrak{S}_5$ had been constructed. Using Łukasiewicz's construction to some considerable extent as a model, I was concerned to employ an 'algorithmic' or 'computative' style in my new system, as opposed to the much more common 'substitution-detachment' style. At this point I would like to mention one or two of the peculiarities of my system which differentiate it from Łukasiewicz's 'Two-valued logic'—apart, of course, from the obvious inequivalence of the two systems, the second being not a system of protothetic at all.

(a) In my system there are no signs 'U' ('I accept') or 'N' ('I reject', 'I deny') which are prefixed to the theses of Łukasiewicz's system.[1]

(b) My system is based on two primitive terms, 'φ' and '$\wedge$' (the logical 'zero'), whereas in Łukasiewicz's system, along with the correlates of these two terms, the logical 'one' appears as a primitive term.[2] (Inasmuch as the four propositions appearing in Łukasiewicz's system which he calls definitions, as well as

---

[1] See Jan Łukasiewicz, 'Logika dwuwartościowa', offprint from the volume in memory of Twardowski (*Przegląd Filozoficzny* 23), Lemberg, 1921, pp. 4 and 5.
[2] Op. cit., pp. 4 and 16.

the author's axioms at the beginning of the system, were introduced with no reference to any directives,[1] I feel compelled to regard these 'definitions' simply as axioms of a special kind, and also to include among the primitive terms of 'Two-valued logic', besides the three terms mentioned above, the four 'defined' terms, as well as the equality sign used by Łukasiewicz in definitions. Cf. the résumé of § 1 of my article.)

(c) While Łukasiewicz introduces three theses as axioms of his system, of which two are formulated with the help of variables, my system is based on one axiom alone, and in this axiom no variables appear.

(d) One of the directives of my system makes possible the addition to the system of any number of definitions.

(e) My system has no substitution directive.[2]

The system of prototheric in question, which I have formalized with the same degree of precision with which I formulated the directives of the system $\mathfrak{S}_5$ in § 11 of my article, could be summarized in the following sketchy and inexact fashion.

Axiom

   A. ╛-(∧∧)

Directives

Directive a: if any expressions 'p' and 'q' are already theses of the system, the corresponding expression '╛-(pq)' may be added.

Directive b: if any expressions 'p' and '╛-$(q∧)$' are already theses of the system, the corresponding expression '╛-(╛-$(pq)∧)$' may be added.

Directive c: if any expressions '╛-$(p∧)$' and 'q' are already theses of the system, the corresponding expression '╛-(pq)' may be added.

Directive d: if any expressions '╛-$(p∧)$' and '╛-$(q∧)$' are already theses of the system, the corresponding expression '╛-(pq)' may be added.

Directive e: this directive permits the addition to the system of definitions having the form of expressions of the type

$$╛\big(╛\text{-}(╛\text{-}(pq)╛\text{-}(╛\text{-}(qp)∧))∧\big),$$

---

[1] Op. cit., pp. 15 and 16.   [2] Cf. op. cit., the 'said rule' a on p. 11.

or of the same type preceded by universal quantifiers. The well-formed propositions represented here by the signs '*p*' and '*q*' constitute, respectively, the *definiens* and the *definiendum* of the definition.

Directive *f*: if the *definiens* of any definition already belonging to the system (or of some substitution instance of such a definition) is a thesis of the system, then the *definiendum* of the definition in question (or of its substitution instance) may be added to the system.

Directive *g*: if in some expression '$\diamond$-(*p*∧)', which is already a thesis of the system, the expression '*p*' is the *definiens* of some definition already belonging to the system (or of some substitution instance of such a definition), then the expression '$\diamond$-(*q*∧)' may be added to the system, where '*q*' is the *definiendum* of the definition in question (or of its substitution instance).

Directive *h*: this directive, which is a correlate of the directive *η* of the system $\mathfrak{S}_2$, generalized for the semantic category of propositions, presupposes (as does the last directive *i*) certain constant terms that may be called basic constants, the latter including in particular the primitive terms '∧' and '$\diamond$-'. A finite number of these basic constants should be defined for each semantic category occurring in the system, according to a scheme inductively characterized in advance. For the semantic category of propositions there are two such basic constants. For the semantic category of signs of propositional functions of one propositional argument, there are four basic constants. For the semantic category of signs of propositional functions of one argument, the latter being itself of the semantic category of such function signs, there are sixteen basic constants, etc. This accords with the observations on p. 37 of my article. The directive in question states that an expression beginning with a universal quantifier can be added to the system as a new thesis, if all those expressions that can be obtained from the given expression by substituting for its variables the basic constants of the same semantic categories are already theses of the system.

Directive *i*: if '*q*' is an expression beginning with a universal quantifier, if an expression '*p*' can be obtained from the expression 'q' by the substitution for its variables of basic constants of the same semantic categories, and if the expression '$\diamond\text{-}(p\wedge)$' is already a thesis of the system, the corresponding expression '$\diamond\text{-}(q\wedge)$' may be added to the system.

In order to illustrate procedures valid in the system of protothetic described here, I deduce from the axiom A, as an example, the thesis

$$\llcorner f\lrcorner^{\ulcorner}\diamond\text{-}\big(f(f(f(\llcorner p\lrcorner^{\ulcorner}p^{\urcorner}))\llcorner p\lrcorner^{\ulcorner}f(p)^{\urcorner})\big)^{\urcorner}.$$

Because the scheme for defining basic constants, mentioned above in the discussion of directive h, has not been effectively formulated by me here, I shall assume (truly) in my derivation that the terms defined by means of the definitions $D1$–$D5$ satisfy the scheme. I shall also assume that these terms, together with the primitive term '$\wedge$', exhaust the basic constants belonging to the semantic categories of (i) propositions, and (ii) signs of propositional functions with one propositional argument. The derivation in question proceeds as follows:

$A.$ $\diamond\text{-}(\wedge\wedge)$

$D1.$ $\diamond\text{-}\Big(\diamond\text{-}\big(\diamond\text{-}(\diamond\text{-}(\wedge\wedge)\vee)\diamond\text{-}(\diamond\text{-}(\vee\diamond\text{-}(\wedge\wedge))\wedge)\big)\wedge\Big)$

(on the basis of directive e)

$D2.$ $\llcorner p\lrcorner^{\ulcorner}\diamond\text{-}\Big(\diamond\text{-}\big(\diamond\text{-}(\diamond\text{-}(pp)\mapsto(p))\diamond\text{-}(\diamond\text{-}(\mapsto(p)\diamond\text{-}(pp))\wedge)\big)\wedge\Big)^{\urcorner}$

(Dir. e)

$D3.$ $\llcorner p\lrcorner^{\ulcorner}\diamond\text{-}\Big(\diamond\text{-}\big(\diamond\text{-}(p\dashv(p))\diamond\text{-}(\diamond\text{-}(\dashv(p)p)\wedge)\big)\wedge\Big)^{\urcorner}$ (Dir. e)

$D4.$ $\llcorner p\lrcorner^{\ulcorner}\diamond\text{-}\Big(\diamond\text{-}\big(\diamond\text{-}(\diamond\text{-}(p\wedge)\vdash(p))\diamond\text{-}(\diamond\text{-}(\vdash(p)\diamond\text{-}(p\wedge))\wedge)\big)\wedge\Big)^{\urcorner}$

(Dir. e)

$D5.$ $\llcorner p\lrcorner^{\ulcorner}\diamond\text{-}\Big(\diamond\text{-}\big(\diamond\text{-}(\diamond\text{-}(\diamond\text{-}(pp)\wedge)-(p))\diamond\text{-}\big(\diamond\text{-}(-(p)\diamond\text{-}$

$(\diamond\text{-}(pp)\wedge))\wedge)\big)\wedge\Big)^{\urcorner}$ (Dir. e)

T1. ◊(⌞p⌟⌜p⌝∧)   (A, Dir. i)

T2. ◊(⌞p⌟⌜p⌝⌞p⌟⌜p⌝)   (T1, Dir. d)

T3. ⊣(⌞p⌟⌜p⌝)   (D2, T2, Dir. f)

T4. ◊(⊣(⌞p⌟⌜p⌝)⊣(⌞p⌟⌜p⌝))   (T3, a)

T5. ⊣(⊣(⌞p⌟⌜p⌝))   (D2, T4, f)

T6. ⊣(∧)   (D2, A, f)

T7. ∨   (D1, A, f)

T8. ◊(∨∨)   (T7, a)

T9. ⊣(∨)   (D2, T8, f)

T10. ⌞p⌟⌜⊣(p)⌝   (T6, T9, h)

T11. ◊(⊣(⊣(⌞p⌟⌜p⌝))⌞p⌟⌜⊣(p)⌝)   (T5, T10, a)

T12. ◊(⊣(⌞p⌟⌜p⌝)∧)   (D3, T1, g)

T13. ◊(⊣(⊣(⌞p⌟⌜p⌝))∧)   (D3, T12, g)

T14. ◊(⊣(∧)∧)   (D3, A, g)

T15. ◊(⌞p⌟⌜⊣(p)⌝∧)   (T14, i)

T16. ◊(⊣(⊣(⌞p⌟⌜p⌝))⌞p⌟⌜⊣(p)⌝)   (T13, T15, d)

T17. ⊢(⌞p⌟⌜p⌝)   (D4, T1, f)

T18. ◊(◊(⊢(⌞p⌟⌜p⌝)∧)∧)   (T17, A, b)

T19. ◊(⊢(⊢(⌞p⌟⌜p⌝))∧)   (D4, T18, g)

T20. ◊(◊(∨∧)∧)   (T7, A, b)

T21. ◊(⊢(∨)∧)   (D4, T20, g)

T22. ◊(⌞p⌟⌜⊢(p)⌝∧)   (T21, i)

T23. ◊(⊢(⊢(⌞p⌟⌜p⌝))⌞p⌟⌜⊢(p)⌝)   (T19, T22, d)

T24. ◊(◊(◊(⌞p⌟⌜p⌝⌞p⌟⌜p⌝)∧)∧)   (T2, A, b)

T25. ◊(−(⌞p⌟⌜p⌝)∧)   (D5, T24, g)

T26. ◊(−(⌞p⌟⌜p⌝)−(⌞p⌟⌜p⌝))   (T25, d)

T27. ◊(◊(◊(−(⌞p⌟⌜p⌝)−(⌞p⌟⌜p⌝))∧)∧)   (T26, A, b)

T28. ◊(−(−(⌞p⌟⌜p⌝))∧)   (D5, T27, g)

T29. ◊(◊(◊(∧∧)∧)∧)   (A, b)

T30. ◊(−(∧)∧)   (D5, T29, g)

$T31.$ $\diamondsuit\text{-}(\llcorner p\lrcorner{}^{\ulcorner}-(p)^{\urcorner}\wedge)$ $(T30, i)$

$T32.$ $\diamondsuit\text{-}(-(-(\llcorner p\lrcorner{}^{\ulcorner}p^{\urcorner}))\llcorner p\lrcorner{}^{\ulcorner}-(p)^{\urcorner})$ $(T28, T31, d)$

$T33.$ $\llcorner f\lrcorner{}^{\ulcorner}\diamondsuit\text{-}\big(f(f(\llcorner p\lrcorner{}^{\ulcorner}p^{\urcorner}))\llcorner p\lrcorner{}^{\ulcorner}f(p)^{\urcorner}\big)^{\urcorner}$

$(T11, T16, T23, T32, h)$

On the model of the system of protothetic sketched here, which is based upon the primitive terms '$\wedge$' and '$\diamondsuit$-', I constructed in 1924 nine further systems of this theory, based upon the following combinations of primitive terms: (1) '-$\diamondsuit$' and '$\wedge$'; (2) '$\diamondsuit$' and '$\wedge$'; (3) '-$\diamondsuit$-' and '$\wedge$'; (4) '-$\diamondsuit$-' and 'V'; (5) '-o-' and 'V'; (6) '-o' and 'V'; (7) 'o-' and 'V'; (8) '$\diamondsuit$' and '$\wedge$'; (9) '$\diamondsuit$' and 'V'. The single axiom, forming the correlate of the axiom A, was constituted in each of these systems by the shortest true proposition that could be formulated with the help of the primitive terms of the given system, together perhaps with parentheses of the type '(' and ')'. (From this it can be seen that the axiom of a system in which the sign 'V' appeared as one of the primitive terms was itself the sign 'V'.) Corresponding to the directives a–d there were four analogous directives in each of the new systems. In the four possible cases in which the arbitrary expressions 'p' and 'q', or their negations expressed in the form '$\diamondsuit$-$(p\wedge)$' and '$\diamondsuit$-$(q\wedge)$', appeared as theses in the system, the directives a–d permitted adding to the system, in accordance with the meaning of the function sign '$\diamondsuit$-', the corresponding expressions '$\diamondsuit$-(pq)' or '$\diamondsuit$-$(\diamondsuit$-$(pq)\wedge)$', the latter being the negation of the former. In the four possible cases in which the arbitrary expressions 'p' and 'q', or their negations expressed with the help of the primitive terms of the given system, appeared as theses, analogous directives in each of the new systems permitted adding to the system, in accordance with the meaning of the primitive function sign 'f', the corresponding expression '$f(pq)$' or its negation expressed with the help of the primitive terms. The correlate of the directive e permitted adding to the system definitions formulated according to an analogous general scheme, adapted to the primitive terms of the system in question. To the directives f–i corresponded four wholly analogous directives. Whereas two of the directives belonging to the first group (directives g and i) concerned certain negations taking the form of expressions of the type '⌀-$(p\wedge)$', the correlates of these directives appealed to corresponding negations, formed with the help of the primitive terms of the system to be constructed. The role of the negation of any given expression 'p', expressed with the help of the primitive terms of the system in question, was played, in the systems constructed with the help of the above combinations 1–9 of primitive terms, by expressions of the following types: (1) '-⌀$(\wedge p)$'; (2) '⌀$(p\wedge)$' or, in a second parallel system, '⌀$(\wedge p)$'; (3) '-⌀-(pp)'; (4) '-⌀-(pp)'; (5) '-o-$(p\vee)$'; or, in a second parallel system, '-o-$(\vee p)$'; (6) '-o$(\vee p)$'; (7) 'o-$(p\vee)$'; (8) 'ò(pp)'; (9) 'ò(pp)'.

During my university course in 1933–4 entitled 'Introduction to Mathematical Logic' (see supplementary remark I) I showed that two further systems of prototetic based on the primitive terms '-⌀-' and '$\vee$' could be constructed, using negations formulated with the help of expressions of the type '-⌀-$(p\vee)$' and '-⌀-$(\vee p)$' respectively. I also showed that two other systems based on the primitive terms 'ò' and '$\wedge$' could be constructed, using negations formulated by means of the expressions 'ò$(p\wedge)$' or 'ò$(\wedge p)$'.

During my lectures entitled 'Foundations of the Propositional Calculus' given in the University of Warsaw during the academic year 1934–5, I remarked that those systems of prototetic mentioned above whose axiom is the sign '$\vee$' possess what I would call, with an eye to the harmony of these systems, a rather annoying characteristic, namely that the axiom in these systems is the only thesis which cannot be repeated. In order to eliminate this disharmony I specified two methods:

Method 1. The axiom '$\vee$' is replaced in the systems in question by a new directive which states that an expression may be added to the system if it is equiform with the eliminated axiom. (In systems modified in this way, each thesis can be repeated any number of times.)

Method 2. The directives which are correlates of a–d and f–i are restricted in such a way that only theses not equiform with any of the theses already belonging to the system may be added. (In systems modified in this way, no thesis can be repeated.)

Supplementary remark VI

While editing the 'terminological explanations' of my article, which were discussed above in the résumé of § 11, and which concerned those terms important for the formulation of the directives of prototethic, I often observed how, the directives remaining unaltered, the careless overlooking of this or that condition or restriction contained in these explanations could lead to an unavoidable contradiction in my system. As a result I began to investigate various other familiar deductive systems from this point of view, and to look for contradictions in these systems which could arise from similar carelessness in their formalization. The central point in these considerations lay in the analysis of the differences in directives between my system and the systems of other authors, in particular in the method of definition and, in general, the method of introducing into the system non-primitive constants. Among other systems, I analysed more closely in this respect the system of arithmetic published by von Neumann (see above the résumé of § 11).

The directives of my system of prototethic permit only single signs and functional expressions of the type $'f[kl...]'$, $'f\{xy...\}$ $[kl...]'$, etc., to be used in the system as function signs.[1] As function signs in von Neumann's system (these are 'operations',[2] 'transformations' of operations,[3] transformations of these transformations,[4] etc.) there can appear, according to the directives of this system, any symbols, with the exception of the parentheses '(' and ')', commas, and expressions of the type $'x_m'$, $'C_m'$ and $'A_m'$, where an arbitrary numeral replaces the letter $'m'$.[5] Taking into account the fact that those symbols which

---

[1] Cf. note 1, p. 66, of my article. See also (i) M. Schönfinkel, 'Über die Bausteine der mathematischen Logik', *Mathematische Annalen* 92, no. 3/4, Berlin, 1924, pp. 307–15; (ii) B. Sobociński, 'O kolejnych uproszczeniach aksjomatyki "ontologji"' Prof. St. Leśniewskiego' ('Successive simplifications of the axiom-system of Leśniewski's Ontology'), offprint from the volume in commemoration of fifteen years' teaching in the University of Warsaw by Prof. T. Kotarbiński, Warsaw, 1934, p. 159. [Paper 8 in this volume—Ed.] I did not cite Schönfinkel's work in the above-mentioned footnote to my article because at the time I was not acquainted with it.

[2] Cf. J. von Neumann, 'Zur Hilbertschen Beweistheorie', *Mathematische Zeitschrift* 26, no. 1 (Berlin, 1927), pp. 4 and 5.

[3] Op. cit., pp. 8 and 9.   [4] Op. cit., p. 8.   [5] Op. cit., pp. 4, 8, and 9.

play the role of function signs in the system under consideration
are not limited by this restriction in the number of signs they
contain, new function signs could be introduced into the system
by the process of transformation. These would be composed of
different combinations of arbitrarily many signs, the latter
already appearing in other function signs of the system.
(Examples of sign combinations of this kind would be the expressions '$+$', '1', '$++$', '$1+$', '11', '$+++$', '$++1$', '$+1+$',
etc., which consist of signs appearing in the function sign '$+1$'
introduced explicitly in the system by von Neumann[1] as a transformation of the operation '$O_3^{(1)}$'.) After comparing the two
systems I suspected that von Neumann's rules regulating the
form of function signs might be so liberal as to lead, in the
context of his entire system, to a logical catastrophe. Further
analysis substantiated my suspicion, inasmuch as von Neumann's system is not developed with sufficient care to preclude
the derivation of two mutually contradictory propositions. In
fact I derived them explicitly in § 11 of my article—see the
résumé of § 11 above.

These remarks concerning von Neumann's system of arithmetic incited him to publish his 'Bemerkungen zu den Ausführungen von Herrn St. Leśniewski über meine Arbeit "Zur
Hilbertschen Beweistheorie"'.[2] This article in turn evoked
Lindenbaum's 'Bemerkung zu den vorhergehenden "Bemerkungen . . ." des Herrn J. v. Neumann'.[3]

In his comments on my article,[4] von Neumann introduced
certain changes, backed up by sentences such as 'I am taking
the liberty of changing his argument somewhat, in ways which
I believe are insignificant, but more practical for the following
discussion',[5] and 'I write $\square$, $\bullet$ (instead of 1, $+$ as Leśniewski
does) in order to avoid giving rise to any arithmetical association'.[6] To prevent any misunderstanding on the part of the
reader, may I remind him, in connexion with the second of
the sentences quoted here, that the symbol '$+1$', consisting

---

[1] Op. cit., p. 15.   [2] *Fundamenta Mathematicae* 17 (1931).
[3] *Fundamenta Mathematicae* 17 (1931).
[4] Cf. von Neumann, op. cit., pp. 331 and 332.
[5] Op. cit., p. 331.   [6] Op. cit., p. 332, note 1.

of the signs '$+$' and '1', was introduced into von Neumann's system by the author himself.

In answer to my thesis that the system under consideration is inconsistent, von Neumann has the following to say:[1]

'May I note here that, in my opinion, his objection rests on a misunderstanding of the concept "sign": if mathematics is to be symbolically formulated, it is imperative that the different signs be discriminable.[2] {[2] Compare D. Hilbert, *Hamb. Abh.* I, pages 162–3.} This discriminability requires not only that two differently written signs should be separately distinguishable from each other, but, more important, that each combination (linear succession) of signs should be unambiguously analysable into its constitutent (printed) parts. Leśniewski selected the symbols □●, □, ●□ (which I designate for the moment as α, β, γ) in accordance with my general transformation rules, but at the same time violated an elementary law of every symbolic 'language': the signs α, β, γ, are not sufficiently distinguishable from one another, since $\alpha\beta$ is identical with $\beta\gamma$. That is, it is impossible to determine whether the combination □●□ is $\alpha\beta$ or $\beta\gamma$.[3] {[3] This naturally does not alter the fact that the mathematical formalism is to be regarded as in principle meaningless.}

'Once again it is important to stress: every symbolic system, my own included, must be constructed out of signs such that two sign combinations $\alpha\beta\gamma...\rho$ and $\lambda\mu\nu...\xi$ can have the same appearance only if they consist of the same number of signs, and if α coincides with λ, β with μ, γ with ν,..., ρ with ξ.

'As this has nothing to do with my object, namely proving the consistency of mathematics, but belongs rather to an earlier and in my opinion unmathematical stage of formalism, I didn't feel particularly compelled to refer to it in my work. However, because a misunderstanding arose, I have nevertheless discussed the matter.'[2]

These paragraphs compel me to make a few comments of an interpretative and terminological nature.

---

[1] The footnotes to the paragraphs quoted here are given in the accompanying brackets.   [2] Op. cit., p. 332.

In discussing the construction of his system of arithmetic in *Zur Hilbertschen Beweistheorie*, von Neumann has established quite precisely what he means by 'simple signs'.[1] This he has not done for 'signs' in general. Nor has he given the reader any indication of the sense in which he uses the word 'symbol'. The question of what 'sign' and 'symbol' mean, in the terminology of the author, is left for the reader to decide on the basis of various characteristic contexts. Having seen nothing in these contexts to require making a qualified interpretation, I interpreted, in fact, the words 'sign' and 'symbol' simply as exact correlates of my term 'expression' (compare pages 60 and 61 of my article). Using this terminology, I wrote the comments on von Neumann's system which I published in my article. I saw at the time, and I see today, no traditional meaning of the terms 'sign' and 'symbol' better adapted to the totality of von Neumann's explanations containing these terms in *Zur Hilbertschen Beweistheorie*.

Using this interpretation of the terms, the long quotation above from von Neumann's later publication contains a series of seemingly paradoxical incongruities. If 'sign' means the same as 'expression', each 'linear succession of signs' which contains at least three words† can be decomposed in at least three different ways into disjoint components which are signs. (For the moment I ignore combinations of only two words, in order to avoid borderline questions concerning von Neumann's use of the expressions 'component' and 'succession'.) For example, the linear succession '0+1' of the two signs '0' and '+1', which contains the three words '0', '+' and '1', is decomposable into
(a) the disjoint components '0', '+' and '1', which are signs,
(b) the disjoint components '0' and '+1', which are signs, and
(c) the disjoint components '0+' and '1', which are signs. The postulate formulated by von Neumann can be satisfied by no two 'sign combinations' of the same form each containing at least three words. According to this postulate 'two sign combina-

[1] See von Neumann, *Zur Hilbertschen Beweistheorie*, pp. 4–6.
† [Ed. note. Up to this point *Wort* has been translated as 'sign'. But Leśniewski now begins to use it in his technical sense of 'simple expression', and it will henceforth in this article be translated as 'word'.]

tions $\alpha\beta\gamma \ldots \rho$ and $\lambda\mu\nu \ldots \xi$ can have the same appearance only if they consist of the same number of signs, and if α coincides with λ, β with μ, γ with ν,..., ρ with ξ. (Thus, the sign combination made up of the sign 'Z' and the subsequent sign '$0+1$', and the sign combination made up of the sign '$Z0$' and the subsequent sign '$+1$' look exactly alike, although neither does the sign 'Z' from the first sign combination coincide with the sign '$Z0$' from the second, nor does the sign '$0+1$' from the first coincide with the sign '$+1$' from the second.) These signs, which should be 'discriminable' from other signs in von Neumann's sense, cannot possibly exist. The fact that the signs '$1+$', '1', and '$+1$', which I made use of in my article to derive a contradiction in von Neumann's system, and which as we saw were replaced in von Neumann's reply by the signs '$\square\bullet$', '$\square$', and '$\bullet\square$', are not 'sufficiently distinguishable' from one another, cannot be taken as implying any 'misunderstanding of the concept "sign"'. (I shall not here take up what is for me a very obscure question, what von Neumann means by the 'unmathematical stage of formalism'.)

To one unfamiliar with the editorial details of the directives of von Neumann's system, and the author's commentary on his system, but who has only read von Neumann's answer to my critique cited here, and on this basis interprets the word 'sign' according to the postulated 'elementary law of every symbolic "language"', it might seem that in von Neumann's terminology 'sign' is exactly the same as 'word' in my terminology (compare pages 60 and 61 of my article). Such an opinion would be incorrect, as the following facts show:

(a) In von Neumann's terminology the expression 'x_1' is a 'sign',[1] and in this expression the numerical index is also a 'sign',[2] so that, according to this interpretation of the word 'sign', both the expression 'x_1' and its index would be words. But not according to my terminology, in which no letter or index that is merely part of a word is itself a word.

(b) As we saw above, von Neumann introduced into his system the expression '$+1$' as a transformation of the operation

---

[1] Op. cit., p. 4.   [2] Op. cit., p. 6.

'$O_3^{(1)}$'. According to the directives concerning transformations given by the author, this expression must be a sign.[1] At the same time, in keeping with our interpretation of the word 'sign', it must be a word. This would not be compatible with my terminology, where no expression consisting of two words is a word.

(c) The expression '$Z0$', consisting also of two words, is thus no word according to my terminology, although in the terminology of von Neumann it is a sign.[2]

I am inclined to believe that the interpretation of 'sign' as 'expression' can be carried out in a thoroughly consistent way within the context of *Zur Hilbertschen Beweistheorie*, which work provided the point of departure for my critical remarks about von Neumann's system. (Note that even a thoroughly consistent terminology in no way excludes the possibility of obtaining contradictory theses in a system whose directives, based on this terminology, are not formulated with sufficient care.) This interpretation in turn forces us to reject those postulates of the author which were first published *ex post facto* by him in his reply to my critical remarks. On the other hand, the attempt to correlate the term 'sign' with my term 'word' leads to obvious incompatibilities in interpretation even when we consider the specific contexts of von Neumann's first, fundamental publications. All this says nothing in favour of interpreting 'signs' as 'words'.

The complications and obscurity described here over the meaning of the word 'sign', which arose through conflating different explanations drawn from the author's two different works, require some radical hypotheses to throw light on the true source of the confusion, and to prepare the ground for its more or less reasonable settlement. As the one who to some extent brought on this tangle through my critique, I feel the need of imparting to the reader one or two confidences concerning the heuristic conception which was helpful to me in overcoming some of the wealth of difficulties described above. This heuristic conception could be summarized more or less as follows.

---

[1] Op. cit., p. 9. [2] Op. cit., pp. 41 and 15.

M

In making use of the terms 'sign' and 'symbol' in *Zur Hilbert-schen Beweistheorie*, von Neumann made no preliminary attempt exactly to circumscribe the use of these terms. And, using these terms in a completely unrestricted and intuitive way, he employed them in practice as correlates of my term 'expression', so that they concurred with my original interpretation of them given above. The author first gave a precise statement of the universal postulate, which the 'signs' of every 'symbolic language' or every 'symbolic system' should satisfy, in the publication of the reply to my critique of his system. The reply was carelessly carried out, and the way in which the author used the word 'sign' is conspicuously inconsistent. The explanation given by the author holds good for 'signs' as 'words', but not for 'signs' as 'expressions'. A successful method of eliminating all these inconsistencies would be to consider the paragraph analysed above as non-existent, and obstinately to treat 'signs' and 'symbols' as 'expressions', at least until the author gives some clearly formulated and convincing reasons for doing otherwise. On the basis of a terminology determined in this way, one can anticipate weakening the directives of von Neumann's system to eliminate the contradiction.

Here I would like to draw to the reader's attention that the impossibility of harmonizing von Neumann's standpoint in *Zur Hilbertschen Beweistheorie* with the 'elementary law of every symbolic language', which was formulated in his reply, can also be proved, if I may so express it, in a more 'immanent' way, without recourse to any hypotheses as to interpretation. The argument proceeds as follows. The words '$\sim$', 'Z', and '0' constitute, in von Neumann's system, transformations of the operations '$O_1^{(1)}$' and '$O_2^{(1)}$' and of the constant 'C_1'.[1] According to the author's directives concerning transformations, these words must be signs.[2] In von Neumann's terminology, the expression '$Z0$' is likewise a sign, as we saw in the discussion of 'signs' as 'words'. The last two steps indicate that the sign combination '$\sim Z0$' can be analysed on the one hand into the successive signs '$\sim$', 'Z', and '0', and on the other into '$\sim$'

---

[1] Op. cit., pp. 10 and 15.   [2] Op. cit., pp. 8 and 9.

and 'Z0'. The 'elementary law of every symbolic language', postulated by the author, is obviously incompatible with this fact.

The article containing von Neumann's polemic against my critique also contains one or two theoretical points worth considering. What is essential in the author's arguments can be brought together in the following quotations:

'As the subject has already been raised, I should like to say one more thing on the question of signs, and at the same time to correct a real oversight in my work.'[1]

'That the system may function at all, it is essential that the construction of any formula should be unique, and that, in particular, this formula should not result in two different ways.'[2]

'I have already mentioned (loc. cit.) that the formal system which I gave, without the "transformations" (pages 8–9), satisfies the postulate.'[3] (The page numbers cited in these paragraphs refer to the earlier of von Neumann's two works discussed here.)

'But in order to include the "transformations" and the simplifying convention concerning parentheses (loc. cit., pages 8–9) one must again be convinced of the validity of the principle of univocity. Here indeed I permitted too much. Thus, for example, the three operations

$$'O_r^{(2)}(\cdot\,,\cdot), \qquad O_s^{(1)}(\cdot), \qquad O_t^{(1)}(\cdot),$$

can be transformed into

$$(\cdot\,\bullet\,\cdot), \qquad \bullet(\cdot), \qquad (\cdot)\bullet,$$

and then (with some constant C_p) the formulae

$$O_r^{(2)}(C_p, O_s^{(1)}(C_p)) \quad \text{and} \quad O_r^{(2)}(O_t^{(1)}(C_p), C_p)$$

can be constructed. Both assume the form

$$(C_p\bullet\,\bullet\,C_p).$$

To avoid such improprieties it suffices, for example, to make the following addition to my "transformation rule": no Γ may

---

[1] Von Neumann, 'Bemerkungen zu den Ausführungen von Herrn St. Leśniewski über meine Arbeit "Zur Hilbertschen Beweistheorie"', p. **333**.
[2] Loc. cit.
[3] Loc. cit.

be applied concurrently in a transformation $\Gamma(:,....,:)$ and in a transformation $(:,....,:)\Gamma$. (The second case could even be limited to the occurrence of a single blank place.)

'It is not difficult to prove the univocity of the systems modified by these precautions. However, with all due respect to the reasons given here, I feel that to carry out the proof is superfluous, particularly as these *pro-domo* discussions already take up too much space.'[1]

My observations on his article, of which a significant part has been included here for the convenience of the reader, already occupy far more space than von Neumann's *'pro-domo'* discussions. Unfortunately, however, I cannot yet conclude these remarks, for I am as sceptical of von Neumann's system, reformed by these precautions, as I was earlier of the author's original system.

I am inclined to believe that, in making no attempt to produce an explicit proof of univocity for his new system because of its alleged simplicity, the author once again was too careless. Contrary to his opinion on this question, I shall attempt to prove that the restriction on the directives, as he introduced it, renders his system in no way univocal; that is to say, the reformed system is as inconsistent as the first one.

This would be a banal task had von Neumann not questioned the correctness of the proof, presented in my article, that his original system was inconsistent. The whole proof, as well as the simple considerations designed to show that the formula '$1+1$' appearing in the proof can be 'constructed' in von Neumann's sense in at least two different ways, would hold true also for the second of his systems. In reply to my critique, von Neumann complicated the situation by charging me, as we know, with proceeding incorrectly (never mind in what respect) in treating the signs '$1+$' and '1' as transformations, along with the sign '$+1$' which was introduced by himself. As I don't wish to make the validity of my argument depend upon whether or not this debatable reproach is justified, I am replacing my original proof by a new construction, in which the transforma-

[1] Op. cit., p. 334.

tions in question no longer appear. To fend off all similar reproofs I introduce, in accordance with the directives of both von Neumann's systems, only those transformations of signs which have already been explicitly introduced by the author himself.

In von Neumann's systems, two mutually contradictory propositions can be obtained in the following way:

From 'schema' 1 of Group III[1] we know that

(α) $Z0$.

Using the transformation rules[2] we transform the symbol '$\Gamma_{1,1}$'[3] into the symbol 'Z'. Schema 3 of Group III,[4] in conjunction with the rule b concerning the omission of parentheses,[5] shows us that

(β) $\sim (Z+1 = 0)$.

Considering the sequences of formulae $b_{u,v}^{(n)}$ [6] discussed by von Neumann we can assert[7] the existence of a natural number l such that the formula '$x_1+1 = 0$' is identical with the formula $b_{l,1}^{(1)}$, the latter being one of the elements of the sequence $b_{u,1}^{(1)}$. This, together with schema 2 of Group VI[8] and rule b concerning the omission of parentheses, allows us to establish that

(γ) $\Omega_{l,1}^{(1)} Z = (Z+1 = 0)$.

According to the rules,[9] we transform the symbol '$\Omega_{l,1}^{(1)}$' in such a way that expressions of the type '$\Omega_{l,1}^{(1)} a_1$' are changed into corresponding expressions of the type '$(a_1)0$'. We can then assert, in conformity with γ and with the rule b, that

(δ) $Z0 = (Z+1 = 0)$.

Schema 1 of group II[10] allows us to assert that

(ϵ) $Z0 = Z0$,

and schema 2' of group II,[11] with the substitution of '$(x_1 = Z0)$' for 'c', that

(ζ) $(Z0 = (Z+1 = 0)) \rightarrow ((Z0 = Z0) \rightarrow ((Z+1 = 0) = Z0))$.

[1] Von Neumann, *Zur Hilbertschen Beweistheorie*, p. 15.
[2] See op. cit., p. 8.   [3] See op. cit., p. 20.   [4] Op. cit., p. 15.
[5] Op. cit., p. 9.   [6] Op. cit., p. 20.
[7] See op. cit., p. 21.   [8] Op. cit., p. 20.
[9] See op. cit., pp. 8 and 9, and the above-mentioned precautions.
[10] Op. cit., p. 15.   [11] Loc. cit.

From δ and ζ we conclude that

(η) $(Z0 = Z0) \rightarrow ((Z+1 = 0) = Z0),$[1]

and from ϵ and η, that

(θ) $(Z+1 = 0) = Z0.$

From schema 2' of Group II, when the formula '$\sim x_1$' is substituted for 'c', it follows that

(ι) $((Z+1 = 0) = Z0) \rightarrow (\sim (Z+1 = 0) \rightarrow \sim Z0).$

From θ and ι we conclude that

(κ) $\sim (Z+1 = 0) \rightarrow \sim Z0,$

and from β and κ that

$\sim Z0,$

which contradicts α.

We may assure ourselves that von Neumann's system, even after the above restrictions of the author, by no means becomes an univocal system, by establishing that the formula '$Z0$' in the proof just given can be derived, using appropriate transformations, on the one hand from the formula '$\Omega^{(1)}_{l,1} \Gamma^{(1)}_{1,1}$', and on the other hand from the formula '$O^{(1)}_2 C^{(1)}_1$'.[2]

Lindenbaum, in his above-mentioned paper, compares my proof of the inconsistency of von Neumann's first system with von Neumann's proof of the 'equivocity' of this system given in one of the paragraphs quoted above, and states the following conclusion:

'Now the entire trouble derives wholly from a different source —again in both constructions in the same way—so that, for example, the rules concerning the elimination of parentheses[1] {[1] *'Zur Hilbertschen Beweistheorie'*, § 4, rule III.} often permit too much.[2] {[2] Nevertheless—in an entirely different treatment —parentheses are completely dispensable.} Should parentheses always be retained, such constructions would be quite impossible.'[3]

It seems to me that Lindenbaum's diagnosis doesn't hit the nail on the head. If I somewhat modified the deduction with the help of which I proved the inconsistency of von Neumann's

---

[1] Op. cit., p. 11.   [2] See op. cit., p. 15.
[3] Lindenbaum, op. cit., p. 336.

system in my article, I could obtain the same result in the following way without using the rules a–c[1] concerning the omission of parentheses.

On the basis of considerations analogous to those in the commentary to thesis α on page 81 of my [original] article, we can add to the system the thesis which states that

(a) $\left(\Omega_{k,1}^{(1)}(0) = 0\right).$

The transformation of the symbol '$\Omega_{k,1}^{(1)}$', according to which expressions of the type '$\Omega_{k,1}^{(1)}(a_1)$' transform into corresponding expressions of the type '$(a_1)+0$', allows us to infer from a that

(b) $((0)+0 = 0).$

According to schema 3 of group III,

(c) $\sim ((0)+1 = 0).$

The transformation of the symbol '0' into the symbol '1' transforms the formulae b and c into formulae which state respectively that

$$((1)+1 = 1)$$

and $\sim ((1)+1 = 1).$

Supplementary remark VII

While the already published part of my article contains the axioms and the directives of the system $\mathfrak{S}_5$, formulated as precisely as possible, and hence contains *in potentia* the entire formalized system of protothetic, my somewhat later abovementioned publication entitled *Über die Grundlagen der Ontologie* gives the axioms and directives (and thus implicitly the whole formalized system) of ontology. Sobociński's basic monograph of 1934, also mentioned above, entitled 'Successive simplifications of the axiom-system of Leśniewski's Ontology'† reports on the results of axiomatic research in this field by myself, by Tarski, and by the author. As for mereology, the third of the theories discussed above in the résumé of the introduction to my article, I have devoted to it the major part of the already published sections of my work 'On the Foundations of Mathematics'.[2]

[1] See von Neumann, op. cit., p. 9.

† [Paper 8 of this volume—Ed.]

[2] S. Leśniewski, '*O podstawach matematyki*' ('On the Foundations of Mathematics'), *Przegląd Filozoficzny*, vols. 30–34.

This work has been appearing since 1927 in *Przegląd Filozoficzny*, and to date comprises the following parts (altogether 171 pages):

1. *Introduction. Section I.* On certain questions concerning the meaning of 'logistic' theses.

Section II. On Russell's 'antinomy' concerning the 'class of classes which are not elements of themselves'.

Section III. On different ways of understanding the words 'class' and 'set'.[1]

2. *Section IV.* On 'Foundations of general set theory. I.'.[2]†

3. *Section V.* Further theorems and definitions of 'general set theory' from the period up to 1920 inclusive.[3]

4. *Section VI.* Axiomatization of 'general set theory' from the year 1918.

Section VII. Axiomatization of 'general set theory' from the year 1920.

Section VIII. On certain conditions, established by Kuratowski and Tarski, necessary and sufficient for P to be the class of a.

Section IX. Further theorems of 'general set theory' from the years 1921–3.[4]

5. *Section X.* Axiomatization of 'general set theory' from the year 1921.

Section XI. On 'singular propositions' of the type '$A \in b$'.[5]

In connexion with the expression 'general set theory', which appears here in the titles of different individual sections, I should like to mention that the theory which I now call 'mereology' I formerly called 'general set theory', or 'general class theory'. I ceased using these two names long ago because, in order not to arouse needless misunderstanding, I wanted to distinguish my theory clearly from the various 'set theories' and 'class theories' which, if I may so express myself, possess an 'official' character. As I began to use the word 'class' in mereology in

---

[1] Vol. 30, 1927.   [2] Vol. 31, 1928.

† [Ed. note. This is the title of Leśniewski's 1916 paper referred to in the résumé of the introduction to his article.]

[3] Vol. 32, 1929.   [4] Vol. 33, 1930.   [5] Vol. 34, 1931.

a way incompatible with the tradition of these theories, I made an effort to rely on the most precise analysis possible of the meaning which in practice I employed, employ, and will continue to employ in the future. As it happened, the meaning I used would seem to harmonize to a great extent with common intuition. I used this word in particular in discussions concerning the 'evidence' or 'non-evidence' of various theses which play a part in the different 'antinomies' constructed by class theorists. I was never able to conceive of a sense of the word 'class' in which I should be at all inclined to ascribe to classes the totality of the properties postulated in these theses. Expressions of the type 'class of objects a' are, on the basis of my mereology, names denoting definite and quite ordinary objects. These expressions naturally have nothing in common either with any mythology of 'classes', considered as objects of some 'higher type' or 'higher order', or with a use of the word 'class' in which the latter is not the name of any object(s), but rather a surrogate *façon de parler* of some entirely different syntactical type, as for example in the system of Whitehead and Russell.[1] The totality of theorems of my system of the foundations of mathematics, which in practice can be handled as theoretical correlates of this or that thesis of these authors' 'theory of classes', forms a proper part of my ontology.†

[1] See Alfred North Whitehead and Bertrand Russell, *Principia Mathematica*, vol. i, 2nd edition (Cambridge, 1925), pp. 71 and 72.

† [Ed. note: The two pages of bibliographical abbreviations and typographical corrections to his original article of 1929 with which Leśniewski concluded his paper are here omitted.]

7

ON DEFINITIONS IN THE SO-CALLED
THEORY OF DEDUCTION†

STANISŁAW LEŚNIEWSKI

THIS paper is a résumé of the course of lectures (in Polish) 'On foundations of the "theory of deduction" ' that I delivered in Warsaw University in the academic year 1930–1. My main task here is to formulate a directive permitting addition, to the system of the 'theory of deduction', of theses of the special kind that I call *definitions*, as distinguished from *axioms* and *theorems*, and codifying as precisely as possible conditions to be satisfied by such definitions.

The problem of definition in the theory of deduction lies quite outside my system of foundations of mathematics, which I have begun publishing in the last few years.[1] What interested me in this problem, if I may so express myself, was its own constructive appeal—in view of the still rather stepmotherly treatment of it even in the current scientific trend in theory of deduction and theory of theory of deduction.

I base my directive for definition formulated below on the well-known bracketless and dotless notation devised for mathematical logic by Jan Łukasiewicz in 1924[2] and since adopted

---

† This paper originally appeared under the title 'Über Definitionen in der sogenannten Theorie der Deduktion' in *Comptes rendus des séances de la Société des Sciences et des Lettres de Varsovie*, Cl. iii, 24 (1931), pp. 289–309, and was presented to the Society by Jan Łukasiewicz on 21 November 1931. Translated by E. C. Luschei.

[1] See Stanisław Leśniewski: (1) 'O podstawach matematyki' (On the foundations of mathematics), *Przegląd Filozoficzny* 30 (1927), pp. 164–206; 31 (1928), pp. 261–91; 32 (1929), pp. 60–101; 33 (1930), pp. 77–105; 34 (1931), pp. 142–70. (2) 'Grundzüge eines neuen Systems der Grundlagen der Mathematik', *Fundamenta Mathematicae* 14 (1929), pp. 1–81. (3) 'Über die Grundlagen der Ontologie', *Comptes rendus des séances de la Société des Sciences et des Lettres de Varsovie*, Cl. iii, 23 (1930), pp. 111–32. Presented by Jan Łukasiewicz on 22 May 1930.

[2] See (1) Jan Łukasiewicz, *Elementy logiki matematycznej* (Elements of

by several others.[1] In terms of this notation—the simplest (though by no means the clearest) symbolism I know for the theory of deduction—the problems of introducing definitions, which if brackets are retained could be resolved by simply adapting my directive for definition in 'protothetic',[2] lose much theoretical banality.

Although my chief problem, in the foundations of the theory of deduction, concerned the directive for definition, naturally I could not carry out my investigations in complete abstraction from other directives of the theory; so, for example, introducing definitions into the theory, I felt compelled to give the 'directive for substitution' too a form permitting replacement of variables even by formulae containing defined terms of the theory. All these considerations have led me to present here a complete system of directives for the theory of deduction.

I base this system of the theory of deduction with definitions on the familiar 33-word axiom quoted below, formulated by Łukasiewicz in terms of negation and implication, which, as he has shown, forms, together with the directives for detachment and substitution, an axiomatic foundation adequate for the ordinary theory of deduction. The directives I give here for a system based on these two primitive terms can very easily be transposed to a system based on others, in particular to the familiar system of Nicod.[3]

Mathematical Logic), Wydawnictwa Koła Matematyczno-Fizycznego Słuchaczów Uniwersytetu Warszawskiego, vol. 18 (1929), pp. 37–40, 45, 154–6, 158–9, 171–2. [English translation, Warsaw 1963—Ed.] (2) Jan Łukasiewicz, 'O znaczeniu i potrzebach logiki matematycznej', *Nauka Polska* 10 (1929), pp. 610–12. (3) Jan Łukasiewicz and Alfred Tarski, 'Untersuchungen über den Aussagenkalkül', *Comptes rendus des séances de la Société des Sciences et des Lettres de Varsovie*, Cl. iii, 23 (1930), pp. 31–32.

[1] See (1) Leon Chwistek, 'Neue Grundlagen der Logik und Mathematik', *Mathematische Zeitschrift*, 30 (1929), p. 713. (2) M. Presburger, 'Über die Vollständigkeit eines gewissen Systems der Arithmetik ganzer Zahlen, in welchem die Addition als einzige Operation hervortritt', *Sprawozdanie z I Kongresu Matematyków Krajów Słowiańskich (Comptes rendus du Ier Congrès des Mathématiciens des pays slaves)*, Warsaw, 1929, pp. 92–93.

[2] See Leśniewski, 'Grundzüge eines neuen Systems der Grundlagen der Mathematik', pp. 70–72, 76.

[3] See Jean G. P. Nicod, 'A reduction in the number of the primitive propositions of logic', *Proceedings of the Cambridge Philosophical Society* 19 (1917), pp. 32–41.

Łukasiewicz's[1] Axiom (L):†

$$CCC\alpha C\beta \alpha CCCN\gamma C\delta N\epsilon CC\gamma C\delta\zeta CC\epsilon\delta C\epsilon\zeta\eta C\theta\eta.$$

Before proceeding to formulate the directives of this system of the theory of deduction based on Axiom L, I give the following series of *terminological explanations* of the technical expressions peculiar to these directives.[2]

Terminological explanation I. I say of object A that it is (the) complex of (the) a^3 if and only if the following conditions are fulfilled:

(1) A is an expression;

(2) if any object is a word that belongs to A, then it belongs to a certain a;

(3) if any object B is a, any object C is a, and some word that belongs to B belongs to C, then B is the same object as C;

(4) if any object is a, then it is an expression that belongs to A.[4]

Examples (I have composed pertinent examples to show the mutual independence of individual conditions of the relevant terminological explanations):

(1) Axiom L is the complex of words that belong to Axiom L.

(2) The first word of Axiom L is not the complex of words that belong to Axiom L [Conditions 1–3 are here fulfilled, Condition 4 is not fulfilled (the 2nd word of Axiom L is a word that belongs to Axiom L, but it is not an expression that belongs to the first word of Axiom L)].

(3) Axiom L is not the complex of expressions that belong to Axiom L [Conditions 1, 2, 4 fulfilled (f.), Condition 3 not fulfilled (n.f.) (Axiom L is an expression that belongs to Axiom L, the

---

[1] See Łukasiewicz and Tarski, op. cit., pp. 36–37.

† [*Translator's note*: Henceforth in this translation called *Axiom L*. Leśniewski used the designation '*Axiom*' in italics.]

[2] To understand the significance of such technical expressions and to preclude possible misinterpretations of these terminological explanations and of the directives, see Leśniewski, op. cit., pp. 59–62.

[3] The uncapitalized variable was here used in the plural.

[4] See Leśniewski, op. cit., p. 63, *T.E. VII*.

first word of Axiom L is an expression that belongs to Axiom L, some word that belongs to Axiom L belongs to the first word of Axiom L, but Axiom L is not the same object as the first word of Axiom L)].

(4) Axiom L is not the complex of expressions that belong to Axiom L and are equiform to the first word of Axiom L [C. 1, 3, 4 f., C. 2 n. f. (the 4th word of Axiom L is a word that belongs to Axiom L, but it belongs to no expression that both belongs to Axiom L and is equiform to the first word of Axiom L)].

(5) The class of expressions[1] that belong to Axiom L and are equiform to the first word of Axiom L is not the complex of expressions that belong to Axiom L and are equiform to the first word of Axiom L [C. 2–4 f., C. 1 n. f.].

Terminological explanation II. I say of object A that it is (the) negate of B if and only if the following conditions are fulfilled:

(1) A is an expression;
(2) B is the complex of objects that are either A or the first of the words that belong to B;
(3) B is not a word;
(4) the first of the words that belong to B is an expression equiform to the 11th word of Axiom L.

Examples.

(1) The 12th word of Axiom L is the negate of the class of objects that are either the 11th or the 12th word of Axiom L.

(2) The class of words of Axiom L that follow the first word of Axiom L is not the negate of Axiom L [C. 1–3 f., C. 4 n. f.].

(3) The 11th word of Axiom L is not the negate of the 11th word of Axiom L [C. 1, 2, 4 f., C. 3 n. f.].

---

[1] Expressions of the form 'class of a' as used here always mean class (i.e. totality) in the *collective* sense of my 'general set theory', which I have come to call mereology. (See Leśniewski: (1) 'O podstawach matematyki', *Przegląd Filozoficzny* 30, pp. 185–206, and 31, pp. 261–5; (2) 'Grundzüge eines neuen Systems der Grundlagen der Mathematik', p. 5.) So, for example, the class of expressions that belong to Axiom L and are equiform to the first word of Axiom L is an object that *consists* of all expressions that belong to Axiom L and are equiform to the first word of Axiom L, just as an orchestra consists of all its members. Expressions of the form 'class of a' occur here only in examples.

(4) Axiom L is not the negate of the class of words of Axiom L that follow the 10th word of Axiom L [C. 1, 3, 4 f., C. 2 n. f.].

(5) It-is-not-true-that[1] the word of Axiom L following the 11th word of Axiom L is the negate of the class of words of Axiom L that follow the 10th word of Axiom L[2] [C. 2–4 f., C. 1 n. f.].

Terminological explanation III. I say of object A that it is (the) implicant of B in C if and only if the following conditions are fulfilled:

(1) C is the complex of objects that are either A, B, or the first of the words that belong to C;

(2) the first of the words that belong to C is an expression equiform to the first word of Axiom L;

(3) A follows the first of the words that belong to C;

(4) B follows A.

Examples.

(1) The 2nd word of Axiom L is the implicant in Axiom L of the class of words of Axiom L that follow the 2nd word of Axiom L.

(2) The class of words of Axiom L that follow the first word of Axiom L is not the implicant in Axiom L of the class of words of Axiom L that follow the first word of Axiom L [C. 1–3 f., C. 4 n. f.].

(3) The first word of Axiom L is not the implicant in Axiom L of the class of words of Axiom L that follow the first word of Axiom L [C. 1, 2, 4 f., C. 3 n. f.].

(4) The 5th word of Axiom L is not the implicant, in the class

---

[1] I use this phrase hyphenated as a 'colloquial' representative of the ordinary propositional negation sign of mathematical logic.

[2] In my terminological explanations and examples 'singular propositions' of the form 'A is b' are used in accordance with the axiom of my 'ontology'. (See Leśniewski, 'Über die Grundlagen der Ontologie', pp. 114–15, 129–31.) It follows, since there is more than one word of Axiom L following the 11th word of Axiom L, that for no a can it be true that the word of Axiom L following the 11th word of Axiom L is a; consequently it cannot be that the word of Axiom L following the 11th word of Axiom L is the negate of the class of words of Axiom L that follow the 10th word of Axiom L, nor even an expression. (See Leśniewski, 'O podstawach matematyki', *Przegląd Filozoficzny* 31, pp. 263–4.)

of words of Axiom *L* that follow the 3rd word of Axiom *L*, of the class of words of Axiom *L* that follow the 5th word of Axiom *L* [C. 1, 3, 4 f., C. 2 n. f.].

(5) The 2nd word of Axiom *L* is not the implicant in Axiom *L* of the 3rd word of Axiom *L* [C. 2–4 f., C. 1 n. f.].

Terminological explanation IV. I say of object *A* that it is subordinate to *B* with respect to (the) *a*,[1] in particular to (the) *b*,[1] relative to *C* if and only if the following conditions are fulfilled:

(1) *B* is an expression that belongs to *C*;

(2) *B* is not a variable;[2]

(3) if any object is a word that belongs to *C* and follows *B*, then it is a variable;

(4) if any object is a word that belongs to some thesis of this system which thesis precedes *C*,[3] then it is not an expression equiform to *B*;

(5) *A* is the complex of objects that are either *b* or the first of the words that belong to *A*;

(6) the first of the words that belong to *A* is an expression equiform to *B*;

(7) if any object is *b*, then it is *a*;

---

[1] The uncapitalized variable was here used in the plural.

[2] I give here no special terminological explanation of the word 'variable'. The details of fixing the denotation of this word are relatively immaterial. But I have to presuppose here that (1) the 4th, 6th, 12th, 14th, 16th, 22nd, 30th, and 32nd words of Axiom *L* are variables; (2) neither the 1st nor the 11th word of Axiom *L* is a variable; (3) if *A* is an expression equiform to *B*, then *A* is a variable if and only if *B* is a variable; (4) any variable is a word; and (5) it is always possible to form new variables (i.e. variables equiform to no variable already used) in the same general sense as it is to form new expressions. A concrete definition, if I had to give one here, might well be to the effect that any object is a variable if and only if it is a word consisting solely of small Greek letters. I could not accept a convention that confined variables to letters of this or that alphabet, since such a convention would preclude forming a proposition containing more nonequiform variables than there are nonequiform letters of the alphabet. Cf. Łukasiewicz and Tarski, op. cit., p. 31.

[3] As theses of this system in addition to Axiom *L* I count only those 'definitions' and 'theorems' *effectively* added to the system, not various other expressions that might be added according to its directives. So the extent of the expression 'thesis of this system' is by no means univocally determined in advance, but rather is conceived as 'growing' by stages. Axiom *L* is the only expression already a thesis of this system.

(8) if any object is b, then it follows the first of the words that belong to A;

(9) there are exactly as many b as words that belong to C and follow B.

Examples.

(1) The class of words of Axiom L that follow the 30th word of Axiom L is subordinate to the 31st word of Axiom L with respect to variables, in particular to objects that are either the 32nd or the 33rd word of Axiom L, relative to Axiom L.

(2) The class of words of Axiom L that follow the 30th word of Axiom L is not subordinate to the 31st word of Axiom L with respect to expressions, in particular to the class of objects that are either the 32nd or the 33rd word of Axiom L,[1] relative to Axiom L [C. 1–8 f., C. 9 n. f.].

(3) The class of objects that are either the 31st or the 32nd word of Axiom L is not subordinate to the 31st word of Axiom L with respect to words, in particular to objects that are either the 31st or the 32nd word of Axiom L, relative to Axiom L [C. 1–7, 9 f., C. 8 n. f. (the 31st word of Axiom L is an object that is either the 31st or the 32nd word of Axiom L, but it does not follow the first of the words that belong to the class of objects that are either the 31st or the 32nd word of Axiom L)].

(4) The class of words of Axiom L that follow the 30th word of Axiom L is not subordinate to the 31st word of Axiom L with respect to Axiom L, in particular to objects that are either the 32nd or the 33rd word of Axiom L, relative to Axiom L [C. 1–6, 8, 9 f., C. 7 n. f. (the 32nd word of Axiom L is an object that is either the 32nd or the 33rd word of Axiom L, but it is not Axiom L)].

(5) The class of words of Axiom L that follow the 29th and precede the 33rd word of Axiom L is not subordinate to the 31st word of Axiom L with respect to words, in particular to objects that are either the 31st or the 32nd word of Axiom L, relative to Axiom L [C. 1–5, 7–9 f., C. 6. n. f.].

(6) Axiom L is not subordinate to the 31st word of Axiom L

---

[1] Of course there is only one such class, as there can be at most one class of a, whatever a may be. See Leśniewski, loc. cit., p. 265, *Axiom III*.

with respect to words, in particular to objects that are either the 32nd or the 33rd word of Axiom L, relative to Axiom L [C. 1–4, 6–9 f., C. 5 n. f.].

(7) The first of the words that follow Axiom L and are expressions equiform to the first word of Axiom L is not subordinate to the first of the words that follow Axiom L and are expressions equiform to the first word of Axiom L with respect to nonquadrangular quadrangles, in particular to nonquadrangular quadrangles, relative to the first of the words that follow Axiom L and are expressions equiform to the first word of Axiom L [C. 1–3, 5–9 f., C. 4 n. f. (the first word of Axiom L is a word that belongs to some thesis of this system which thesis precedes the first of the words that follow Axiom L and are expressions equiform to the first word of Axiom L, but it is an expression equiform to the first of the words that follow Axiom L and are expressions equiform to the first word of Axiom L)].

(8) Axiom L is not subordinate to the first word of Axiom L with respect to words, in particular to words of Axiom L that follow the first word of Axiom L, relative to Axiom L [C. 1, 2, 4–9 f., C. 3 n. f. (the 2nd word of Axiom L is a word that belongs to Axiom L and follows the first word of Axiom L, but it is not a variable)].

(9) The class of words of Axiom L that follow the 31st word of Axiom L is not subordinate to the 32nd word of Axiom L with respect to words, in particular to the 33rd word of Axiom L, relative to Axiom L [C. 1, 3–9 f., C. 2 n. f.].

(10) The second word of Axiom L is not subordinate to the second word of Axiom L with respect to Axiom(s) L, in particular to nonquadrangular quadrangles, relative to the first word of Axiom L [C. 2–9 f., C. 1 n. f.].

Terminological explanation V. I say of object A that it is an expression fundamental for (the) a,[1] relative to B, if and only if the following conditions are fulfilled:

(1) A is an expression;

(2) some expression is a;

---

[1] The uncapitalized variable was here used in the plural.

(3) if any object is a, then it is an expression that belongs to A;

(4) if any object is a variable that belongs to A, then it is a;

(5) if any object C is an expression that belongs to A, and the negate of C is a, then C is a;

(6) if any object C is the same object as B or is a thesis of this system which thesis precedes B, and any object D belongs to A and is subordinate to some expression with respect to a, in particular to any arbitrary objects b, relative to C, then D is a.

Examples.

(1) Axiom L is an expression fundamental for expressions that belong to Axiom L, relative to Axiom L.

(2) The class of words of Axiom L that follow the 30th word of Axiom L is not an expression fundamental for objects that are either the 32nd or the 33rd word of Axiom L, relative to Axiom L [C. 1–5 f., C. 6 n. f. (Axiom L is the same object as Axiom L or is a thesis of this system which thesis precedes Axiom L, the class of words of Axiom L that follow the 30th word of Axiom L belongs to the class of words of Axiom L that follow the 30th word of Axiom L and is subordinate to some expression with respect to objects that are either the 32nd or the 33rd word of Axiom L, in particular to objects that are either the 32nd or the 33rd word of Axiom L, relative to Axiom L, but the class of words of Axiom L that follow the 30th word of Axiom L is not an object that is either the 32nd or the 33rd word of Axiom L)].

(3) The class of objects that are either the 11th or the 12th word of Axiom L is not an expression fundamental for the 12th word(s) of Axiom L, relative to Axiom L [C. 1–4, 6 f., C. 5 n. f. (the class of objects that are either the 11th or the 12th word of Axiom L is an expression that belongs to the class of objects that are either the 11th or the 12th word of Axiom L, the negate of the class of objects that are either the 11th or the 12th word of Axiom L is 12th word of Axiom L, but the class of objects that are either the 11th or the 12th word of Axiom L is not 12th word of Axiom L)].

(4) Axiom L is not an expression fundamental for Axiom(s) L, relative to Axiom L [C. 1–3, 5, 6 f., C. 4. n. f. (the 4th word of Axiom L is a variable that belongs to Axiom L, but it is not Axiom L)].

(5) Axiom L is not an expression fundamental for expressions, relative to Axiom L [C. 1, 2, 4–6 f., C. 3 n. f. (the title of this paper is an expression, but is not an expression that belongs to Axiom L)].

(6) The first word of Axiom L is not an expression fundamental for nonquadrangular quadrangles, relative to Axiom L [C. 1, 3–6 f., C. 2 n. f.].

(7) The class of objects that are either the 1st or the 4th word of Axiom L is not an expression fundamental for the 4th word(s) of Axiom L, relative to Axiom L [C. 2–6 f., C. 1 n. f.].

Terminological explanation VI. I say of object A that it is propositional (i.e. belongs to the category of propositions) relative to B if and only if the following conditions are fulfilled:

(1) A is an expression;

(2) some variable belongs to A;

(3) if A is an expression fundamental for any arbitrary objects a, relative to B, then A is a.[1]

---

[1] *Terminological explanation VI* is based on the ideas of 'hereditary class' and 'ancestral relation', well known in mathematical logic. My explanation is a 'generalization', for the theory of deduction including definitions, of the definition of the '*set S of all sentences*' given by Łukasiewicz and Tarski (op. cit., p. 31). [According to Leśniewski's explanation, loosely paraphrased, A is propositional relative to B if and only if expression A belongs to the closure of the one or more variables in A with respect to negation in A or subordination in A to some expression, relative to a thesis relative to B.—Translator.] It could easily be proved that the extent of the expression 'propositional relative to B' would not be altered by omitting Condition 5 from the six defining conditions of *Terminological explanation V* above. I have nevertheless retained it because I favour always being able to confirm that a propositional expression is propositional, relative to a given thesis of the system, by a combinatorial decision procedure referring only to a corresponding specific finite domain of expressions. I know yet another method, quite different from that explained here, for defining propositions of various deductive theories. I first explained this other method, in which 'hereditary classes' and the 'ancestral relation' play no role, and which essentially originated in 1922, in my 'logistic' lectures in the academic year 1924–5 (see Leśniewski, 'Grundzüge eines neuen Systems der Grundlagen der Mathematik', p. 59), in application to my system of proto-thetic (see op. cit., pp. 9–81). In terms of the 'symbolic' abbreviations used in

Examples.

(1) Axiom *L* is propositional relative to Axiom *L*.

(2) The class of words of Axiom *L* that follow the 31st word of Axiom *L* is not propositional relative to Axiom *L* [C. 1, 2 f., C. 3 n. f. (the class of words of Axiom *L* that follow the 31st word of Axiom *L* is an expression fundamental for objects that are either the 32nd or the 33rd word of Axiom *L*, relative to Axiom *L*, but the class of words of Axiom *L* that follow the 31st word of Axiom *L* is not an object that is either the 32nd or the 33rd word of Axiom *L*)].

the terminological explanations and directives for protothetic (see op. cit., pp. 59–76), such a definition for protothetic could be formulated as follows (essentially as I worded it in 1926—regarding the suffix 'p' of 'propp' see op. cit., pp. 68–69):

$[A, B] :\because A \in \text{propp } (B) . = :: B \in \text{thp} ::$

$\qquad\qquad [\exists C] \therefore C \in \text{vrb}. C \in \text{frp } (B) . A \in \text{cnf}$

$(C) \therefore [D, E] : D \in \text{thp } (B) . E \in \text{ingr } (D) . \supset . C \in N \text{ (cnvar } (C, E)) \therefore \vee.$

$\quad [\exists C] . C \in \text{frp } (B) . A \in \text{genfnct } (C) . \vee . A \in \text{gnrl} ::$

$\qquad\qquad [C] \therefore C \in \text{trm}. C \in \text{ingr } (A) . \supset : C \in \text{Id}$

$(A) . \vee . [\exists D] . D \in \text{qntf}. D \in \text{ingr } (A) . C \in \text{int } (D) . \vee . [\exists D, E] . D \in \text{ingr}$

$\quad (A) . C \in \text{var } (E, D) . \vee . C \in \text{constp } (B, A) ::$

$\qquad\qquad [C, D] : D \in \text{qntf}. D \in \text{ingr } (A) . C \in \text{int}$

$(D) . \supset . [\exists E, F] . E \in \text{ingr } (A) . F \in \text{var } (C, E) ::$

$\qquad\qquad [C, D, E] \therefore E \in \text{ingr } (A) . C \in \text{cnvar } (D,$

$E) . \supset : C \in \text{Id } (D) . \vee . [\exists F, G] . C \in \text{quasihomosemp } (D, B, A, F, G) ::$

$\qquad\qquad [C] \therefore C \in \text{gnrl}. C \in \text{ingr } (A) . \supset : C \in \text{Id}$

$(A) . \vee . [\exists D, E, F, G] . D \in \text{thp } (B) . E \in \text{ingr } (D) . F \in \text{ingr } (A) . G \in \text{homo-}$

semp $(B, B) . G \in \text{Anarg } (C, E, F) ::$

$\qquad\qquad [C, D] \therefore C \in \text{gnrl}. C \in \text{ingr } (A) . D \in$

Essnt $(C) . \supset : D \in \text{vrb} . \vee . [\exists E] . E \in \text{frp } (B) . D \in \text{genfnct } (E) ::$

$\qquad\qquad [C] \therefore C \in \text{fnct}. C \in \text{ingr } (A) . \supset : C \in \text{Id}$

$(A) . \vee . [\exists D] . D \in \text{gnrl}. D \in \text{ingr } (A) . C \in \text{Essnt } (D) . \vee . [\exists D, E] . C \in \text{fnctp}$

$\quad (B, A, D, E)$

It is not difficult to formulate analogous definitions for further theories belonging to my system of foundations of mathematics. In one of the first lectures of my above-mentioned university course 'On foundations of the "theory of deduction"', I remarked that the same scheme of definition could very easily be adapted to the theory of deduction, if brackets were used in this theory. At the same time I mentioned that, for the bracketless symbolism of Łukasiewicz, I did not know whether or how one could find such a definition equivalent to *Terminological explanation VI* but essentially independent of the idea of 'hereditary class' and that of the 'ancestral relation'.

(3) The first word of Axiom L is not propositional relative to Axiom L [C. 1, 3 f., C. 2 n. f.].

(4) The class of variables that belong to Axiom L is not propositional relative to Axiom L [C. 2, 3 f., C. 1 n. f.].

Terminological explanation VII. I say of object A that it is a consequence of B, relative to C, by substitution of (the) a^1 if and only if the following conditions are fulfilled:

(1) A is the complex of (the) a;

(2) there are exactly as many a as words that belong to B;

(3) if any object D is a word that belongs to B, any object E is a, and there are exactly as many a that precede E as words that belong to B and precede D, then D is a variable or is an expression equiform to E;

(4) if any object D is a variable that belongs to B, any object E is a, and there are exactly as many a that precede E as words that belong to B and precede D, then E is propositional relative to C;

(5) if any object D is a word that belongs to B, any object E is an expression that belongs to B and is equiform to D, any object F is a, any object G is a, there are exactly as many a that precede F as words that belong to B and precede D, and there are exactly as many a that precede G as words that belong to B and precede E, then G is an expression equiform to F.

Examples.

(1) Axiom L is a consequence of Axiom L, relative to Axiom L, by substitution of the words that belong to Axiom L.

(2) The class of words between the 18th and the 23rd word of Axiom L (i.e. words of Axiom L which follow the 18th and precede the 23rd word of Axiom L) is not a consequence of the class of words between the 3rd and the 8th word of Axiom L (i.e. words of Axiom L which follow the 3rd and precede the 8th word of Axiom L), relative to Axiom L, by substitution of the words between the 18th and the 23rd word of Axiom L [C. 1–4 f., C. 5 n. f. (the 4th word of Axiom L is a word that belongs

---

1 The uncapitalized variable was here used in the plural.

to the class of words between the 3rd and the 8th word of Axiom L, the 7th word of Axiom L is an expression that belongs to the class of words between the 3rd and the 8th word of Axiom L and is equiform to the 4th word of Axiom L, the 19th word of Axiom L is a word between the 18th and the 23rd word of Axiom L, the 22nd word of Axiom L is a word between the 18th and the 23rd word of Axiom L, there are exactly as many words between the 18th and the 23rd word of Axiom L that precede the 19th word of Axiom L as words that belong to the class of words between the 3rd and the 8th word of Axiom L and precede the 4th word of Axiom L, and there are exactly as many words between the 18th and the 23rd word of Axiom L that precede the 22nd word of Axiom L as words that belong to the class of words between the 3rd and the 8th word of Axiom L and precede the 7th word of Axiom L, but the 22nd word of Axiom L is not an expression equiform to the 19th word of Axiom L)].

(3) The 2nd word of Axiom L is not a consequence of the 4th word of Axiom L, relative to Axiom L, by substitution of the 2nd word(s) of Axiom L [C. 1–3, 5 f., C. 4 n. f. (the 4th word of Axiom L is a variable that belongs to the 4th word of Axiom L, the 2nd word of Axiom L is 2nd word of Axiom L, there are exactly as many 2nd words of Axiom L that precede the 2nd word of Axiom L as words that belong to the 4th word of Axiom L and precede the 4th word of Axiom L, but the 2nd word of Axiom L is not propositional relative to Axiom L)].

(4) Axiom L is not a consequence of the first word of Axiom L, relative to Axiom L, by substitution of Axiom(s) L [C. 1, 2, 4, 5 f., C. 3 n. f. (the first word of Axiom L is a word that belongs to the first word of Axiom L, Axiom L is Axiom L, there are exactly as many Axiom(s) L that precede Axiom L as words that belong to the first word of Axiom L and precede the first word of Axiom L, but the first word of Axiom L neither is a variable nor is an expression equiform to Axiom L)].

(5) The 2nd word of Axiom L is not a consequence of the class of words of Axiom L that precede the 3rd word of Axiom L, relative to Axiom L, by substitution of the 2nd word(s) of Axiom L [C. 1, 3–5 f., C. 2 n. f.].

(6) The first word of Axiom L is not a consequence of Axiom L, relative to Axiom L, by substitution of the words that belong to Axiom L [C. 2–5 f., C. 1 n. f.].

Terminological explanation VIII. I say of object A that it is a consequence of B by substitution, relative to C, if and only if, for some a,[1] A is a consequence of B, relative to C, by substitution of the a.

Examples.

(1) Axiom L is a consequence of Axiom L by substitution, relative to Axiom L.[2]

(2) The first word of Axiom L is not a consequence of Axiom L by substitution, relative to Axiom L.

Terminological explanation IX. I say of object A that it is a consequence of B by detachment, relative to C, with respect to D and to E if and only if the following conditions are fulfilled:

(1) D is implicant of E in B;

(2) C is an expression equiform to D;

(3) A is an expression equiform to E.

Examples.

(1) The 33rd word of Axiom L is a consequence, by detachment, of the class of words of Axiom L that follow the 30th word of Axiom L, relative to the 32nd word of Axiom L, with respect to the 32nd and to the 33rd word of Axiom L.

(2) Axiom L is not a consequence, by detachment, of the class of words of Axiom L that follow the 30th word of Axiom L, relative to the 32nd word of Axiom L, with respect to the 32nd and to the 33rd word of Axiom L [C. 1, 2 f., C. 3 n. f.].

(3) The 33rd word of Axiom L is not a consequence, by detachment, of the class of words of Axiom L that follow the 30th word of Axiom L, relative to Axiom L, with respect to the 32nd and to the 33rd word of Axiom L [C. 1, 3 f., C. 2 n. f.].

(4) Axiom L is not a consequence of Axiom L by detach-

---

[1] The expression 'for some a' here corresponds to the *particular* quantifier '[∃a]' of my 'symbolic' language.

[2] See example 1 of the preceding terminological explanation.

ment, relative to Axiom L, with respect to Axiom L and to Axiom L [C. 2, 3, f., C. 1 n. f.].

Terminological explanation X. I say of object A that it is a consequence of B by detachment, relative to C, if and only if A is a consequence of B by detachment, relative to C, with respect to some expression and to some expression.

Examples.

(1) The 33rd word of Axiom L is a consequence, by detachment, of the class of words of Axiom L that follow the 30th word of Axiom L, relative to the 32nd word of Axiom L.[1]

(2) Axiom L is not a consequence of Axiom L by detachment, relative to Axiom L.

Terminological explanation XI. I say of object A that it is a definition of B, relative to C, by means of D, and with respect to E if and only if the following conditions are fulfilled:

(1) D is propositional relative to C;
(2) the first of the words that belong to B is not a variable;
(3) if any object F is the same object as C or is a thesis of this system which thesis precedes C, and any object G is a word that belongs to F, then the first of the words that belong to B is not an expression equiform to G;
(4) if any object F is a word that belongs to B, any object G is a word that belongs to B, and F is an expression equiform to G, then F is the same object as G;
(5) if any object is a variable that belongs to D, then it is an expression equiform to some word that belongs to B;
(6) if any object is a word that belongs to B and follows the first of the words that belong to B, then it is an expression equiform to some variable that belongs to D;
(7) the implicant of B in the negate of E is an expression equiform to D;
(8) the implicant of D in the implicant of E in the negate of A is an expression equiform to B.

---

[1] See example 1 of the preceding terminological explanation.

Examples.

(1) If any object A is one of the equiform expressions '$NCCF\alpha\alpha NC\alpha F\alpha$', then it is a definition of the class of words of A that follow the 9th word of A, relative to Axiom L, by means of the 6th word of A, and with respect to the class of words of A that follow the 6th word of A.

(2) If any object A is one of the equiform expressions '$NC\alpha F\alpha$', then it is not a definition of the class of words of A that follow the 3rd word of A, relative to Axiom L, by means of the 3rd word of A, and with respect to A [C. 1–7 f., C. 8 n. f.].

(3) If any object A is one of the equiform expressions '$NCCF\alpha\alpha$', then it is not a definition of the class of objects that are either the 4th or the 5th word of A, relative to Axiom L, by means of the 6th word of A, and with respect to the 7th word of A [C. 1–6, 8 f., C. 7 n. f.].

(4) If any object A is one of the equiform expressions '$NCCFN\alpha\alpha NC\alpha FN\alpha$', then it is not a definition of the class of words of A which follow the 10th word of A, relative to Axiom L, by means of the 7th word of A, and with respect to the class of words of A which follow the 7th word of A [C. 1–5, 7, 8 f., C. 6 n. f. (the 12th word of A is a word that belongs to the class of words of A which follow the 10th word of A and follows the first of the words that belong to the class of words of A which follow the 10th word of A, but it is an expression equiform to no variable that belongs to the 7th word of A)].

(5) If any object A is one of the equiform expressions '$NCCF\alpha NC\alpha F$', then it is not a definition of the 9th word of A, relative to Axiom L, by means of the 5th word of A, and with respect to the class of words of A that follow the 5th word of A [C. 1–4, 6–8 f., C. 5 n. f. (the 5th word of A is a variable that belongs to the 5th word of A, but it is an expression equiform to no word that belongs to the 9th word of A)].

(6) If any object A is one of the equiform expressions '$NCCF\alpha\alpha NC\alpha F\alpha\alpha$', then it is not a definition of the class of words of A which follow the 10th word of A, relative to Axiom L, by means of the 7th word of A, and with respect to the class of words of A which follow the 7th word of A [C. 1–3, 5–8 f.,

C. 4 n. f. (the 12th word of A is a word that belongs to the class of words of A which follow the 10th word of A, the 13th word of A is a word that belongs to the class of words of A which follow the 10th word of A, the 12th word of A is an expression equiform to the 13th word of A, but the 12th word of A is not the same object as the 13th word of A)].

(7) If any object A is one of the equiform expressions '$NCCN\alpha\alpha NC\alpha N\alpha$', then it is not a definition of the class of words of A which follow the 9th word of A, relative to Axiom L, by means of the 6th word of A, and with respect to the class of words of A which follow the 6th word of A [C. 1, 2, 4–8 f., C. 3 n. f. (Axiom L is the same object as Axiom L or is a thesis of this system which thesis precedes Axiom L, the 11th word of Axiom L is a word that belongs to Axiom L, but the first of the words that belong to the class of words of A which follow the 9th word of A is an expression equiform to the 11th word of Axiom L)].

(8) If any object A is one of the equiform expressions '$NCCuNCu$', then it is not a definition of the 9th word of A, relative to Axiom L, by means of the 5th word of A, and with respect to the class of words of A that follow the 5th word of A [C. 1, 3–8 f., C. 2 n. f.].

(9) If any object A is one of the equiform expressions '$NCCFFNCFF$', then it is not a definition of the 9th word of A, relative to Axiom L, by means of the 5th word of A, and with respect to the class of words of A that follow the 5th word of A [C. 2–8 f., C. 1 n. f.].

Terminological explanation XII. I say of object A that it is a definition, relative to C, if and only if A is a definition of some expression, relative to C, by means of some expression, and with respect to some expression.[1]

Examples.

(1) If any object A is one of the equiform expressions '$NCCF\alpha\alpha NC\alpha F\alpha$', then it is a definition, relative to Axiom L.[2]

(2) Axiom L is not a definition, relative to Axiom L.

[1] Regarding this definition of definition cf. op. cit., p. 11.
[2] See example 1 of the preceding terminological explanation.

I add further theses to this system of the theory of deduction, of which the first thesis is Axiom L, only if at least one of the three following conditions is fulfilled:

(1) the added thesis is a consequence, by substitution, of one of the preceding theses of this system, relative to the last of the preceding theses of this system;

(2) the added thesis is a consequence, by detachment, of one of the preceding theses of this system, relative to one of the preceding theses of this system;

(3) the added thesis is a definition, relative to the last of the preceding theses of this system.

The directives for constructing this system of the theory of deduction inclusive of definitions are thus complete.

8

SUCCESSIVE SIMPLIFICATIONS OF THE AXIOM-SYSTEM OF LEŚNIEWSKI'S ONTOLOGY†

BOLESŁAW SOBOCIŃSKI

THE aim of this paper is to provide a detailed account of the successive steps by which the axiom of Ontology was simplified.[1] I originally intended to present only the results that I obtained myself. However, in my research I took advantage of various contributions, published and unpublished, made by others, namely by Professor Stanisław Leśniewski and Dr. Alfred Tarski. If no detailed account of their results were given, the reader could have difficulties in following my own proofs. Moreover, this omission could obscure the development of all these investigations, and, contrary to my wishes, belittle the share of some contributors or their contributions. To avoid such misunderstandings and to present a complete picture of the results obtained in this field of research, I decided to give an account of all of them, for which the persons concerned granted me their kind permission.[2]

† This paper appeared originally under the title 'O kolejnych uproszczeniach aksjomatyki "ontologji" prof. St. Leśniewskiego' in *Fragmenty Filozoficzne*, a volume in commemoration of fifteen years' teaching in the University of Warsaw by Professor T. Kotarbiński, Warsaw, 1934, pp. 143–60. Translated by Z. Jordan.

[1] For the sake of conciseness, in this paper the term 'Ontology' is always used instead of the expression 'Ontology of Stanisław Leśniewski'. Concerning the question what the system of 'Ontology' is, see Leśniewski [4], Leśniewski [3], pp. 153–70, and Kotarbiński [1], pp. 227–47. [Bibliography at end of paper—Ed.]

[2] Editorial and space-saving considerations oblige me to use in this paper a slightly modified Peano–Russellian symbolic notation instead of the original symbolism of Ontology (which can be learned from Leśniewski [4], see pp. 114, 115, and 129–32). The notation should be comprehensible to everybody who is acquainted with the ordinary Peano–Russell symbolism and with Leśniewski

The stages of successive simplifications of the single axiom of Ontology were as follows.[1]

1. The original axiom of Ontology, formulated by Leśniewski in 1920, is the theorem:[2]

$$A. \ [Aa] :: \epsilon\{Aa\} . \ \equiv \ \therefore \ [\exists B] . \ \epsilon\{BA\} \ \therefore \ [BC] : \epsilon\{BA\} . \ \epsilon\{CA\}.$$
$$\supset . \ \epsilon\{BC\} \ \therefore \ [B] : \epsilon\{BA\} . \supset . \ \epsilon\{Ba\}.$$

[1] and Leśniewski [4]. For this reason detailed explanations can be dispensed with. To avoid misunderstandings, all the more important theorems occurring in this paper are given in the footnotes in the original symbolism of Ontology. The above-mentioned editorial and space-saving considerations are also responsible for the proofs not being given in full. They have the character of abbreviations of complete proofs, expressed in the system of notation already described and established in accordance with the rules of the system of Ontology formulated in Leśniewski [4] (pp. 115–27). The reader who is not particularly interested in the problems concerning the rules of Ontology may simply consider the incomplete proofs given in this paper 'intuitively' (incidentally, they are modelled on the form of proof to be found in Leśniewski [5] or Leśniewski [6]), as this is the case with the majority of mathematical proofs, including those in the investigations concerned with the axiomatic basis of particular theories. If the reader adopts this 'intuitive' approach to the proofs given below, he need not refer to the rules of Ontology, unless the author of this paper does it explicitly. While verifying the proofs he can take full advantage of the results of the propositional calculus and apply to the theorems of Ontology the 'commonly accepted' methods of operating with quantifiers. As far as the existential quantifier is concerned (see Leśniewski [4], pp. 114–15) it should be remembered that in Leśniewski's theories, that is, in Protothetic (see Leśniewski [1]) and Ontology, there is only the universal quantifier; there is no possibility of introducing the existential quantifier into the system. For this reason the latter will be made use of solely as an abbreviation, to make it possible to write down formulae in a way which is shorter and more easy to grasp. Of course, all the proofs in which formulae involving the existential quantifier occur can easily be given by means of formulae involving only the universal quantifier. Every reader familiar with the laws connecting the universal and the existential quantifiers can easily do it himself.

Apart from Leśniewski [1] and Leśniewski [4], the way in which Leśniewski formulates inference rules in deductive systems can be learned from Leśniewski [2].

[1] A concise description of these successive simplifications can be found in Leśniewski [4], pp. 114–15 and pp. 129–32.

[2] For the way in which Leśniewski obtained this thesis and the intuitive meaning which he attached to it, see Leśniewski [3], pp. 153–70 (and also Kotarbiński [1], pp. 227–9). Theorem A corresponds to the following thesis when expressed in the original symbolism of Ontology (see Leśniewski [4], p. 115):

$$\llcorner Aa \lrcorner^{\ulcorner} \Phi \Big(\epsilon\{Aa\} \varphi \big(\vdash (\llcorner B \lrcorner^{\ulcorner} \vdash (\epsilon\{BA\}) \urcorner) \llcorner BC \lrcorner^{\ulcorner} \Diamond{-}$$
$$\big(\varphi (\epsilon\{BA\} \ \epsilon\{CA\}) \ \epsilon\{BC\} \big)^{\urcorner} \llcorner B \lrcorner^{\ulcorner} \Diamond{-} (\epsilon\{BA\} \ \epsilon\{Ba\}) \urcorner \big) \Big)^{\urcorner}.$$

2. Theorem A consists of the following theses:[1]

$B.$ $[Aa]: \epsilon\{Aa\} . \supset . [\exists B] . \epsilon\{BA\}$[2]

$C.$ $[AaBC]: \epsilon\{Aa\} . \epsilon\{BA\} . \epsilon\{CA\} . \supset . \epsilon\{BC\}$[3]

$D.$ $[AaB]: \epsilon\{Aa\} . \epsilon\{BA\} . \supset . \epsilon\{Ba\}$[4]

$E.$ $[AaB] :: \epsilon\{BA\} \therefore [BC] : \epsilon\{BA\} . \epsilon\{CA\} . \supset . \epsilon\{BC\} \therefore [B]$
$: \epsilon\{BA\} . \supset . \epsilon\{Ba\} \therefore \supset . \epsilon\{Aa\}.$[5]

3. In 1921 Tarski proved that thesis C can be deduced from thesis D.[6]

4. Having learned of Tarski's result, Leśniewski showed in 1921 that theorem A is inferentially equivalent to a shorter one:[7]

In Leśniewski [4], p. 114, the above thesis is expressed in another symbolism:

$$(A,a) :: \epsilon\{Aa\} . \equiv \therefore \sim \big((B) . \sim (\epsilon\{BA\})\big) \therefore (B, C) : \epsilon\{BA\} .$$

$$\epsilon\{CA\} . \supset . \epsilon\{BC\} \therefore (B) : \epsilon\{BA\} . \supset . \epsilon\{Ba\}.$$

In Kotarbiński [1], p. 227 (cf. Leśniewski [3], p. 164) theorem A is expressed in yet another symbolism:

$$\Pi\,A, B\,\{A\ est\ B = [\Pi X(X\ est\ A < X\ est\ B) . \Sigma\,X(X\ est\ A) .$$

$$\Pi\,X, Y\ (X\ est\ A . Y\ est\ A < X\ est\ Y)]\}.$$

---

[1] See Leśniewski [4], p. 129.

[2] Thesis B corresponds to the following theorem when expressed in the original symbolism of Ontology (see Leśniewski [4], p. 129):

$$\llcorner Aa \lrcorner {}^{\ulcorner} \diamondsuit \big(\epsilon\{Aa\} \vdash \big(\llcorner B \lrcorner {}^{\ulcorner} \vdash (\epsilon\{BA\}){}^{\urcorner}\big){}^{\urcorner}\big){}^{\urcorner}.$$

[3] Thesis C corresponds to the following theorem when expressed in the original symbolism of Ontology (see Leśniewski [4], p. 129):

$$\llcorner AaBC \lrcorner {}^{\ulcorner} \diamondsuit \big(\varrho(\epsilon\{Aa\}\ \epsilon\{BA\}\ \epsilon\{CA\})\ \epsilon\{BC\}\big){}^{\urcorner}.$$

[4] Thesis D corresponds to the following theorem when expressed in the original symbolism of Ontology (see Leśniewski [4], p. 129):

$$\llcorner AaB \lrcorner {}^{\ulcorner} \diamondsuit \big(\varrho(\epsilon\{Aa\}\ \epsilon\{BA\})\ \epsilon\{Ba\}\big){}^{\urcorner}.$$

[5] Thesis E corresponds to the following theorem when expressed in the original symbolism of Ontology (see Leśniewski [4], p. 129):

$$\llcorner AaB \lrcorner {}^{\ulcorner} \diamondsuit \big(\varrho(\epsilon\{BA\}\llcorner BC \lrcorner {}^{\ulcorner} \diamondsuit \big(\varrho(\epsilon\{BA\}\ \epsilon\{CA\})\ \epsilon\{BC\}\big){}^{\urcorner}$$

$$\llcorner B \lrcorner {}^{\ulcorner} \diamondsuit (\epsilon\{BA\}\ \epsilon\{Ba\}){}^{\urcorner}\big)\ \epsilon\{Aa\}\big){}^{\urcorner}.$$

[6] See Leśniewski [4], p. 131.   [7] Ibid.

$F.$ $[Aa] :: \epsilon\{Aa\} . \equiv \therefore [\exists B] . \epsilon\{BA\} . \epsilon\{Ba\} \therefore [BC] : \epsilon\{BA\} .$
$\epsilon\{CA\} . \supset . \epsilon\{BC\}.$[1]

5. By making use of Tarski's result, mentioned above in 3, and by applying the rule of extensionality, which in the years 1920–1 had been no part of Ontology, I proved in 1929 that thesis F is inferentially equivalent to a shorter thesis:[2]

$G.$ $[Aa] :: \epsilon\{Aa\} . \equiv \therefore [\exists B] . \epsilon\{BA\} . \epsilon\{Ba\} \therefore [B] : \epsilon\{BA\} .$
$\supset . \epsilon\{AB\}.$[3]

6. In connexion with the above result I observed at the same time that thesis F is inferentially equivalent to the expression consisting of theses B and D.[4]

7. Having acquainted himself with these results, Leśniewski proved in 1929 that the expression consisting of theses B and D is inferentially equivalent to:[5]

$H.$ $[Aa] : \epsilon\{Aa\} . \equiv . [\exists B] . \epsilon\{AB\} . \epsilon\{Ba\}.$[6]

It can be seen from the above that each of theses A, F, G, or H may be adopted as the single axiom of Ontology. It should be noted that the results given in sections 5–7 were obtained in

---

[1] Thesis F corresponds to the following theorem when expressed in the original symbolism of Ontology (see Leśniewski [4], p. 131):

$$\llcorner Aa \lrcorner \ulcorner \phi \Big(\epsilon\{Aa\}\varphi \big(\vdash \big(\llcorner B \lrcorner \ulcorner \vdash \big(\varphi(\epsilon\{BA\} \ \epsilon\{Ba\}) \big) \urcorner \big) \llcorner BC \lrcorner$$
$$\ulcorner \phi \cdot \big(\varphi(\epsilon\{BA\} \ \epsilon\{CA\}) \ \epsilon\{BC\} \big) \urcorner \big) \Big) \urcorner .$$

[2] See Leśniewski [4], p. 131.

[3] Thesis G corresponds to the following theorem when expressed in the original symbolism of Ontology (see Leśniewski [4], p. 131):

$$\llcorner Aa \lrcorner \ulcorner \phi \Big(\epsilon\{Aa\} \ \varphi \big(\vdash \big(\llcorner B \lrcorner \ulcorner \vdash \big(\varphi(\epsilon \{BA\} \ \epsilon\{Ba\}) \big) \urcorner \big) \llcorner B \lrcorner$$
$$\ulcorner \phi \cdot (\epsilon\{BA\} \ \epsilon\{AB\}) \urcorner \big) \Big) \urcorner .$$

[4] Ibid.   [5] Ibid., p. 132.

[6] Thesis H corresponds to the following theorem when expressed in the original symbolism of Ontology (ibid.):

$$\llcorner Aa \lrcorner \ulcorner \phi \Big(\epsilon\{Aa\} \vdash \big(\llcorner B \lrcorner \ulcorner \vdash \big(\varphi(\epsilon\{AB\} \ \epsilon\{Ba\}) \big) \urcorner \big) \urcorner .$$

virtue of the fact that I managed to prove that K can be deduced from thesis D alone:

K. $[AaB] :: \epsilon\{BA\} . \epsilon\{Ba\} \therefore [BC] : \epsilon\{BA\} . \epsilon\{CA\} . \supset . \epsilon\{BC\}$
$\therefore \supset . \epsilon\{Aa\}.$[1]

It also should be emphasized that the proofs establishing that theses G and F follow from theorem A, that is, from the original axiom of Ontology, were given by Leśniewski.

The account of the results just reviewed will proceed in the following way:

§ 1 shows how Leśniewski obtained some theorems, necessary for further deductions, from theorem A, that is, from the original axiom of Ontology;

§ 2 gives the proof that thesis C follows from thesis D;

§ 3 gives the proof that theses F, G, H, and K follow from theorem A;

§ 4 gives the proof that theses B, C, D, and E follow from theorem F;

§ 5 gives the proof that thesis K follows from thesis D;

§ 6 gives the proof that thesis F follows from theses B and D;

§ 7 gives the proof that theses B and D follow from thesis G;

§ 8 gives the proof that theses B and D follow from thesis H.

In this way all the results obtained so far and concerned with the subject of this paper will be taken into account and presented to the reader.

§ 1. Leśniewski's proof that some theorems necessary for further deductions can be derived from theorem A

A. $[Aa] :: \epsilon\{Aa\} . \equiv . \therefore [\exists B] . \epsilon\{BA\} \therefore [BC] : \epsilon\{BA\} . \epsilon\{CA\}$
$. \supset . \epsilon\{BC\} \therefore [B] : \epsilon\{BA\} . \supset . \epsilon\{Ba\}$

B. $[Aa] : \epsilon\{Aa\} . \supset . [\exists B] . \epsilon\{BA\}$ [follows from A]

C. $[AaBC] : \epsilon\{Aa\} . \epsilon\{BA\} . \epsilon\{CA\} . \supset . \epsilon\{BC\}$ [from A]

D. $[AaB] : \epsilon\{Aa\} . \epsilon\{BA\} . \supset . \epsilon\{Ba\}$ [A]

---

[1] In the original symbolism of Ontology thesis K corresponds to the following theorem:

$$\llcorner AaB \lrcorner \ulcorner \Leftrightarrow \left(\varphi \left(\epsilon\{BA\} \ \epsilon\{Ba\} \llcorner BC \lrcorner \ulcorner \Leftrightarrow \left(\varphi(\epsilon\{BA\} \ \epsilon\{CA\}) \ \epsilon\{BC\} \right) \urcorner \right) \epsilon\{Aa\} \right) \urcorner.$$

E. $[AaB] :: \epsilon\{BA\} \therefore [BC] : \epsilon\{BA\} . \epsilon\{CA\} . \supset . \epsilon\{BC\} \therefore [B]$
   $: \epsilon\{BA\} . \supset . \epsilon\{Ba\} \therefore \supset . \epsilon\{Aa\}$ [A]

I. $[AB] :: \epsilon\{BA\} \therefore [BC] : \epsilon\{BA\} . \epsilon\{CA\} . \supset . \epsilon\{BC\} \therefore \supset . \epsilon$
   $\{AA\}$ [E]

II. $[AB] :: \epsilon\{BA\} \therefore [BC] : \epsilon\{BA\} . \epsilon\{CA\} . \supset . \epsilon\{BC\} \therefore \supset . \epsilon$
   $\{AB\}$

Proof.

$[AB] ::$

(α) $\epsilon\{BA\} \therefore$
(β) $[BC] : \epsilon\{BA\} . \epsilon\{CA\} . \supset . \epsilon\{BC\} \therefore \supset .$
(γ) $\epsilon\{AA\}.$ [I, α, β]
     $\epsilon\{AB\}$ [β, γ, α]

III. $[AaB] : \epsilon\{Aa\} . \epsilon\{BA\} . \supset . \epsilon\{AB\}^1$

Proof.

$[AaB] ::$

(α) $\epsilon\{Aa\} .$
(β) $\epsilon\{BA\} . \supset \therefore$
(γ) $[BC] : \epsilon\{BA\} . \epsilon\{CA\} . \supset . \epsilon\{BC\} \therefore$ [C, α]
     $\epsilon\{AB\}$ [II, β, γ]

IV. $[Aa] : \epsilon\{Aa\} . \supset . \epsilon\{AA\}^2$

Proof.

$[Aa] ::$

(α) $\epsilon\{Aa\} . \supset \therefore$
(β) $[\exists B] . \epsilon\{BA\} \therefore [BC] : \epsilon\{BA\} . \epsilon\{CA\} . \supset . \epsilon\{BC\} \therefore$
          [A, α]
     $\epsilon\{AA\}$ [β, I]

[1] In the original symbolism of Ontology thesis III corresponds to the following theorem: $\llcorner AaB \lrcorner \ulcorner \Diamond\!\!-\!(\wp(\epsilon\{Aa\} \, \epsilon\{BA\}) \, \epsilon\{AB\})\urcorner.$

Expressed in a different symbolism, this important and characteristic thesis of Ontology is to be found, together with elucidating remarks, in Kotarbiński [1], p. 246, n. 38:

$$A \ est \ B . B \ est \ C < B \ est \ A.$$

[2] Thesis IV corresponds to the following theorem when expressed in the original symbolism of Ontology (see Leśniewski [4], p. 129):

$$\llcorner Aa \lrcorner \ulcorner \Diamond\!\!-\!(\epsilon\{Aa\} \, \epsilon\{AA\})\urcorner.$$

This thesis, called by Leśniewski the 'ontological law of identity', is discussed in Leśniewski [4], pp. 129 and 130.

Df I. $[AB] : \epsilon\{AB\} . \epsilon\{BA\} . \equiv . = \{AB\}^1$

Df II. $[A\phi] : \epsilon\{AA\} . \phi\{A\} . \equiv . \epsilon\{A \text{ stsf}\langle\phi\rangle\}^2$

V. $[AB\phi] : = \{AB\} . \phi\{A\} . \supset . \phi\{B\}^3$

Proof.

$[AB\phi]$:

(α)	$= \{AB\} .$	
(β)	$\phi\{A\} . \supset .$	
(γ)	$\epsilon\{AB\} . \epsilon\{BA\} .$	$[Df\ I, \alpha]$
(δ)	$\epsilon\{AA\} .$	$[IV, \gamma]$
(ϵ)	$\epsilon\{A \text{ stsf}\langle\phi\rangle\} .$	$[Df\ II, \delta, \beta]$
(ζ)	$\epsilon\{B \text{ stsf}\langle\phi\rangle\} .$	$[D, \epsilon, \gamma]$
	$\phi\{B\}$	$[Df\ II, \zeta]$

§ 2. Tarski's proof that thesis *C* is derivable from thesis *D*

D. $[AaB] . \epsilon\{Aa\} . \epsilon\{BA\} . \supset . \epsilon\{Ba\}$

Df A. $[AaB] : \epsilon\{Aa\} . \epsilon\{BA\} . \equiv . \epsilon\{A * [aB]\}^4$

C. $[AaBC] : \epsilon\{Aa\} . \epsilon\{BA\} . \epsilon\{CA\} . \supset . \epsilon\{BC\}$

---

[1] This thesis, which is here in the form of a definition, corresponds to the following theorem when expressed in the original symbolism of Ontology (see Leśniewski [4], p. 130):

$$\llcorner AB \lrcorner \ulcorner \phi\big(\wp(\epsilon\{AB\}\ \epsilon\{BA\}) = \{AB\}\big)\urcorner.$$

Expressed in a different symbolic notation, the thesis which corresponds to *Df.* I is to be found, together with elucidating remarks, in Kotarbiński [1], p. 237, Df. 11:

$$\Pi\ A, B\ [A\ id\ B = (A\ est\ B . B\ est\ A)].$$

[2] In the original symbolism of Ontology, *Df.* II corresponds to the following theorem:
$$\llcorner A\phi \lrcorner \ulcorner \phi\big(\wp(\epsilon\{AA\}\phi\{A\})\ \epsilon\{A \text{ stsf}\langle\phi\rangle\}\big)\urcorner.$$

[3] In the original symbolism of Ontology, thesis V corresponds to the following theorem:
$$\llcorner AB\phi \lrcorner \ulcorner \diamondsuit\big(\wp(= \{AB\}\phi\{A\})\phi\{B\}\big)\urcorner.$$

The thesis $\llcorner AB\phi \lrcorner \ulcorner \diamondsuit\big(= \{AB\}\diamondsuit(\phi\{A\}\phi\{B\})\big)\urcorner,$

to be found in Leśniewski [4], p. 130, slightly strengthens the preceding theorem. Every reader will be able to prove it easily. Thesis V is the 'law of extensionality for identities'. It should be noted that thesis V is proved without resorting to the rule of extensionality.

[4] In the original symbolism of Ontology *Df.* A corresponds to the following theorem:
$$\llcorner AaB \lrcorner \ulcorner \phi\big(\wp(\epsilon\{Aa\}\ \epsilon\{BA\})\ \epsilon\{A * [aB]\}\big)\urcorner.$$

Proof.

$[AaBC]$:

(α) $\epsilon\{Aa\}$.

(β) $\epsilon\{BA\}$.

(γ) $\epsilon\{CA\}$. ⊃ .

(δ) $\epsilon\{A * [aB]\}$. $[Df\,A, \alpha, \beta]$

(ε) $\epsilon\{C * [aB]\}$. $[D, \delta, \gamma]$

        $\epsilon\{BC\}$ $[Df\,A, \epsilon]$

§ 3. Leśniewski's proof that theses F, G, H, and K are derivable from thesis A

VI. $[Aa] : \epsilon\{Aa\}$. ⊃ . $[\exists B]$. $\epsilon\{BA\}$. $\epsilon\{Ba\}$

Proof.

$[Aa] \therefore$

(α) $\epsilon\{Aa\}$. ⊃ :

(β) $\epsilon\{AA\}$: $[IV, \alpha]$

        $[\exists B]$. $\epsilon\{BA\}$. $\epsilon\{Ba\}$ $[\beta, \alpha]$

K. $[AaB] :: \epsilon\{BA\}$. $\epsilon\{Ba\} \therefore [BC] : \epsilon\{BA\}$. $\epsilon\{CA\}$. ⊃ . ϵ $\{BC\} \therefore \supset . \epsilon\{Aa\}$

Proof.

$[AaB] ::$

(α) $\epsilon\{BA\}$.

(β) $\epsilon\{Ba\} \therefore$

(γ) $[BC] : \epsilon\{BA\}$. $\epsilon\{CA\}$. ⊃ . $\epsilon\{BC\} \therefore \supset .$

(δ) $\epsilon\{AB\}$. $[II, \alpha, \gamma]$

        $\epsilon\{Aa\}$ $[D, \beta, \delta]$

F. $[Aa] :: \epsilon\{Aa\}$. $\equiv \therefore [\exists B]$. $\epsilon\{BA\}$. $\epsilon\{Ba\} \therefore [BC] : \epsilon$ $\{BA\}$. $\epsilon\{CA\}$. ⊃ . $\epsilon\{BC\}$ $[VI, C, K]$

VII. $[AaB] :: \epsilon\{BA\}$. $\epsilon\{Ba\} \therefore [B] : \epsilon\{BA\}$. ⊃ . $\epsilon\{AB\} \therefore \supset .$ $\epsilon\{Aa\}$

Proof.

$[AaB] ::$

(α) $\epsilon\{BA\}$.

(β) $\epsilon\{Ba\} \therefore$

(γ) $[B] : \epsilon\{BA\}$. ⊃ . $\epsilon\{AB\} \therefore \supset ,$

(δ) $\epsilon\{AB\}$.   [γ, α]

           $\epsilon\{Aa\}$ [D, β, δ]

G.   [Aa] :: $\epsilon\{Aa\}$. ≡ ∴ . [∃B] . $\epsilon\{BA\}$. $\epsilon\{Ba\}$ ∴ [B] : $\epsilon\{BA\}$.

      ⊃ . $\epsilon\{AB\}$ [VI, III, VII]

VIII. [Aa] : $\epsilon\{Aa\}$. ⊃ . [∃B] . $\epsilon\{AB\}$. $\epsilon\{Ba\}$

Proof.

   [Aa] ∴

(α) $\epsilon\{Aa\}$. ⊃ :

(β) $\epsilon\{AA\}$:   [IV, α]

           [∃B] . $\epsilon\{AB\}$. $\epsilon\{Ba\}$ [β, α]

H.   [Aa] : $\epsilon\{Aa\}$. ≡ . [∃B] . $\epsilon\{AB\}$. $\epsilon\{Ba\}$ [VIII, D]

§ 4. Leśniewski's proof that theses B, C, D, E, and, consequently, A, are derivable from thesis F

F.   [Aa] :: $\epsilon\{Aa\}$. ≡ ∴ . [∃B] . $\epsilon\{BA\}$. $\epsilon\{Ba\}$ ∴ [BC] : ϵ
      $\{BA\}$. $\epsilon\{CA\}$. ⊃ . $\epsilon\{BC\}$

B.   [Aa] : $\epsilon\{Aa\}$. ⊃ . [∃B] . $\epsilon\{BA\}$ [F]

C.   [AaBC] : $\epsilon\{Aa\}$. $\epsilon\{BA\}$. $\epsilon\{CA\}$. ⊃ . $\epsilon\{BC\}$ [F]

D.   [AaB] : $\epsilon\{Aa\}$. $\epsilon\{BA\}$. ⊃ . $\epsilon\{Ba\}$

Proof.

   [AaB] ::

(α) $\epsilon\{Aa\}$.

(β) $\epsilon\{BA\}$. ⊃ ∴

(γ)   [BC] : $\epsilon\{BA\}$. $\epsilon\{CA\}$. ⊃ . $\epsilon\{BC\}$ ∴   [F, α]

(δ) $\epsilon\{AA\}$.   [F, β, γ]

(ε) $\epsilon\{AB\}$ ∴   [γ, δ, β]

(ζ)   [CD] : $\epsilon\{DB\}$. $\epsilon\{CB\}$. ⊃ . $\epsilon\{DC\}$ ∴   [F, β]

           $\epsilon\{Ba\}$ [F, ε, α, ζ]

E.   [AaB] :: $\epsilon\{BA\}$ ∴ [BC] : $\epsilon\{BA\}$. $\epsilon\{CA\}$. ⊃ . $\epsilon\{BC\}$ ∴
      [B] : $\epsilon\{BA\}$. ⊃ . $\epsilon\{Ba\}$ ∴ ⊃ . $\epsilon\{Aa\}$

Proof.

   [AaB] ::

(α) $\epsilon\{BA\}$ ∴

(β)   [BC] : $\epsilon\{BA\}$. $\epsilon\{CA\}$. ⊃ . $\epsilon\{BC\}$ ∴

(γ)   [B] : $\epsilon\{BA\}$. ⊃ . $\epsilon\{Ba\}$ ∴ ⊃ .

(δ) $\quad \epsilon\{Ba\}$. $\qquad\qquad\qquad\qquad$ [γ, α]

$\quad\quad \epsilon\{Aa\}$ $\qquad\qquad\qquad\qquad\qquad$ $[F, \alpha, \delta, \beta]^1$

§ 5. Sobociński's proof that thesis K is derivable from thesis D

$D.$ $\quad [AaB] : \epsilon\{Aa\} . \epsilon\{BA\} . \supset . \epsilon\{Ba\}$

$Df\Omega.$ $\; [Aa] . \epsilon\{Aa\} . \equiv . \epsilon\{a\}\{A\}^2$

$\Omega.$ $\quad [ab] \therefore [A] : \epsilon\{Aa\} . \equiv . \epsilon\{Ab\} : \equiv : [\phi] : \phi\{a\} . \equiv . \phi\{b\}^3$

$K.$ $\quad [AaB] :: \epsilon\{BA\} . \epsilon\{Ba\} \therefore [BC] : \epsilon\{BA\} . \epsilon\{CA\} . \supset . \epsilon\{BC\}$

$\quad\quad \therefore \supset . \epsilon\{Aa\}$

---

[1] As can be seen from §§ 3 and 4, Leśniewski did not take advantage of Tarski's result, given in § 2, in the proof of the inferential equivalence of theses A and F. The knowledge of this result only prompted the search for a thesis shorter than and equivalent to A. Only the proofs given in §§ 6 and 8 make essential use of Tarski's result.

[2] In the original symbolism of Ontology $Df. \Omega$ corresponds to the following theorem:
$$\llcorner Aa \lrcorner \ulcorner \phi(\epsilon\{Aa\}\ \epsilon\{a\}\{A\})\urcorner .$$

The functor 'ϵ' occurring in the *definiendum* of the above theorem is a quite different functor from 'ϵ' occurring in the *definiens*. For the question as to when in Leśniewski's theories two equiform symbols may be applied in the same theory to denote two different functors, see Leśniewski [1] (p. 76, point 1 of the rule, and p. 70, terminological explanation T.E. XLIV), as well as Leśniewski [4] (p. 127, points 1 and 2 of the rule, p. 118, terminological explanation T.E. XLIV°, and p. 123, terminological explanation T.E. LVI°).

Leśniewski used definition Df. Ω before I established the proof given above. This definition is a typical instance of definitions in which 'many-link functions' occur. As far as many-link functions are concerned, see Leśniewski [1], p. 66, footnote 1. Leśniewski drew my attention to the fact that Moses Schönfinkel discussed similar functions in Schönfinkel [1], pp. 307–15.

[3] In the original symbolism of Ontology, Ω corresponds to the following theorem:
$$\llcorner ab \lrcorner \ulcorner \phi\big(\llcorner A \lrcorner \ulcorner \phi(\epsilon\{Aa\}\ \epsilon\{Ab\})\urcorner \llcorner \phi \lrcorner \ulcorner \phi(\phi\{a\}\phi\{b\})\urcorner \big)\urcorner .$$

Thesis Ω, which was known to Leśniewski earlier, is obtained by means of the rule of extensionality. For this rule see Leśniewski [4], p. 127, points 6 and 7 of the directive. At the time when the results given in §§ 1–4 were established, there was no rule of extensionality in the system of Ontology. The only 'law of extensionality' which was proved at that time was the 'law of extensionality for identities'. See Leśniewski [4], pp. 130–1, and also § 1 and footnote 3, p. 194, above. Only later did Leśniewski add the rule of extensionality to the system of Ontology and he did it for different reasons, unrelated to the possibility of simplifying the axiom-system of Ontology by means of this rule. It appears that I was the first to draw attention to the fact that the introduction of the rule of extensionality creates new logical relations holding between various elementary propositions of Ontology. The discovery of these relations made it possible to simplify the axiom-system.

Proof.

[AaB] ::

(α)	$\epsilon\{BA\}$.	
(β)	$\epsilon\{Ba\}$ $\therefore$	
(γ)	$[BC]: \epsilon\{BA\} . \epsilon\{CA\} . \supset . \epsilon\{BC\} \therefore \supset \therefore$	
(δ)	$[C]: \epsilon\{CB\} . \supset . \epsilon\{CA\} \therefore$	$[D, \alpha]$
(ϵ)	$[C]: \epsilon\{CA\} . \supset . \epsilon\{CB\} \therefore$	$[\gamma, \alpha]$
(ζ)	$[C]: \epsilon\{CB\} . \equiv . \epsilon\{CA\} \therefore$	$[\delta, \epsilon]$
(η)	$\epsilon\{a\}\{B\}$.	$[Df\,\Omega, \beta]$
(θ)	$\epsilon\{a\}\{A\}$.	$[\Omega, \zeta, \eta]$
	$\epsilon\{Aa\}$	$[Df\,\Omega, \theta]$

§ 6. Sobociński's proof that thesis *F* is derivable from theses *B* and *D*

B. $[Aa]: \epsilon\{Aa\} . \supset . [\exists B] . \epsilon\{BA\}$

D. $[AaB]: \epsilon\{Aa\} . \epsilon\{BA\} . \supset . \epsilon\{Ba\}$

VI. $[Aa]: \epsilon\{Aa\} . \supset . [\exists B] . \epsilon\{BA\} . \epsilon\{Ba\}$

Proof.

[Aa] $\therefore$

(α)	$\epsilon\{Aa\} . \supset :$ $[\exists B]$.	
(β)	$\epsilon\{BA\}$.	$[B, \alpha]$
(γ)	$\epsilon\{Ba\}:$	$[D, \alpha, \beta]$
	$[\exists B]: \epsilon\{BA\} . \epsilon\{Ba\}$	$[\beta, \gamma]$

C. $[AaBC]: \epsilon\{Aa\} . \epsilon\{BA\} . \epsilon\{CA\} . \supset . \epsilon\{BC\}$

  [Follows from *D*, as in § 2]

K. $[AaB] :: \epsilon\{BA\} . \epsilon\{Ba\} \therefore [BC]: \epsilon\{BA\} . \epsilon\{CA\} . \supset . \epsilon\{BC\}$
  $\therefore \supset . \epsilon\{Aa\}$ [Follows from *D*, as in § 5]

F. $[Aa] :: \epsilon\{Aa\} . \equiv \therefore [\exists B] . \epsilon\{BA\} . \epsilon\{Ba\} \therefore [BC]: \epsilon\{BA\} .$
  $\epsilon\{CA\} . \supset . \epsilon\{BC\}$ [VI, *C*, *K*]

§ 7. Sobociński's proof that theses *B* and *D* are derivable from thesis *G*

G. $[Aa] :: \epsilon\{Aa\} . \equiv \therefore [\exists B] . \epsilon\{BA\} . \epsilon\{Ba\} \therefore [B]: \epsilon\{BA\} .$
  $\supset . \epsilon\{AB\}$

B. $[Aa]: \epsilon\{Aa\} . \supset . [\exists B] . \epsilon\{BA\}$ $[G]$

D. $[AaB]: \epsilon\{Aa\} . \epsilon\{BA\} . \supset . \epsilon\{Ba\}$

Proof.

$[AaB]$:

(α)	$\epsilon\{Aa\}$.	
(β)	$\epsilon\{BA\}$. $\supset$ $\therefore$	
(γ)	$[C] : \epsilon\{CB\}$. $\supset$. $\epsilon\{BC\}$ $\therefore$	$[G, \beta]$
(δ)	$\epsilon\{AB\}$.	$[G, \alpha, \beta]$
	$\epsilon\{Ba\}$	$[G, \delta, \alpha, \gamma]$

§ 8. Leśniewski's proof that theses B and D are derivable from thesis H

H. $\quad [Aa] : \epsilon\{Aa\}$. $\equiv$. $[\exists B]$. $\epsilon\{AB\}$. $\epsilon\{Ba\}$

D. $\quad [AaB] : \epsilon\{Aa\}$. $\epsilon\{BA\}$. $\supset$. $\epsilon\{Ba\}$ $\qquad\qquad [H]$

C. $\quad [AaBC] : \epsilon\{Aa\}$. $\epsilon\{BA\}$. $\epsilon\{CA\}$. $\supset$. $\epsilon\{BC\}$

$\qquad\qquad\qquad\qquad$ [Follows from *D*, as in § 2]

B. $\quad [Aa] : \epsilon\{Aa\}$. $\supset$. $[\exists B]$. $\epsilon\{BA\}$

Proof.

$[Aa]$ $\therefore$

(α)	$\epsilon\{Aa\}$. $\supset$:	
	$[\exists B]$.	
(β)	$\epsilon\{AB\}$. $\epsilon\{Ba\}$:	$[H, \alpha]$
(γ)	$\epsilon\{AA\}$:	$[C, \beta]$
	$[\exists B]$. $\epsilon\{BA\}$	$[\gamma]$

BIBLIOGRAPHICAL ABBREVIATIONS

Kotarbiński [1]   Tadeusz Kotarbiński, *Elementy teorji poznania, logiki formalnej i metodologji nauk* (*Elements of Epistemology, Formal Logic and Methodology of Science*), Lwów, 1929.

Leśniewski [1]   Stanisław Leśniewski, 'Grundzüge eines neuen Systems der Grundlagen der Mathematik', Introduction and §§ 1–11, *Fundamenta Mathematicae* 14 (1929), pp. 1–81.

Leśniewski [2]   Stanisław Leśniewski, 'Über Definitionen in der sogenannten Theorie der Deduktion', *Comptes rendus des séances de la Société des Sciences et des Lettres de Varsovie*, Cl. iii, 24 (1931), pp. 289–309. [Translated as paper 7 of this volume. Ed.]

Leśniewski [3]   Stanisław Leśniewski, 'O podstawach matematyki',

(On the Foundations of Mathematics), chapters x and xi, *Przegląd Filozoficzny* 34 (1931), pp. 142–70.

Leśniewski [4]   Stanisław Leśniewski, 'Über die Grundlagen der Ontologie', *Comptes rendus des séances de la Société des Sciences et des Lettres de Varsovie*, Cl. iii, 23 (1930), pp. 111–32.

Leśniewski [5]   Stanisław Leśniewski, 'Über Funktionen, deren Felder Abelsche Gruppen in Bezug auf diese Funktionen sind', *Fundamenta Mathematicae* 14 (1929), pp. 242–51.

Leśniewski [6]   Stanisław Leśniewski, 'Über Funktionen, deren Felder Gruppen mit Rücksicht auf diese Funktionen sind', *Fundamenta Mathematicae* 13 (1929), pp. 319–33.

Schönfinkel [1]   Moses Schönfinkel, 'Über die Bausteine der mathematischen Logik', *Mathematische Annalen*, vol. 92, no. 3/4 (1924), pp. 305–16.

9

AN INVESTIGATION OF PROTOTHETIC†

BOLESŁAW SOBOCIŃSKI

WHILE investigating various problems of protothetic, I observed that a number of theorems which, as far as I could ascertain, remained unknown at that time (December 1935), were theses of protothetic.[1] As this finding is closely associated with the theorem of Dr. Alfred Tarski concerning the definability, in protothetic, of conjunction in terms of equivalence,[2] I have decided to publish the theses discovered by myself.

Editorial and space-saving considerations compel me to use a slightly modified form of the symbolism of Peano and Russell instead of the original symbolism of protothetic. This notation should be easy to follow for anyone familiar with the usual symbolism of Peano and Russell and with the contributions Leśniewski [1], Leśniewski [3], Sobociński [1], Tarski [1], Tarski [2], and Tarski [3]. However, to avoid any misunderstanding, all the more important theorems which occur in this contribution will be given in footnotes in the original symbolism of protothetic.[3] For the reasons already mentioned I have also decided to present the discovered theorems without proofs, for I assume that every reader familiar with elementary logic can easily prove them by the truth table method.[4]

† This paper was intended to appear, under the title 'Z badań nad protototetyką', in vol. 1 of the periodical *Collectanea Logica* (Warsaw, 1939), pp. 171–6. (For the fate of this periodical, see the bibligrapical footnote to paper 5 of this volume.) An English translation of the paper, made by Dr. Sobociński, appeared as no. 5 of the *Cahiers de l'Institut d'Études polonaises en Belgique* (Brussels, 1949). This version is translated anew from the Polish by Z. Jordan.

---

[1] For information about protothetic see Leśniewski [1], Leśniewski [2], and Leśniewski [3]. The expression 'thesis' as used in this paper means 'theorem valid in protothetic'.

[2] See Tarski [1], Tarski [2], and Tarski [3], and also Leśniewski [1], pp. 11 ff.

[3] Cf. Sobociński [1], p. 155, footnote 2.

[4] See Tarski [1], pp. 75 ff., and Tarski [2].

As we know, Alfred Tarski proved the following two theses of protothetic:

$A.$ $[pq] :: p \cdot q \cdot \equiv \therefore [f] \therefore p \equiv : [r] \cdot p \equiv f(r) \cdot \equiv \cdot [r] \cdot q$
$\equiv f(r)$

$B.$ $[pq] \therefore p \cdot q \cdot \equiv : [f] : p \equiv \cdot f(p) \equiv f(q),$[1]

each of which can be accepted in protothetic as the definition of the connective '.' by means of the connective '$\equiv$'. In their efforts to establish and simplify the axiom system of protothetic, Professor Stanisław Leśniewski and Dr. Mordchaj Wajsberg discovered numerous modifications of theses A and B; but these modifications, which, with one exception, remained unpublished, keep the characteristic features of Tarski's theorem.[2] On the other hand, although the theorems I have discovered are based on Tarski's idea that in protothetic conjunction can be defined by means of equivalence, the universal quantifier, and variable propositional functors, they differ significantly from theorems A and B as well as from all their known modifications. The theorems discovered by myself are as follows:

$C1.$ $[pq] : p \cdot q \cdot \equiv \cdot [f] \cdot f(pq) \equiv f(q1)$[3]

$C2.$ $[pq] : p \cdot q \cdot \equiv \cdot [f] \cdot f(pq) \equiv f(1p)$

$C3.$ $[pq] : p \cdot q \cdot \equiv \cdot [f] \cdot f(qp) \equiv f(p1)$

$C4.$ $[pq] : p \cdot q \cdot \equiv \cdot [f] \cdot f(qp) \equiv f(1q)$

$C5.$ $[pq] : p \cdot q \cdot \equiv \cdot [f] \cdot f(pp) \equiv f(q1)$

$C6.$ $[pq] : p \cdot q \cdot \equiv \cdot [f] \cdot f(pp) \equiv f(1q)$

$C7.$ $[pq] : p \cdot q \cdot \equiv \cdot [f] \cdot f(qq) \equiv f(p1)$

[1] See Tarski [1], pp. 73 and 75, and Tarski [3], pp. 199–200.

[2] The thesis which was published is to be found in Leśniewski [1], p. 32.

[3] For the sake of clarity I use the symbol '1' for '$[u] . u \equiv u$', and the symbol '0' for '$[u] . u$' (see, for instance, Leśniewski [1], pp. 12–13). It should be observed that if in the system of protothetic which Leśniewski called $\mathfrak{S}_5$ one of the theses given here were to be accepted as a definition, then, in accordance with the directive accepted in system $\mathfrak{S}_5$, the *definiens* should be written on the left-hand side and the *definiendum* on the right-hand side of the equivalence. See Leśniewski [1], pp. 54 and 76, point 1 of the directive, and p. 70, terminological explanation T.E. XLIV.

In the original symbolism of protothetic the thesis corresponding to $C1$ is written as follows:

$$\llcorner pq \lrcorner ^{\ulcorner}\varphi\big(\varphi(pq)_\llcorner f \lrcorner ^{\ulcorner}\varphi\big(f(pq)f(q_\llcorner u \lrcorner ^{\ulcorner}\varphi(uu)^\urcorner)\big)^\urcorner\big)^\urcorner.$$

$C8.\ \ [pq]:p\cdot q\cdot \equiv \cdot [f]\cdot f(qq)\equiv f(1p)$

$C9.\ \ [pq]:p\cdot q\cdot \equiv \cdot [f]\cdot f(pq)\equiv f(11)^1$

$C10.\ \ [pq]:p\cdot q\cdot \equiv \cdot [f]\cdot f(qp)\equiv f(11)$

$C11.\ \ [pq]:p\cdot q\cdot \equiv \cdot [f]\cdot f(1p)\equiv f(q1)^2$

$C12.\ \ [pq]:p\cdot q\cdot \equiv \cdot [f]\cdot f(1q)\equiv f(p1)$

It should be noted that theorems $C1$–$C12$ can be written in a modified form by substituting everywhere for the sign '1' one of the expressions: '$p\equiv p$', '$q\equiv q$', '$p\equiv q$', or '$q\equiv p$'. It can be seen that the fundamental difference between theorems A and B on the one hand and theorems $C1$–$C12$ on the other is twofold. First, in theorems $C1$–$C12$ the variable functor 'f' is a propositional functor of two arguments instead of one. Second, while in each of theorems $C1$–$C12$ the expression in the scope of the inside quantifier is an equivalence, neither of whose arguments is an equivalence, in Tarski's theorems one of the arguments of the equivalence occurring after the inside quantifier is also an equivalence.

Having discovered theorems $C1$–$C12$, I noted that in protothetic we are able to construct in a similar fashion theorems which within this system can be accepted as the definitions of three other propositional functors of two arguments, namely the functors defined by means of the expressions '$\sim (p\vee q)$', '$\sim (p\supset q)$', and '$\sim (p\vee \sim q)$'.

If the symbolic expression '$p\,\dot{6}\,q$', equivalent to '$\sim (p\vee q)$', is introduced, so that

$$[pq]:p\,\dot{6}\,q\cdot \equiv \cdot \sim (p\vee q)^3$$

---

[1] In the original notation of protothetic the thesis corresponding to $C9$ is written as follows:

$$\llcorner pq\lrcorner \ulcorner \phi\big(\varphi(pq)\llcorner f\lrcorner \ulcorner \phi\big(f(pq)f(\llcorner u\lrcorner \ulcorner \phi(uu)\urcorner \llcorner u\lrcorner \ulcorner \phi(uu)\urcorner)\big)\urcorner\big)\urcorner.$$

[2] In the original notation of protothetic the thesis corresponding to $C11$ is written as follows:

$$\llcorner pq\lrcorner \ulcorner \phi\big(\varphi(pq)\llcorner f\lrcorner \ulcorner \phi\big(f(\llcorner u\lrcorner \ulcorner \phi(uu)\urcorner p)f(q\llcorner u\lrcorner \ulcorner \phi(uu)\urcorner)\big)\big)\urcorner\big)\urcorner.$$

[3] The sign '$\dot{6}$' is taken from the original notation of protothetic. The functor '$\dot{6}$' is one of the two propositional functors of two arguments, each of which is sufficient to define in the theory of deduction all the remaining functors. This proposition has been proved by Henry M. Sheffer; see Sheffer [1]. Moreover, Professor Eustachy Żyliński has proved that out of sixteen possible binary

becomes a valid theorem, the following theses of protothetic can be established:

$D1.$ $[pq]: p \, \mathbf{\phi} \, q \, . \equiv . \, [f] \, . f(pq) \equiv f(q0)^{1}$

$D2.$ $[pq]: p \, \mathbf{\phi} \, q \, . \equiv . \, [f] \, . f(0q) \equiv f(p0)$

$D3.$ $[pq]: p \, \mathbf{\phi} \, q \, . \equiv . \, [f] \, . f(0p) \equiv f(q0)$

$D4.$ $[pq] \, . \, p \, \mathbf{\phi} \, q \, . \equiv . \, [f] \, . f(qp) \equiv f(00)$

$D5.$ $[pq]: p \, \mathbf{\phi} \, q \, . \equiv . \, [f] \, . f(pq) \equiv f(00)$

$D6.$ $[pq]: p \, \mathbf{\phi} \, q \, . \equiv . \, [f] \, . f(qq) \equiv f(0p)$

$D7.$ $[pq]: p \, \mathbf{\phi} \, q \, . \equiv . \, [f] \, . f(qq) \equiv f(p0)$

$D8.$ $[pq]: p \, \mathbf{\phi} \, q \, . \equiv . \, [f] \, . f(pp) \equiv f(0q)$

$D9.$ $[pq]: p \, \mathbf{\phi} \, q \, . \equiv . \, [f] \, . f(pp) \equiv f(q0)$

$D10.$ $[pq]: p \, \mathbf{\phi} \, q \, . \equiv . \, [f] \, . f(qp) \equiv f(0q)$

$D11.$ $[pq]: p \, \mathbf{\phi} \, q \, . \equiv . \, [f] \, . f(qp) \equiv f(p0)$

$D12.$ $[pq]: p \, \mathbf{\phi} \, q \, . \equiv . \, [f] \, . f(pq) \equiv f(0p)$

If we regard the new symbolic expression '$p \, \text{-o} \, q$' as equivalent to '$\sim (p \supset q)$' and accept the proposition

$$[pq]: p \, \text{-o} \, q \, . \equiv . \sim (p \supset q)^{2}$$

connectives only these two propositional functors of two arguments have the characteristic in question. See Żyliński [1] and Żyliński [2].

The truth-functional behaviour of this functor can be described by means of the matrix method in the following way:

$\mathbf{\phi}$	0	1
0	1	0
1	0	0

For the matrix method see Łukasiewicz–Tarski [1], §1, Definition 3, p. 4.

[1] In the original symbolism of protothetic the thesis corresponding to $D1$ is written as follows:

$$\llcorner pq \lrcorner \ulcorner \phi \big(\phi(pq) {\llcorner} f {\lrcorner} \ulcorner \phi(f(pq)f(q {\llcorner} u {\lrcorner} \ulcorner u \urcorner)) \urcorner \big) \urcorner.$$

[2] The sign '-o' is taken from the original notation of protothetic. With the aid of the matrix method the truth table of '-o' can be presented as follows:

-o	0	1
0	0	0
1	1	0

the following theorems become theses of protothetic:

E1. $[pq] : p \multimap q . \equiv . [f] . f(pq) \equiv f(10)$[1]

E2. $[pq] : p \multimap q . \equiv . [f] . f(qp) \equiv f(01)$

E3. $[pq] : p \multimap q . \equiv . [f] . f(1q) \equiv f(p0)$

E4. $[pq] : p \multimap q . \equiv . [f] . f(0p) \equiv f(q1)$

If the new symbolic expression '$p \circ\!\!- q$', equivalent to '$\sim (p \vee \sim q)$', and the proposition

$$[pq] : p \circ\!\!- q . \equiv . \sim (p \vee \sim q)$$[2]

are accepted, the following theses should be included in protothetic:

F1. $[pq] : p \circ\!\!- q . \equiv . [f] . f(pq) \equiv f(01)$[3]

F2. $[pq] : p \circ\!\!- q . \equiv . [f] . f(qp) \equiv f(10)$

F3. $[pq] : p \circ\!\!- q . \equiv . [f] . f(1p) \equiv f(q0)$

F4. $[pq] : p \circ\!\!- q . \equiv . [f] . f(0q) \equiv f(p1)$

It should be observed that a modified form of theorems E1–F4 can be obtained by the substitution of either '$p \equiv p$' or '$q \equiv q$' for '1'.

BIBLIOGRAPHICAL ABBREVIATIONS

Leśniewski [1]   Stanisław Leśniewski, 'Grundzüge eines neuen Systems der Mathematik', Introduction and §§ 1–11, *Fundamenta Mathematicae* 14 (1929), pp. 1–81.

[1] The thesis corresponding to E1 is written in the original symbolism of protothetic as follows:

$$\llcorner pq \lrcorner \ulcorner \phi \big(\multimap(pq) \llcorner f \lrcorner \ulcorner \phi \big(f(pq) f(\llcorner u \lrcorner \ulcorner \phi(uu) \urcorner \llcorner u \lrcorner \ulcorner u \urcorner) \big) \urcorner \big) \urcorner .$$

[2] The sign '$\circ\!\!-$' is taken from the original notation of protothetic. The truth table of '$\circ\!\!-$', presented by means of the matrix method, is as follows:

$\circ\!\!-$	0	1
0	0	1
1	0	0

[3] In the original symbolism of protothetic the thesis corresponding to F1 is written as follows:

$$\llcorner pq \lrcorner \ulcorner \phi \big(\circ\!\!-(pq) \llcorner f \lrcorner \ulcorner \phi \big(f(pq) f(\llcorner u \lrcorner \ulcorner u \urcorner \llcorner u \lrcorner \ulcorner \phi(uu) \urcorner) \big) \urcorner \big) \urcorner .$$

Leśniewski [2]   Stanisław Leśniewski, 'O podstawach matematyki' (On the foundations of Mathematics),
chapter 1, *Przegląd Filozoficzny* 30 (1927), pp.
169–81.

Leśniewski [3]   Stanisław Leśniewski, 'Über die Grundlagen
der Ontologie', *Comptes rendus des séances de la
Société des Sciences et des Lettres de Varsovie*,
Cl. iii, 23 (1930), pp. 111–32.

Łukasiewicz–Tarski [1]   J. Łukasiewicz and A. Tarski, 'Untersuchungen
über den Aussagenkalkül', *Comptes rendus des
séances de la Société des Sciences et des Lettres
de Varsovie*, Cl. iii, 23 (1930), pp. 30–50.

Sheffer [1]   Henry Maurice Sheffer, 'A set of five independent postulates for Boolean Algebras, with
applications to logical constants', *Transactions
of the American Mathematical Society* 14 (1913),
pp. 481–8.

Sobociński [1]   Bolesław Sobociński, 'O kolejnych uproszczeniach aksjomatyki "ontologji"' prof. St. Leśniewskiego' (Successive simplifications of the
axiom-system of Leśniewski's Ontology), *Fragmenty Filozoficzne* (Volume in commemoration
of 15 years' teaching in the University of
Warsaw by Professor T. Kotarbiński), Warsaw,
1934. [Translated as paper 8 of this volume
—Ed.]

Tarski [1]   Alfred Tajtelbaum-Tarski, 'O wyrazie pierwotnym logistyki' (On the primitive term of logistic), Doctoral thesis, *Przegląd Filozoficzny* 26
(1923), pp. 68–89.

Tarski [2]   Alfred Tajtelbaum-Tarski, 'Sur les truth-
functions au sens de MM. Russell et Whitehead', *Fundamenta Mathematicae* 5 (1924), pp.
59–74.

Tarski [3]   Alfred Tajtelbaum-Tarski, 'Sur le terme primitif
de la Logistique', *Fundamenta Mathematicae* 4
(1923), pp. 196–200.

Żyliński [1]   Eustachy Żyliński, 'O przedstawialności funkcyj
prawdziwościowych jednych przez drugie'
(On the mutual representability of truth-
functions by one another), *Przegląd Filozoficzny*
30 (1927), p. 290 ff.

Żyliński [2]   Eustachy Żyliński, 'Some remarks concerning
the theory of deduction', *Fundamenta Mathematicae* 7 (1925), p. 203 ff.

10

SYNTACTIC CONNEXION†

KAZIMIERZ AJDUKIEWICZ

I

1. THE discovery of the antinomies, and the method of their resolution, have made problems of linguistic syntax the most important problems of logic (provided this word is understood in a sense that also includes metatheoretical considerations). Among these problems that of syntactic connexion is of the greatest importance for logic. It is concerned with the specification of the conditions under which a word pattern, constituted of meaningful words, forms an expression which itself has a unified meaning (constituted, to be sure, by the meaning of the single words belonging to it). A word pattern of this kind is *syntactically connected*.

The word pattern *'John loves Ann'*, for instance, is composed of words of the English language in syntactic connexion, and is a significant expression in English. However, the expression *'perhaps horse if will however shine'* is constructed of meaningful English words, but lacks syntactic connexion, and does not belong to the meaningful expressions of the English language.

There are several solutions to this problem of syntactic connexion. Russell's theory of types, for example, offers a solution. But a particularly elegant and simple way of grasping the concept of syntactic connexion is offered by the theory of semantic categories developed by Professor Stanislaw Leśniewski. We shall base our work here on the relevant results of Leśniewski,[1]

† This paper originally appeared under the title 'Die syntaktische Konnexität' in *Studia Philosophica* 1 (1935), pp. 1–27. Translated by H. Weber.

[1] Stanislaw Leśniewski, *Grundzüge eines neuen Systems der Grundlagen der Mathematik.* (Reprinted from *Fundamenta Mathematicae* 14 (Warsaw, 1929),

adding on our part a symbolism, in principle applicable to almost all languages, which makes it possible to formally define and examine the syntactic connexion of a word pattern.

2. Both the concept and the term 'semantic category' (*Bedeutungskategorie*) were first introduced by Husserl. In his *Logische Untersuchungen*[1] Husserl mentions that single words and complex expressions of a language can be divided into classes such that two words or expressions belonging to the same class can be substituted for one another, in a context possessing unified meaning, without that context becoming an incoherent word pattern and losing unified sense. On the other hand, two words or expressions belonging to different classes do not possess this property. Take the sentence '*the sun shines*' as an example of a context having a coherent meaning. If we substitute for the word '*shines*' the word '*burns*' or the word '*whistles*' or the word '*dances*', we obtain from the sentence '*the sun shines*' other true or false sentences which have coherent meaning. If we replace '*shines*' by '*if*' or '*green*' or '*perhaps*' we obtain incoherent word patterns. Husserl terms these classes of words or expressions semantic categories.

We want to define this concept a little more precisely. The word or expression A, taken in sense x, and the word or expression B, taken in sense y, belong to the same semantic category if and only if there is a sentence (or sentential function) S_A, in which A occurs with meaning x, and which has the property that if S_A is transformed into S_B upon replacing A by B (with meaning y), then S_B is also a sentence (or sentential function). (It is understood that in this process the other words and the structure of S_A remain the same.)

The system of semantic categories is closely related to the simplified hierarchy of logical types—although ramified to a

pp. 13 ff., 67 ff.) We borrow from Leśniewski only the basic idea of semantic categories and their type. Leśniewski cannot be held responsible for the wording of the definitions and explanations we offer, nor for the details of the content we assign to this term, since his definitions are not general, but apply only to his special symbolism, in a quite distinct, highly precise, and purely structural sense.

[1] Edmund Husserl, *Logische Untersuchungen*, vol. ii, part 1 (2nd rev. ed. Halle/S., 1913), pp. 294, 295, 305–12, 316–21, 326–42.

much higher degree than the latter—and basically constitutes its grammatical semantic counterpart.[1]

Among all semantic categories two kinds can be distinguished, *basic categories* and *functor categories*. (The term 'functor' derives from Kotarbiński; that of 'basic category' is my own.) Unfortunately we are unable to define these terms with any tolerable precision. An understanding, however, of what is meant by them can be reached easily. A 'functor' is the same as a 'functional sign', or an 'unsaturated symbol' with 'brackets following it'. Functor categories are the semantic categories to which functors belong. I shall call a semantic category a basic category when it is not a functor category.

From the above definition of a semantic category it follows at once that any two sentences belong to the same semantic category. Sentences are, of course, not functors, consequently the semantic category of sentences is among the basic categories. Besides the sentence category there can also be other basic categories. Leśniewski has only one basic category besides the sentence category: the name category, to which belong singular as well as general names. If one compared the simplified theory of types with the theory of semantic categories, the sentence type and the proper-name type would have to be counted among the basic categories. All other types would belong to the functor categories. It seems that not all names form a single semantic category in ordinary language. In our view, at least two semantic categories can be distinguished among names in ordinary speech: first, the semantic category to which belong the singular names of individuals, and the general names of individuals in so far as these are taken *in suppositione personali*; secondly, the semantic category of general names in so far as they occur *in suppositione simplici* (i.e. as the names of universals).

If the concept of syntactic connexion were to be defined in strict generality, nothing could be decided about the number and kind of basic semantic and functor categories, since these

---

[1] R. Carnap, *Abriss der Logistik* (Vienna, 1929), p. 30; A. Tarski, *Pojęcie prawdy w językach nauk dedukcyjnych* (The concept of truth in formalized deductive sciences) (Warsaw, 1933), p. 67.

may vary in different languages. For the sake of simplicity, however, we shall restrict ourselves (like Leśniewski) to languages having only two basic semantic categories, that of sentences and that of names. Apart from these two basic semantic categories, we shall assume, again in accordance with Leśniewski, an unbounded and ramified ascending hierarchy of functor categories characterized in two ways: first by the number and semantic category of their arguments taken in order; second by the semantic category of the whole composite expression formed by them together with their arguments. Thus, for example, the functors which form a sentence with *one* name as argument would represent a distinct semantic category; the functors which form a sentence with *two* names as arguments would form a different semantic category, and so forth. Functors that form a *name* with one name as their argument would again constitute a different semantic category. Also, sentence-forming functors which allow a sentence as their only argument (e.g. the '$\sim$' sign in mathematical logic) would be assigned to a separate semantic category, and so forth.

3. We assume that the semantic category of a single word is defined by its meaning. We shall give single words an index according to their semantic category, viz. the simple index 's' to single words belonging to the sentence category; the simple index 'n' to words belonging to the category of names. To single words belonging not to a basic category but to functor categories we shall assign a fractional index, formed of a numerator and a denominator. In the numerator will be the index of the semantic category to which the whole expression composed of the functional sign plus its arguments belongs, while in the denominator appear, one after the other, the indices of the semantic categories of the arguments, which together with the functor combine into a significant whole. For example, a word forming a sentence with two names as its arguments gets the fractional index s/nn. In this way each semantic category receives a characteristic index. The hierarchy of semantic categories would be reflected in a series of indices (far from being complete) of the following form:

$$s,\ n,\ \frac{s}{n},\ \frac{s}{nn},\ \frac{s}{nnn},\ldots,\ \frac{s}{s},\ \frac{s}{ss},\ \frac{s}{sss},\ldots,\ \frac{s}{ns},\ \frac{s}{sn},\ldots,$$

$$\frac{s}{\dfrac{s}{n}},\ \frac{s}{\dfrac{s}{n}\dfrac{s}{n}}\ldots,\ \frac{n}{n},\ \frac{n}{nn},\ \frac{n}{sn},\ldots,\ \frac{\dfrac{s}{n}}{\dfrac{s}{n}},\ \text{etc.}$$

To illustrate this index symbolism with a proposition from mathematical logic, e.g.

$$\sim p \supset p \,.\, \supset \,.\, p$$

we add indices to the single words and obtain:

$$\sim p \supset p \,.\, \supset \,.\, p$$
$$\frac{s}{s}\ \frac{s}{ss}\ \frac{s}{ss}\ \frac{s}{ss}\ \ s\ \ s$$

If one applies this index symbolism to ordinary language, the semantic categories which we have assumed (in accordance with Leśniewski) will not always suffice, since ordinary languages apparently are richer in semantic categories. Furthermore, the decision to which semantic category a word belongs is rendered difficult by fluctuations in the meaning of words. At times it is uncertain what should be considered as *one* word. In simple and favourable cases, however, the index apparatus cited above will be quite suitable for linguistic use, as is illustrated by the following example:

The lilac smells very strongly and the rose blooms.

$\frac{n}{n}$	n	$\frac{s}{n}$	$\frac{s}{n}$	$\frac{s}{n}$	$\frac{s}{ss}$	$\frac{n}{n}$	n	$\frac{s}{n}$
			$\frac{s}{n}$	$\frac{s}{n}$				

$$\frac{\dfrac{s}{n}}{\dfrac{s}{n}}$$

4. In every significant composite expression it is indicated, in one way or another, which expressions occur as arguments,

and to what expressions, appearing as functors, they belong. Whenever the functor has several arguments, it must be indicated which of these arguments is the first, which is the second, etc. The order of arguments plays an important role; the difference between subject and predicate, or between antecedent and consequent of a hypothetical proposition, are special cases of the important differences made by the order of arguments. Generally speaking, this order is not identical with the sequence in which arguments occur in expressions. In fact, the order is not a purely structural, i.e. purely external affair, but is based on the semantic qualities of the whole expression. Only in symbolic languages and in some ordinary languages does the order of arguments correspond to their sequential ordering.

In order to indicate the manifold interrelationships of the parts of an expression, symbolic languages use conventions concerning the 'scope' of various functors, as well as employing brackets and making use of word order. In ordinary speech these interrelations are indicated by means of word order, inflexion, prepositions, and punctuation marks. A semantic pattern in which these interrelations are not indicated at all, or only incompletely, is devoid of any coherent meaning.

In every significant composite expression the relations of functors to their arguments have to be such that the entire expression may be divided into parts, of which one is a functor (possibly itself a composite expression) and the others are its arguments. This functor we call the *main functor* of the expression. (The concept of the main functor and the basic notion of its definition we owe to Leśniewski.) In the logical example given above, the second implication sign is the main functor of the whole proposition; in the example taken from ordinary language, the word 'and' is the main functor. When it is possible to divide a composite expression into a main functor and its arguments, we call such an expression *well articulated*. The main functor of an expression and its arguments we call *first-order parts* of this expression. If the first-order parts of an expression *A* either consist of single words, or, being composite, are themselves well articulated; and if, descending to the parts of parts, and to the parts of parts of parts, etc., i.e., i.e. to the nth order parts, we always meet with single words or well articulated expressions, we call the expression A *well articulated throughout*.

It should be mentioned that ordinary language often admits elliptical expressions, so that sometimes a significant composite expression cannot be well articulated throughout on the sole basis of the words explicitly contained within it. But a good overall articulation can be easily established by introducing the words omitted but implicit. The difficulties are greater if the language, e.g. German, admits separable words. In this case the criterion for a *single word* cannot be stated purely structurally.

5. Good articulation throughout is a necessary but not sufficient condition for a composite expression's having a unified sense, and hence being a meaningful expression. This condition must be supplemented by others. In order that an expression which is well articulated throughout may be meaningful, all its parts of the same order (which are related as functor and arguments) must fit together. That is, to each part of the nth degree, which occurs as the main functor either of the entire expression or of an $(n-1)$th-order part of it, and which requires—according to its semantic category—so-and-so many arguments belonging to certain semantic categories, there must correspond as arguments just this number of parts, of the nth degree and of the required semantic categories. To a main functor of the semantic category with the index s/ns, for instance, two arguments have first of all to correspond. Secondly, the first argument must belong to the semantic category of names, and the second to that of sentences. An expression well articulated throughout, and complying with both the above-mentioned conditions, we shall call *syntactically connected*.

These conditions can be formulated in another way, and more precisely, by means of our index symbolism. For this purpose we introduce the concept of the *exponent* of an expression, and illustrate this with an example. If we take the expression

$$p \lor p . \supset . p$$

and add indices to each word, we obtain

$$p \vee p \,.\, \supset \,.\, p \qquad\qquad (A)$$

$$\frac{s \quad s \quad s \quad s \quad s}{ss \qquad ss}$$

We arrange the parts of this expression according to the following principle. First we write down the main functor of the entire expression, and then, successively, its first, second (and if necessary third, fourth, etc.) arguments. We thus obtain:

$$\supset, \; p \vee p, \; p. \qquad\qquad (B)$$

$$\frac{s \quad s \quad s \quad s \quad s}{ss \quad ss}$$

If any part, appearing in this sequence, is still an expression composed of a main functor and its arguments, we separate this part into parts of the next higher degree and order them by the same principle, putting first the main functor and then its first, second, etc., arguments. Hence we obtain:

$$\supset, \; \vee, \; p, \; p, \; p. \qquad\qquad (C)$$

$$\frac{s \quad s \quad s \quad s \quad s}{ss \quad ss}$$

Any composite part left in this sequence is decomposed according to the same principle, and this procedure is continued to the point where only simple, single words appear as parts of the sequence. A sequence ordered in this way, and consisting of the words of an expression, we call the *proper word sequence* of the expression. In our example the proper word sequence has already been obtained with the second step; i.e. (C) is the proper word sequence of (A). If we now separate the indices from the words of the proper word sequence of an expression, writing down the indices in the same order, we obtain what is called the *proper index sequence* of the expression.

The proper index sequence of expression (A) takes the following form:

$$\frac{s \quad s}{ss \quad ss} \; s \; s \; s. \qquad\qquad (1)$$

We now see whether we can find in this index sequence, reading from left to right, a combination of indices with a fractional index in the initial position, followed immediately by exactly the same indices that occur in the denominator of the fractional index. If we find one or more of these combinations of indices, we cancel the first one (again reading from left to right) from the index sequence, and replace it by the numerator of the fractional index. The new index sequence obtained by this procedure we call the *first derivative* of the proper index sequence of the expression A. It looks like this:

$$\frac{s}{ss}\ s\ s. \tag{2}$$

This first derivative consists of a fractional index followed directly by the same index combination that forms the denominator of the fractional index. We can therefore transform it once more in the manner described above, forming its second derivative, which consists of the single index

$$s. \tag{3}$$

This we shall call the final derivative, as no further derivations are possible.

The final derivative of the proper index sequence of an expression we call the *exponent* of the expression.

Let us now determine the exponent of the ordinary language proposition given on page 211. Its proper index sequence and its successive derivatives appear as follows:

$$
\begin{array}{ccc}
\dfrac{s}{n} & & \\[2ex]
\dfrac{\frac{s}{n}}{\frac{s}{n}} & \dfrac{s}{n} & \\[2ex]
\dfrac{s}{ss}\quad \dfrac{\frac{s}{n}}{\frac{s}{n}} & \dfrac{\frac{s}{n}}{\frac{s}{n}}\ \dfrac{s}{n}\ \dfrac{n}{n}\ n\ \dfrac{s}{n}\ \dfrac{n}{n}\ n & \text{(proper index sequence)} \\[2ex]
\dfrac{s}{n} & \\
\end{array}
$$

$$\frac{s}{ss} \qquad \frac{\dfrac{s}{n}}{\dfrac{s}{n}}\, n \qquad \frac{s}{n}\,\frac{n}{n}\, n \qquad\qquad \text{(1st derivative)}$$

$$\frac{s}{ss}\,\frac{s}{n}\,\frac{n}{n}\, n \quad \frac{s}{n}\,\frac{n}{n}\, n \qquad\qquad \text{(2nd derivative)}$$

$$\frac{s}{ss}\,\frac{s}{n}\, n \quad \frac{s}{n}\,\frac{n}{n}\, n \qquad\qquad \text{(3rd derivative)}$$

$$\frac{s}{ss}\, s\, \frac{s}{n}\,\frac{n}{n}\, n \qquad\qquad \text{(4th derivative)}$$

$$\frac{s}{ss}\, s\, \frac{s}{n}\, n \qquad\qquad \text{(5th derivative)}$$

$$\frac{s}{ss}\, s\, s \qquad\qquad \text{(6th derivative)}$$

$$s \qquad\qquad \text{(7th and final derivative)}$$

We are now able to make a definition. An expression is *syntactically connected* if, and only if, (1) it is well articulated throughout; (2) to every functor that occurs as a main functor of any order, there correspond exactly as many arguments as there are letters in the denominator of its index; (3) the expression possesses an exponent consisting of a single index.[1]

---

[1] Compliance with the first and third conditions alone does not guarantee syntactic connexion. For example,

$$\sim (\phi,\, x)$$
$$\frac{s}{s}\;\frac{s}{n}\; n$$

is not syntactically connected, although the expression is well articulated throughout, and its exponent, which we find by the following procedure:

$$\frac{s}{s}\;\frac{s}{n}\; n$$

$$\frac{s}{s}\; s$$

$$s$$

is a single index.

This index may take the form of a single letter, but it may also be a fraction. For instance, the expression

$$
\begin{array}{ccc}
smells & very & nice \\[4pt]
\dfrac{s}{n} & \dfrac{s}{n} & \dfrac{s}{n} \\[10pt]
 & \dfrac{s}{n} & \dfrac{s}{n} \\[10pt]
 & \dfrac{s}{n} & \\[10pt]
 & \dfrac{s}{n} &
\end{array}
$$

which has the proper index sequence

$$
\begin{array}{ccc}
\dfrac{s}{n} & & \\[8pt]
\dfrac{s}{n} & \dfrac{s}{n} & \\[8pt]
\dfrac{s}{n} & \dfrac{s}{n} & \dfrac{s}{n}, \\[8pt]
\dfrac{s}{n} & &
\end{array}
$$

has as its exponent the fractional index

$$
\frac{s}{n}.
$$

As an example of an expression not syntactically connected, consider the following word pattern:

$$
\begin{array}{cccccc}
F & (\phi) & : \equiv & : \sim & \phi & (\phi) \\[4pt]
\dfrac{s}{\dfrac{s}{n}} & \dfrac{s}{n} & \dfrac{s}{ss} & \dfrac{s}{s} & \dfrac{s}{n} & \dfrac{s}{n}.
\end{array}
$$

Its proper index sequence and its derivatives are:

$$
\begin{array}{cccccc}
\dfrac{s}{ss} & \dfrac{s}{\dfrac{s}{n}} & \dfrac{s}{n} & \dfrac{s}{s} & \dfrac{s}{n} & \dfrac{s}{n}
\end{array}
$$

$$
\frac{s}{ss} \quad s \quad \frac{s}{s} \frac{s}{n} \frac{s}{n}.
$$

The first derivative, being also the final one in this instance, forms the exponent, which patently consists of several indices. Therefore the expression is not syntactically connected. (The word pattern examined in this last example forms the well-known 'definition' which leads to Russell's antinomy of the class of classes not containing themselves as members.)

The exponent of a syntactically connected expression gives the semantic category to which the whole composite expression belongs.

6. A symbolism which appended indices to single words would need no parentheses or other means to show the articulation of its syntactically connected expressions; i.e. to denote the mutual relations of functors and their arguments. A strict adherence, in ordering words, to the same principle that governs the order of indices in the proper index sequence of an expression would suffice. That is, the words of each composite expression would have to be arranged in such a way that the sign for the main functor would always be followed by its first, its second, etc., arguments. The proposition, for example, which reads in Russell's symbolism:

$$p \cdot q \cdot \supset r : \equiv : \sim r \cdot q \cdot \supset \sim p \qquad (A)$$

would be written by this principle as follows:

$$(B)$$

We call a functor n-adic, when the denominator of its index contains n indices. Then we can say, the expression A is the kth argument of the n-adic functor F in the expression B, if and only if, (i) an uninterrupted part T, following directly to the right of F, can be lifted out of the expression B, the exponent of T being the same as the denominator of the index of F; (ii) this part T can be completely decomposed into n uninterrupted subparts in such a way that the exponents of the subparts are, taken successively, the same as the indices contained in the denominator of the index of F; (iii) A is the kth of these subparts; and (iv) F together with T constitutes either the entire expression B or a part of B. (Strictly speaking, this explanation should be replaced by a recursive definition.)

For example, according to this explanation, the part of expression B designated as 3 is the first argument, and that designated as 4 the second argument, of the implication sign designated as 5 in expression B. For (i) the uninterrupted part 1, following directly to the right of 5, and having an exponent identical with the denominator of the index of 5, can be lifted out of the expression B; (ii) part 1 can be completely reduced to two uninterrupted subparts, their exponents being, in order, the same as the indices contained in the denominator of the index of 5; (iii) 3 is the first of these subparts, 4 the second; and (iv) 5 together with 1 constitutes a part of B.

The advantage of the index symbolism, which makes parentheses dispensable, may perhaps seem insignificant if one considers only examples from the propositional calculus. In fact, for the propositional calculus Łukasiewicz has introduced a symbolism that needs no parentheses or other signs in order to signify the articulation of syntactically connected expressions, and which involves no indices.[1] The possibility of dispensing with parentheses without introducing indices is explained by the fact that the propositional calculus is concerned with only very few (in practice only three) semantic categories. All variables belong to one semantic category, and the number of constants is limited, thus making it possible to indicate the semantic category of an expression by the form of the symbols themselves. In this case, the rules of formation are easily enumerated. But in the case of a large, theoretically unlimited number of different semantic categories, some systematic mode of differentiation like our index symbolism has to be employed.

[1] Cf. Jan Łukasiewicz, 'Philosophische Bemerkungen zu mehrwertigen Systemen des Aussagenkalküls', *Comptes rendus des séances de la Société des Sciences et des Lettres de Varsovie* 13, Cl. iii (Warsaw, 1930). [Appearing as paper 3 in this volume.—Ed.]

So far our inquiries have concerned only expressions containing no operators (see below, section 7). We will now consider expressions that contain them.

II

7. It has been assumed above that any word of a language can be assigned to a certain semantic category by its meaning, and can consequently be furnished with an index. Composite expressions may be analysed according to the pattern, 'functors and their arguments', only if this assumption holds. For some languages it may hold; it does not, however, seem to be applicable to certain symbolic languages. We are referring here to languages that make use of what are called operators. By this term we understand such signs as the universal quantifier in logic, '(Πx)' or '(x)', which is also called the all-operator (cf. Carnap, *Abriss der Logistik* (Vienna, 1929), p. 13); the existential operator '$(\exists x)$'; the algebraic summation sign

$$\sum_{k=1}^{10};$$

the product sign
$$\prod_{x=1}^{100};$$

the definite integral sign $\int_{0}^{1} dx$, etc.

All of these signs have the property of applying to an expression that contains one or more variables, and reducing one or more of these variables to the role of an apparent variable. For instance, when an operator is applied to an expression which contains only one variable, a composite expression of constant value is created. Thus the expressions '$(\exists x).x$ is human' and

$$\sum_{x=1}^{10} x^2$$

have constant values, although variables occur in them. These variables are reduced to apparent variables by the operator; or, in other words, they are 'bound' by the operator.

The analysis of an expression that contains an operator, e.g. the general proposition '$(\Pi x).fx$', into functors and arguments

with appropriate semantic categories, seems to meet with in-
superable difficulties.

Disregarding the inner structure of the composite operator
'(Πx)', we shall first reject the obvious interpretation of the syntactical structure of the general proposition '$(\Pi x) \cdot fx$'; to wit,
that which implies that in a proposition of this kind the operator
'(Πx)' plays the role of the main functor and the propositional
function the role of its argument. If this were the correct syntactical analysis of a general proposition, the universal operator
'(Πx)' would have to be counted among the functors which form
a sentence with one sentence as their argument, and hence
belong to the $\frac{s}{s}$ category. But, on the other hand, it is to be
considered that an $\frac{s}{s}$ functor must be a truth functor in an extensionalistic logic. Therefore it has to correspond to one of the
following four tables:

p	$f_1 p$		p	$f_2 p$		p	$f_3 p$		p	$f_4 p$
0	0		0	1		0	1		0	0
1	1		1	0		1	1		1	0

In other words, if the universal operator were an $\frac{s}{s}$ functor,
the proposition '$(\Pi x) \cdot fx$' would have to be either (1) equivalent
to 'fx'; or (2) equivalent to '$\sim fx$'; or (3) always true, independent of 'x'; or (4) always false. All these cases are, however,
excluded. Hence, in an extensionalistic logic, the '(Πx)'-opera-
tor may not be interpreted as an $\frac{s}{s}$ functor. Yet, since it forms
a proposition with the proposition 'fx', it cannot be any other
kind of functor.

The thought, however, comes to mind that the syntactical
structure of a general proposition

$$(\Pi x) \cdot fx$$

could also be interpreted in a different way. Perhaps 'Πx' is
not the main functor in this proposition and 'fx' not its argument, but the 'Π'-sign might be the main functor and 'x' its

first, 'fx' its second argument. Then the general proposition would have to be written in the following form in order to be correct: $\Pi(x, fx)$.

Since 'x' can belong to different semantic categories, 'Π' would have to be of various types. If, for example, 'x' belongs to the name category and 'f' to the $\frac{s}{n}$ category, Π would have to belong to the $\frac{s}{ns}$ category. But if 'x' belonged to the category of sentences, and 'f' to the $\frac{s}{s}$ category, 'Π' would have to belong to the $\frac{s}{ss}$ category for '$\Pi(x, fx)$' to be a proposition. In this latter case, 'Π' would have to be a dyadic truth functor in an extensionalistic logic; it would therefore correspond to one of the sixteen known tables for dyadic truth functors. But it is easily demonstrated that this too is incompatible with the meaning of the general proposition '$(\Pi x) . fx$'.

Thus both the first and the second way of interpreting the syntactical structure of general propositions as composed of functors and arguments have proved unsuccessful.

8. Nothing may be substituted for a variable, in a legitimate proposition, which is bound by an operator (assuming that the latter is not a universal operator forming a main part of the entire proposition). This is the meaning of the 'apparency' or 'boundness' of the variables. Functors, on the other hand, behave in the opposite way in this respect. If, therefore, functors are conceived of as non-binding and operators as binding, it becomes evident that an operator may not be counted as a functor. As a secondary difference between a functor and an operator, it should be noted that a functor can appear as the argument of another functor, while an operator can never be the argument of a functor.

Despite these differences, there is a similarity between an operator and a functor. An operator can form a composite, coherent whole with the expression to which it applies, just as a functor can with its arguments. Therefore, one should also be able to assign indices to operators. These would have to be different from the indices assigned to functors, as they cannot be treated in the same manner as the functor indices in determining the exponent of an expression. Since an operator can never be an argument, its index must not be combined with an index preceding it in the proper index sequence or its derivatives, but must always be combined with the indices following it. We therefore propose for operator indices a fraction with a vertical dash to its left. The universal operator '(Πx)' would thus receive the index $\left|\dfrac{s}{s}\right.$, since it forms a sentence out of a sentence.

We assign *one* index to the entire operator, although the operator appears to consist of several words. By this we do not abandon the principle that from the start an index should be assigned only to a single word, and that indices of composite expressions are considered to be their exponents (i.e. the final derivatives of their index sequences). For an operator must not be treated as an expression composed of several words. An operator is basically a single word composed of several letters. There are ways of writing operators which make this clear. Professor Scholz, for instance, writes '$\tilde{x}$' for '(Πx)'. And the usual notation '(x)' instead of '(Πx)', or 'Π_x' instead of '(Πx)', makes the single-word character of operators evident.

9. The exponent of an expression containing an operator has to be determined in a way other than the one given above. If we proceeded with operator indices as with functor indices, it might happen that the operator index was combined with an index preceding it, and this, as we have seen, is not permissible.

Let us take the following expression as an example:

$$F \cdot \Pi x \cdot x. \qquad\qquad (A)$$
$$\frac{s}{n} \quad \left|\frac{s}{s}\right. \quad n$$
$$\frac{s}{s}$$

If we formed its exponent by the rule given earlier, we would obtain the following derivatives:

$$(1^0) \quad \dfrac{\dfrac{s}{n}}{\dfrac{s}{s}} \quad \left|\begin{array}{c} s \\ \hline s \end{array}\right. n, \qquad (2^0) \quad \dfrac{s}{n} \; n, \qquad (3^0) \quad s.$$

According to this the exponent would prove to be a sentence index, although the expression (A) is obviously syntactic nonsense.

Our new rule for the formation of the exponent of an expression requires that the part of its proper index sequence which begins with a vertical dash on the extreme left must be treated separately. For this part, which contains a dashed index only at the beginning, the final derivative must be determined by the original method. In this process the dashed index is treated exactly like an undashed one; thus, for example,

$$\left| \begin{array}{c} {}' s \\ \dfrac{}{s} \ s \end{array} \right.{}' \quad \text{as well as} \quad \begin{array}{c} {}' s \\ \dfrac{}{s} \ s \end{array}{}'$$

is replaced by the index 's', and similarly in other cases.

When the final derivative of this partial index sequence has been determined, it is substituted for the partial index sequence in the complete index sequence. Two possibilities must now be distinguished. Either the initial dashed index of the partial index sequence has disintegrated in the course of determining its final derivative (i.e. in forming the nth derivative from the n—1th, the initial dashed index, together with the indices following it, was replaced by its numerator); or it has remained intact.

In the second case, where the dashed index heading the partial index sequence remains intact, we stop and declare the complete index sequence (which results from replacing the separated partial index sequence by its final derivative) to be the final derivative, i.e. the exponent, of the original proper index sequence of the expression.

In the first case, where the dashed index heading the partial index sequence disintegrates, its vertical dash disappears from the whole index sequence, diminishing the number of vertical dashes by one. In this case the procedure is repeated, and continued either until some dashed index does not disintegrate, or until all dashed indices have vanished and we have arrived at an index sequence without dashes and with no further derivatives. The index sequence obtained is taken as the final derivative of the proper index sequence of the original expression, and hence as its exponent.

We shall demonstrate this new procedure by the following example:

$$(\Pi fg):.(\Pi x).f\ \ x \supset\ g\ \ x :\supset: (\Pi x).f\ \ x .\supset. (\Pi x).\ g\ \ x. \quad (A)$$

$$\left|\frac{s}{s}\quad\right|\frac{s}{s}\ \ \frac{s}{n}\ \ n\ \ \frac{s}{ss}\ \ \frac{s}{n}\ \ n\ \ \frac{s}{ss}\quad\left|\frac{s}{s}\ \ \frac{s}{n}\ \ n\ \ \frac{s}{ss}\quad\right|\frac{s}{s}\ \ \frac{s}{n}\ \ n$$

The proper index sequence of this expression is

$$\left|\frac{s}{s}\ \frac{s}{ss}\ \right|\frac{s}{s}\ \frac{s}{ss}\ \frac{s}{n}\ n\ \frac{s}{n}\ n\ \frac{s}{ss}\ \left|\frac{s}{s}\ \frac{s}{n}\ n\ \right|\frac{s}{s}\ \frac{s}{n}\ n. \quad (I)$$

First we form the final derivative of the part to the right of the last vertical dash:

$$(1)\quad\left|\frac{s}{s}\ \frac{s}{n}\ n,\quad\right.\qquad(2)\quad\left|\frac{s}{s}\ s\quad\right.\qquad(3)\quad s.$$

We now replace in I this part by its final derivative, thus reducing the number of dashes by one:

$$\left|\frac{s}{s}\ \frac{s}{ss}\ \right|\frac{s}{s}\ \frac{s}{ss}\ \frac{s}{n}\ n\ \frac{s}{n}\ n\ \frac{s}{ss}\ \left|\frac{s}{s}\ \frac{s}{n}\ n\ s.\right.\quad (II)$$

Proceeding with II in the same manner as with I we get

$$\left|\frac{s}{s}\ \frac{s}{ss}\ \right|\frac{s}{s}\ \frac{s}{ss}\ \frac{s}{n}\ n\ \frac{s}{n}\ n\ \frac{s}{ss}\ s\ s.\quad (III)$$

which is again to be dealt with in the same way. What must be determined is the final derivative of the part separated by the last dash. Since this is a lengthy procedure, we demonstrate it here:

$$\left|\frac{s}{s}\ \frac{s}{ss}\ \frac{s}{n}\ n\ \frac{s}{n}\ n\ \frac{s}{ss}\ s\ s\right. \quad (1)$$

$$\left|\frac{s}{s}\ \frac{s}{ss}\ s\ \frac{s}{n}\ n\ \frac{s}{ss}\ s\ s\right. \quad (2)$$

$$\left| \frac{s}{s} \ \frac{s}{ss} \ s \ s \ \frac{s}{ss} \ s \ s \right. \tag{3}$$

$$\left| \frac{s}{s} \ s \ \frac{s}{ss} \ s \ s \right. \tag{4}$$

$$\left| \frac{s}{s} \ s \ s \right. \tag{5}$$

$$s \ s \tag{6}$$

We replace the part separated by the last dash in (III) by this quantity and obtain

$$\left| \frac{s}{s} \ \frac{s}{ss} \ s \ s. \right. \tag{IV}$$

The final derivative of the remaining index sequence (IV) we determine without difficulty to be

$$s.$$

This final derivative of the original index sequence is the exponent of the expression (A).

As another example we shall examine a case in which not all dashed indices disintegrate. Take for instance the expression

$$(\Pi x) . f \quad x : \supset : (\Pi x) . \quad g \quad (x, z). \tag{B}$$
$$\left| \frac{s}{s} \quad \frac{s}{n} \quad n \quad \frac{s}{ss} \right| \frac{s}{s} \quad \frac{n}{nn} \quad n \quad n$$

Its proper index sequence is

$$\frac{s}{ss} \left| \frac{s}{s} \ \frac{s}{n} \ n \right| \frac{s}{s} \ \frac{n}{nn} \ n \ n. \tag{I}$$

Forming the final derivative of the part separated by the last dash we obtain

$$\left| \frac{s}{s} \ n. \right.$$

Since the dashed index has not disintegrated, the last dash is not omitted, and the final derivative of I, which is also the exponent of B, takes the form of

$$\frac{s}{ss} \left| \frac{s}{s} \ \frac{s}{n} \ n \right| \frac{s}{s} \ n.$$

Hence the expression B does not have a single index as exponent.

We have now become familiar with the method of forming the exponent of expressions containing operators. Clearly the earlier method for operator-free expressions is contained in it as a special case. (In its formulation mention would merely have to be made of possibly appearing dashed indices.) We could now repeat word for word the definition of syntactic connexion given earlier; it would now also apply to expressions containing operators.

10. The syntactic connexion of operator-free expressions coincides with their syntactical correctness. However, besides being syntactically connected, expressions containing operators have to comply with a further condition. This condition stipulates that to each variable contained in the operator there must correspond, in the argument of the operator (i.e. in the expression to which the operator applies),[1] a variable of the same form which is not bound within this argument. Only when this condition has been fulfilled will a syntactically connected expression containing operators be syntactically correct.

III

11. We have taken the binding role of operators to be their characteristic property, distinguishing them from functors. Binding one or more variables is a property common to all operators, although some operators distinguish themselves by playing other roles as well. There is, however, one species of operator which is confined to binding one or more variables, to which category the circumflex sign ' ^ ', introduced by Whitehead and Russell, appears to belong. Russell employs the circumflex in order to distinguish between what he calls the 'ambiguous value of a function' and 'the function itself'. If 'fx' symbolizes the ambiguous value of a function, '$f\hat{x}$' denotes the function itself. Closer examination reveals that Russell's 'ambiguous value of a function' is identical with what is commonly termed 'value of the dependent variable'. What Russell calls

---

[1] Strictly speaking one should not speak of the 'argument' of an operator, but perhaps use the word 'operand'. Our earlier remarks about 'good articulation' of an expression must obviously be extended to the operator-operand relationship.

'the function itself' is, on the other hand, not a variable, but a constant. A further study of the explanations by which Russell clarifies the notion of the 'function itself' leads us to surmise that by this term he means what we would call the objective correlate of a functor. Therefore $f\hat{x}$ is the same as f and the symbols '$f\hat{x}$' and 'f' denote the same thing. If this supposition is correct, the circumflex sign can be counted as an operator, for its function is to 'cancel' or 'bind' a variable. In addition it should be mentioned that several variables can be bound simultaneously in an expression by the circumflex sign. Thus '$f\hat{x}\hat{y}$' represents the functor 'f' of two arguments.

In the simplest cases, where the circumflex sign is placed over all the arguments of the main functor of an expression, as in the schematic examples '$f\hat{x}$' or '$f\hat{x}\hat{y}$', the circumflex sign has the same effect as a stroke cancelling the circumflexed variable, which is thus eliminated. If, however, not all arguments of the main functor of the whole expression are circumflexed, the role of the circumflex sign can no longer be compared to that of an ordinary cancellation mark. For example, '$\hat{p} \supset . a . \sim a$' ('$a$' being a constant proposition) represents an $\frac{s}{s}$ functor 'f', for which the following equivalence holds:

$$fp . \equiv . p \supset . a . \sim a$$

But it is obvious that the negation sign will do in place of the 'f' of this equivalence. Therefore '$\hat{p} \supset . a . \sim a$' means the same as '$\sim$'; while '$\supset . a . \sim a$', which is the same as '$p \supset . a . \sim a$' with 'p' cancelled, is not an $\frac{s}{s}$ functor, and is in fact not even a syntactically connected expression.

12. If an entire expression, in which the circumflex sign is applied to a variable, belongs to the category of sentences, we find in Russell's symbolism another sign with which the circumflex sign can be equated. This is the prefix $(\hat{x})$ used in class symbols, or the prefixes $(\hat{x}, \hat{y})$, used in symbols for relations. If 'fx' stands for a propositional function, and if

we disregard certain complications arising from the admission of intensional functions, dropped in any case by Russell in the second edition of *Principia*, the symbol '$(\hat{x})\,.\,fx$' designates the same thing as the functor 'f', and hence the same thing as '$f\hat{x}$'. The same is true with regard to the symbols '$(\hat{x}\hat{y})\,.\,fxy$' and '$f\hat{x}\hat{y}$'.

We wish also to employ the prefixes $(\hat{x})$ or $(\hat{x}\hat{y})$ when the expression to which they refer does not belong to the category of sentences, so that we may then in general write '$(\hat{x})\,.\,fx$' for '$f\hat{x}$'-symbols and '$(\hat{x}\hat{y})\,.\,fxy$' for '$f\hat{x}\hat{y}$'-symbols. This revised notation for the circumflex operator has the advantage of allowing the whole expression to which the circumflex operation applies to be clearly marked. This was not possible with the previous notation, and could lead to ambiguities in complicated cases. Apart from this, the new notation permits us to apply the circumflex operator several times successively to the same expression. Thus we may write '$(\hat{x}) : (\hat{y})\,.\,fxy$', which is not the same as '$(\hat{x}\hat{y})\,.\,fxy$' (previously written '$f\hat{x}\hat{y}$'). The operational character of the circumflex symbol is also made more evident by the new notation.

13. In our index symbolism the circumflex symbols $(\hat{x})$, $(\hat{x}\hat{y})$, etc., are assigned, as operators, a dashed index. Since, however, circumflex operators can be applied to expressions of various semantic categories, and can also transform them into expressions of various semantic categories, circumflex symbols do not always receive the same dashed index.

The general definition of the (monadic) circumflex operator would be the following. An operator '$(\hat{x})$' referring to a variable X in an expression A is a circumflex operator, if, together with this expression, it forms a functor, and if the functor, with the variable X as its argument, forms an expression equivalent to the expression A. This can be illustrated by the following example, in which the expression A takes the form 'fx', and the variable X the form 'x':

$$(\hat{x})\,.\,fx : x\,.\,:\,\equiv\,.\,fx.$$

From the above it is clear that, if the expression A to which the operator refers has the exponent 'E_1', and the variable X has the index 'E_2', the operator must have the dashed index

$$\left|\begin{array}{c} E_1 \\ \hline E_2 \\ \hline E_1 \end{array}\right.$$

Depending on the indices to be substituted for 'E_1' and 'E_2', the dashed index of the circumflex operator will assume various forms. Many-place circumflex operators behave similarly.

It was mentioned above that circumflex operators appear to be restricted to binding variables only. Other operators, however, have wider roles. We have seen that the main difference between a functor and an operator lies in the fact that an operator plays the role of binding while a functor does not. This suggests that it might be possible to divide in two the role of those operators not restricted to binding alone; thus the part of binding could be performed by the circumflex operator, the other by a functor. Let us, for example, introduce the functor 'Π', to which we assign the index

$$\frac{s}{\dfrac{s}{n}}$$

Syntactically speaking, 'Π' is a sentence-forming functor with an $\dfrac{s}{n}$ functor as its argument. In addition we make the following definition of this functor: '$\Pi(f)$' is satisfied by all and only those $\dfrac{s}{n}$ functors (in the place of 'f') which form a *true* proposition with every name. Thus we have:

$$\Pi(f) \, . \equiv . \, (\Pi x) \, . fx.$$

We shall call such a functor a universal functor. We can now replace the universal operator by the universal functor in all cases where, in place of the propositional function to which the universal (Πx) operator applies, we can specify a functor which, with x as its argument, forms an expression equivalent to the original propositional function. But this the circumflex operator always permits us to do. For '$(\hat{x})$. fx' is just the functor sought for the propositional function 'fx', no matter what form this propositional function may take. We may therefore always write '$\Pi((\hat{x})$. $fx)$' for '(Πx) . fx'. Thus the role of the universal operator can be filled by a combination of the roles of the universal functor and the circumflex operator. We need not stress the fact that there will be not just *one*, but several universal functors, differing from each other by their semantic category, according to the semantic category of the functors which they take as their arguments.

Owing to the equivalence

$$\Pi(f) \,.\, \equiv \,.\, (\Pi x) \,.\, fx,$$

the universal functor can be easily defined with the assistance of the universal operator. Its definition meets, however, with insurmountable difficulties if one does not want to make use of the universal operator. Nevertheless, in our opinion, a substitute for the definition of the universal functor might be found in a statement of those rules of inference which pertain to its deductive use. This would introduce the 'Π' symbol openly into logic as a primitive sign, which would then have a clearer place in the system of this science than the smuggled-in universal operator, which belongs neither to the defined nor to the primitive signs of logic.

However, either the circumflex operator must be defined, or else it must be smuggled into logic in the manner of the universal operator. We shall refrain from resolving this dilemma here. Should, however, it be decided to smuggle the circumflex operator in, we would permit ourselves the suggestion that this subterfuge might well pay, for it is possible that all other operators (of which there are many in the deductive sciences) might be replaced by the circumflex operator and by corresponding functors. It would be a great advantage, in our opinion, if the employment of operators could in all cases be restricted to one kind, that of circumflex operators.

Allatum est die 15. Iulii 1934.

11

ON THE RULES OF SUPPOSITIONS IN FORMAL LOGIC†

STANISŁAW JAŚKOWSKI

§ 1. The theory of deduction based on the method of suppositions

IN 1926 Professor J. Łukasiewicz called attention to the fact that mathematicians in their proofs do not appeal to the theses of the theory of deduction, but make use of other methods of reasoning. The chief means employed in their method is that of an arbitrary supposition. The problem raised by Mr. Łukasiewicz was to put those methods under the form of structural rules and to analyse their relation to the theory of deduction. The present paper contains the solution of that problem.[1]

Here we consider as structural rules only those which refer to the external appearance of expressions. It is possible to formulate such rules only for a formal system in which all propositions are written in symbols. In the present paper we shall use Mr. Łukasiewicz's[2] bracket-free symbolism. The implication 'if α, then β' will be symbolized by '$C\alpha\beta$' and the negation 'not α' by '$N\alpha$'. In the above, α and β stand for significant expressions of the system. The significant expression built up by means of 'C' and 'N' can be defined by the two following

---

† This paper appeared in English in *Studia Logica* 1 (1934), pp. 5–32.

[1] The first results on that subject obtained by the author in 1926 at Professor Łukasiewicz's seminar were presented at the First Polish Mathematical Congress in Lwów in 1927 and were mentioned in the proceedings of the Congress: *Księga pamiątkowa pierwszego polskiego zjazdu matematycznego*, Kraków, 1929.

[2] J. Łukasiewicz, *Elementy logiki matematycznej*. Opracował M. Presburger. Warszawa, 1929 (lithogr.), p. 38ff. [English translation: *Elements of Mathematical Logic*, Oxford, 1963—Ed.] J. Łukasiewicz und A. Tarski, 'Untersuchungen über den Aussagenkalkül', *Comptes rendus des séances de la Société des Sciences et des Lettres de Varsovie*, Cl. iii, 23 (1930). [English translation in Tarski, *Logic, Semantics, Metamathematics*, Oxford, 1956—Ed.]

conditions: (1) such an expression contains more variables than symbols '*C*', (2) no initial part of this expression contains more variables than symbols '*C*'. In condition (2) we must take 'part' in the sense of *proper* part, i.e. as distinct from the whole. As to the variables, we shall make use of small Latin letters '*p*','*q*', '*r*', etc. In accordance with the above explanations, the symbolic expression '*CpCCpqq*' can be read: 'If *p*, then, if *p* implies *q*, *q*.'

We intend to analyse a practical proof by making use of the method of suppositions. How can we convince ourselves of the truth of the proposition '*CpCCpqq*'? We shall do it as follows:

Suppose p. This supposition being granted, *suppose Cpq.* Thus we have assumed '*p*' and 'if *p*, then *q*'. Hence *q* follows. We then observe that '*q*' is a consequence of the supposition '*Cpq*', and obtain as a deduction: 'if *p* implies *q*, then *q*' i.e. *CCpqq*. Thus having supposed '*p*', we have deduced this last proposition; from this fact, we can infer *CpCCpqq*.

This last proposition does not depend upon any supposition. It would remain true even in case the suppositions used above should be false. All such processes as the above grow clearer when we introduce prefixes denoting which propositions are consequences of a given supposition. These prefixes will contain numbers classifying the suppositions; thus the number '1' will correspond to our first supposition '*p*'. Such a number must be written before the supposition to which it corresponds and before all expressions which are assumed under the conviction that this supposition is true. One of the expressions written within the scope of validity of the first supposition was '*suppose Cpq*'; its prefix therefore must begin with the number '1'. On the other hand '*Cpq*', being a supposition itself, will obtain its own number, which will take the second place in the prefix. '*Cpq*', being the first supposition made in the scope of the former one, will also have '1' as its number, but its prefix will be different, namely '1 . 1 .'. The dots are adjoined to the prefix in order to remove ambiguity. The word '*suppose*' of the usual language will be symbolized by the letter '*S*' written down immediately after the prefix. The above conventions lead us to some new expressions which must be considered as significant ones. In the following explanations, however, we shall retain for the term 'proposition' the meaning already given, namely the significant proposition of the usual theory of deduction as above defined. Thus our sketch of demonstration takes the following form:

$$1.Sp$$
$$1.1.SCpq$$
$$1.1.q$$
$$1.CCpqq$$
$$CpCCpqq$$

The reader will easily understand the following process:

$$2.SCNpNq$$
$$2.1.Sq$$
$$2.1.1.SNp$$
$$2.1.1.Nq$$

The supposition 'Np' with the prefix '2.1.1.' leads us to a contradiction consisting of the simultaneous validity of 'q' and 'Nq'. We can therefore deduce:

$$2.1.p$$
$$2.Cqp$$
$$CCNpNqCqp[1]$$

---

[1] In 1926 the system was expressed in another symbolism, which is shown below for the same examples of reasoning:

$$CpCCpqq$$

We now intend to make possible the assuming of the above expressions for theses of a deductive system whose structural rules we shall formulate. In the formulation of these rules, we shall make use of some abbreviated modes of speaking about the expressions written in the system. In order to avoid any misunderstanding, we must always remember that, by an expression, a thesis etc., we shall treat a given inscription as a material object, just as Professor S. Leśniewski did in the explanations concerning his systems.[1] Thus two inscriptions having the same appearance but written down in different places must never be taken as identical; they can only be said to be *equiform* with each other.

In order to be able to draw examples illustrative of the conventions we shall make, we may suppose that the steps of proofs

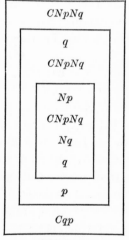

$CNpNq$

q
$CNpNq$

Np
$CNpNq$
Nq
q

p

Cqp

$CCNpNqCqp$

Certain expressions as 'p', 'q', '$CNpNq$' which have been written outside of some rectangles, have been repeated inside of them. In doing so, we obeyed a particular rule which now, through the modification of others, has become superfluous.

[1] S. Leśniewski, 'Grundzüge eines neuen Systems der Grundlagen der Mathematik', *Fundamenta Mathematicae* 14 (1929), pp. 59 ff. S. Leśniewski, 'Über die Grundlagen der Ontologie', *Comptes rendus des séances de la Société des Sciences et des Lettres de Varsovie*, Cl. iii, 23 (1930), pp. 115 ff. We use the terms 'thesis', 'system', with the same meaning as Mr. Leśniewski does.

written above are already theses of the system and that no other thesis exists. Consider a thesis containing a supposition, e.g. the thesis having the form '2.1.Sq'. All other theses having their initial parts equiform with the prefix '2.1.' are those which have the following forms: '2.1.1.SNp', '2.1.1.Nq', and '2.1.p'. The class composed of a supposition α and of all expressions which in other theses are preceded by initial parts equiform to the prefix of α will be called the *domain of the supposition* α. Thus in our example, the domain of the supposition having the form 'q' is a class of which the elements have the following forms: 'q', '1.SNp', '1.Nq', and 'p'. Besides domains of suppositions, we shall give the name 'domain' to the class of all theses belonging to the system. This will be called the absolute domain.[1]

The meaning of the word 'domain' like that of 'system' depends upon the set of theses which are written down up to a given moment. The domains grow wider in conformity with the development of the system, for they obtain new elements. The absolute domain is assumed as existing, though empty, before the first thesis has been written.

In reality, the rules we shall give will enable us to subjoin new theses to the system. Nevertheless, for the sake of brevity, we shall use other words for expressing this fact in the formulation of the rules. So we shall say: 'It is allowed to subjoin every expression satisfying some condition Φ to a given domain D' and by so saying we shall mean what can be expressed more exactly as follows: 'Given a domain D, it is allowed to subjoin to the system a new thesis which contains an expression satisfying the condition Φ and which belongs to a new domain composed of the elements of the former domain D and of this expression'.

This abbreviation will be applied in the following rule I.

---

[1] The class is taken here in the meaning employed by A. N. Whitehead and B. Russell in *Principia Mathematica*, vol. i, Cambridge, 1925. It is possible to understand the domain as a class of expressions in conformity with Mr. Leśniewski's view of a class as a material object, but in that case the subsequent explanations would have to be modified and the formulation of the rules would become more complicated.

Rule I. *To every domain D, it is allowed to subjoin an expression composed successively:*

(1) *of a number not equiform with the initial number of any other element of the domain D,*

(2) *of a dot,*

(3) *of a symbol 'S', and*

(4) *of a proposition.*

In virtue of this rule, we should be able to obtain some expressions among the above-written, for instance those having the forms: '1.Sp', '1.1.$SCpq$', '2.1.Sq', '2.1.1.SNp'.

Given two different domains D and D', where D is the domain of a supposition α and D' either the absolute domain or the domain of a supposition β whose prefix is equiform with an initial part of the prefix of α, we shall say that D is a *subdomain* of D'. Thus the domain of the above-mentioned supposition 'Np' with the prefix '2.1.1.' is a subdomain of the domain of the supposition 'q' with the prefix '2.1.'. A proposition belonging to a domain will be called *valid* in every subdomain of that domain. Thus the proposition '$CpCCpqq$' is valid in the domains of suppositions 'p', 'Cpq', '$CNpNq$', 'q', and 'Np'.

Rule II. *If in the domain D of a supposition α a proposition β is valid, it is allowed to subjoin a proposition of the form '$C\alpha\beta$' to the domain whereof D is an immediate subdomain.*

Here we regard a given subdomain D of a domain D' as an *immediate subdomain* then and only then, when D is not the subdomain of any subdomain of D'. In our given example, the rule II would allow us to obtain expressions which have the forms '1.$CCpqq$' (in reference to '1.1.$SCpq$' and '1.1.q') and '$CCNpNqCqp$' (from '2.$SCNpNq$' and '2.Cqp').

The next rule is a generalization of the usual rule of inference (*modus ponens*):

Rule III. *Given a domain D in which two propositions are valid, one of them being α and the other being composed successively:*

(1) *of a symbol 'C',*

(2) *of a proposition equiform with α,*

(3) *of a proposition β,*

it is allowed to subjoin to the domain D a proposition equiform with β.

As an example of the application of this rule, we may cite the deduction of '1.1.q' from '1.Sp' and '1.1.SCpq'.

It remains to formulate a rule embodying the principle of *reductio ad absurdum*. We can distinguish between two forms of this principle. The first form states: 'Because the supposition α has led us to β and not-β, not-α must be the case'. This form perhaps is more natural than the other, but it has less deductive power, as we shall see in § 3. Therefore we must use the other form of the principle, namely: 'Because the supposition not-α has led us to β and not-β, α must be the case.' This form is embodied in the following structural rule:

Rule IV. *Given a domain D of a supposition composed successively of a symbol 'N' and of a proposition α, if two propositions β and γ are valid in D such that γ is composed successively of a symbol 'N' and of a proposition equiform with β, it is allowed to subjoin a proposition equiform with α to that domain whereof D is an immediate subdomain.*

Example. The conclusion '2.1.p' from the premisses

'2.1.SNp', '2.1.Sq', and '2.1.1.Nq'.

Thus we have formulated all the rules of our system, which has the peculiarity of requiring no axioms. Now we shall give some theses of this system. In order to facilitate reference to various theses, we shall classify them independently of the numbers constituting parts of theses.

Furthermore, since other systems will be developed in subsequent chapters, we shall prefix to the numbers of theses the letters 'td' as abbreviation for 'theory of deduction'. To the right of each thesis, we shall write the number of the rule used in obtaining that thesis and the numbers of theses to which we appeal. None of these numbers are parts of the theses, in contrast to the numbers belonging to the prefixes.

We begin by repeating the theses which were obtained above in an intuitive way.

td 1	1. Sp	I	
td 2	1.1. $SCpq$	I	
td 3	1.1. q	III	2, 1
td 4	1. $CCpqq$	II	2, 3
td 5	$CpCCpqq$	II	1, 4
td 6	2. $SCNpNq$	I	
td 7	2.1. Sq	I	
td 8	2.1.1. SNp	I	
td 9	2.1.1. Nq	III	6, 8
td 10	2.1. p	IV	8, 7, 9
td 11	2. Cqp	II	7, 10
td 12	$CCNpNqCqp$	II	6, 11
td 13	1.2. Sq	I	
td 14	1. Cqp	II	13, 1
td 15	$CpCqp$	II	1, 14
td 16	1.3. SNp	I	
td 17	1.3.1. SNq	I	
td 18	1.3. q	IV	17, 1, 16
td 19	1. $CNpq$	II	16, 18
td 20	$CpCNpq$	II	1, 19
td 21	3. $SCpq$	I	
td 22	3.1. $SCqr$	I	
td 23	3.1.1. Sp	I	
td 24	3.1.1. q	III	21, 23
td 25	3.1.1. r	III	22, 24
td 26	3.1. Cpr	II	23, 25
td 27	3. $CCqrCpr$	II	22, 26
td 28	$CCpqCCqrCpr$	II	21, 27
td 29	4. $SCpCqr$	I	
td 30	4.1. $SCpq$	I	
td 31	4.1.1. Sp	I	
td 32	4.1.1. Cqr	III	29, 31
td 33	4.1.1. q	III	30, 31
td 34	4.1.1. r	III	32, 33
td 35	4.1. Cpr	II	31, 34

td 36	4. $CCpqCpr$	II 30, 35
td 37	$CCpCqrCCpqCpr$	II 29, 36
td 38	5. $SCNpp$	I
td 39	5.1. SNp	I
td 40	5.1. p	III 38, 39
td 41	5. p	IV 39, 40, 39
td 42	$CCNppp$	II 38, 41
td 43	6. $SCCpqp$	I
td 44	6.1. SNp	I
td 45	6.1.1. Sp	I
td 46	6.1.1. $CNpq$	III 20, 45
td 47	6.1.1. q	III 46, 44
td 48	6.1. Cpq	II 45, 47
td 49	6.1. p	III 43, 48
td 50	6. p	IV 44, 49, 44
td 51	$CCCpqpp$	II 43, 50

Although the last thesis does not contain any negation, we made use of the rule of *reductio ad absurdum* in proving it. It can be shown, in virtue of theorem 4 of § 3, that it is impossible to avoid that rule in this proof.

td 52	7. St	I
td 53	7.1. SNu	I
td 54	7.1.1. Sp	I
td 55	7.1. Cpt	II 54, 52
td 56	7.1.2. Su	I
td 57	7.1.2.1. SNp	I
td 58	7.1.2. p	IV 57, 56, 53
td 59	7.1. Cup	II 56, 58

The domain having the prefix '7.1.' can be used for showing an interesting property of systems dealing with suppositions. In the suppositions valid in that domain, two variables appear: '*t*' and '*u*'. We can assign to each of them the meaning of a constant, namely to '*t*' that of the true proposition and to '*u*' that of the false one. A precise analysis would show that in the domain with the prefix '7.1.' we can obtain all those propositions and only those which we are able to deduce in the usual theory of deduction in which '*t*' and '*u*' would be constants and the suppositions belonging to the theses td 52 and td 53 would be taken as supplementary axioms. Thus the domain in question can be considered as a system of the theory of deduction with those two constants. The prefix '7.1.' is analogous to the assertion sign of that system.

The above example shows the analogy between domains and deductive systems. Every domain can be considered as a system having its own axioms and constants, though not every domain gives a complete system, much less an interesting one. Thus the system we were occupied with in the present chapter can be considered as composed of many systems. For this reason this system will be called the *composite system* of the theory of deduction, and, in contrast to it, the usual system will be called the *simple* one.

§ 2. The relation between the composite and the simple system of the theory of deduction

Among the theses obtained in § 1 we find those which, being taken as a set of axioms, give the complete simple system of the theory of deduction. For instance the theses td 28, td 42, and td 20 are equiform with the axioms of Łukasiewicz.[1]

The simple system of the theory of deduction having '*C*' and '*N*' as constant terms will be symbolized with capital letters TD. The axioms of Łukasiewicz are taken as the first theses:

TD 1   *CCpqCCqrCpr*
TD 2   *CCNppp*
TD 3   *CpCNpq*

For purposes of brevity the theses of the composite system will be called theses td, those of the simple system theses TD.

Now let Φ denote the property exhibited by those theses TD for which an equiform thesis td can be obtained. As we have seen above, the axioms TD 1, TD 2, and TD 3 have the property Φ. Hence for proving that all theses TD have the same property, it is sufficient to show that Φ is a hereditary property with

---

[1] Łukasiewicz, op. cit. *Elementy logiki matematycznej*, p. 45.

respect to both rules of the simple system. Suppose that α is thesis TD having the property Φ and $\alpha(p_1/\beta_1, p_2/\beta_2, ..., p_k/\beta_k)$ is the result of replacing the variables by propositions $\beta_1, \beta_2, ..., \beta_k$. We can see that a thesis equiform to $\alpha(p_1/\beta_1, p_2/\beta_2, ..., p_k/\beta_k)$ can be obtained in the composite system as the result of a proof analogous to that of the thesis td equiform to α. Only the following modifications have to be made: (1) instead of the variables p_1, $p_2, ..., p_k$, we must always write the corresponding propositions $\beta_1, \beta_2, ..., \beta_k$, and (2) some numbers in the prefixes must be changed. Thus $\alpha(p_1/\beta_1, p_2/\beta_2, ..., p_k/\beta)$ has the property Φ.

If now two theses TD having the forms '$C\alpha\beta$' and 'α' have the property Φ, the thesis TD of the form 'β' obtained from them in virtue of the rule of inference has also the property Φ: for, an equiform thesis td can be obtained in accordance with the rule III. Thus we see that no rule of the simple system can give us the first thesis TD which has not the property Φ. Hence follows the following theorem 1:

THEOREM 1. *Given any thesis* TD, *we can obtain a thesis* td *equiform with the former.*

It is obvious that such a theorem cannot be inverted, for some theses td, not being 'propositions', are meaningless in the simple system. Their intuitive meaning can, however, be interpreted by means of a proposition which will be called the development of the given thesis td. Suppose that we have the following theses containing suppositions:

$$n_1 \, . \, S\alpha_1$$
$$n_1 \, . \, n_2 \, . \, S\alpha_2$$
$$\cdot \quad \cdot \quad \cdot \quad \cdot \quad \cdot$$
$$n_1 \, . \, n_2 \, . \, ... \, n_k \, . \, S\alpha_k$$

where $k \geqslant 1$. Let β be an element of the domain of the last supposition α_k. It can happen that β is identical with α_k or not. In the first case the thesis containing β is written above, in the second case this thesis has the form

$$n_1 \, . \, n_2 \, . \, ... \, n_k \, . \, \beta$$

In both cases any expression of the form

$$C\alpha_1 \, C\alpha_2 \, ... \, C\alpha_k \, \beta$$

will be called the *development* of the thesis containing β. Thus we have defined the development of any thesis possessing a prefix. In the case of a thesis which is a proposition β having no prefix, the development proceeds by considering every proposition equiform with that thesis. In that case, the development can be represented by the above-given general schema by putting $k = 0$. Now we are able to explain what it means for a simple system and a composite one to be two formal systems of the same theory. It occurs then and only then, when (1) for every thesis of the simple system an equiform thesis can be obtained in the composite one and (2) for every thesis of the composite system we can obtain a development which is a thesis of the simple system. Two such systems will be called correspondent to each other. The two systems of theory of deduction TD and td are correspondent systems, as is shown by theorem 1 and the following theorem 2.

THEOREM 2. *Given any thesis* td, *we can obtain a thesis* TD *which is its development.*

It is known that, for any expression satisfying the method of verification through substitution of the values 1 and 0 for variables, we can obtain an equiform thesis TD. Hence for proving the theorem 2, it is sufficient to show that all developments of theses td satisfy this method of verification. Consider an arbitrary substitution, for variables, of the values: 1 standing for the truth, 0 standing for the falsehood. The development '$C\alpha_1 C\alpha_2 \ldots C\alpha_k \beta$' receives the value 1, if at least one of the propositions $\alpha_1, \alpha_2, \ldots, \alpha_k$ has the value 0 or if β has the value 1. Now it can be easily shown that the property of having 1 as value of the developments is hereditary with respect to all the rules of the composite system, whence it follows that all theses td have this property.

§ 3. Regarding some incomplete systems of the theory of deduction

The first question which we shall answer in the present chapter is: which is the simple system correspondent to that composite one in which no other rule than I, II, III holds, and in which 'C' is the only constant term of the propositions? We shall show that this simple system is the incomplete system known as the 'positive logic', based on Hilbert's[1] four axioms containing no negation. Here another set of axioms will be taken, namely the following two taken from among those of Frege,[2] since the equivalence between the two sets of axioms has been proved by Mr. Łukasiewicz.

PTD 1 $CpCqp$
PTD 2 $CCpCqrCCpqCpr$

In the composite system with which we shall now deal, the rules I, II, and III are valid, but in rule I we must give another meaning to the term 'proposition', for propositions cannot contain the letter 'N'. This system will be symbolized by 'ptd' as an abbreviation for the 'positive theory of deduction'.

ptd 1	1. Sp	I
ptd 2	1.1. Sq	I
ptd 3	1. Cqp	II 2, 1
ptd 4	$CpCqp$	II 1, 3
ptd 5	2. $SCpCqr$	I
ptd 6	2.1. $SCpq$	I
ptd 7	2.1.1. Sp	I
ptd 8	2.1.1. Cqr	III 5, 7
ptd 9	2.1.1. q	III 6, 7
ptd 10	2.1.1. r	III 8, 9
ptd 11	2.1. Cpr	II 7, 10
ptd 12	2. $CCpqCpr$	II 6, 11
ptd 13	$CCpCqrCCpqCpr$	II 5, 12

Having obtained the theses ptd 4 and ptd 13, equiform to the axioms PTD 1 and PTD 2, we can prove the following theorem by means of a demonstration analogous to that of the theorem 1.

THEOREM 3. *Given any thesis* PTD, *we can obtain an equiform thesis* ptd.

[1] D. Hilbert, 'Die logischen Grundlagen der Mathematik', *Math. Annalen* 88 (1922), p. 153.
[2] G. Frege, *Begriffsschrift* (Halle, 1879), p. 26.

We shall be able to state that the systems PTD and ptd are correspondent, if we prove the following theorem.

THEOREM 4. *Given any thesis* ptd, *it is possible to obtain a thesis* PTD *which is its development.*

The proof requires reference to some theses PTD which will be obtained for that purpose below. As in Mr. Łukasiewicz's works,[1] all the theses with exception of the axioms will be preceded by the proof lines which will indicate the applications of the rules. The numbers in those lines are the numbers of theses PTD.

$$2\ q/Cqp,\ r/p\ *\ C1\ q/Cqp—C1—3$$

PTD 3 Cpp

$$1\ p/CCpCqrCCpqCpr,\ q/s\ *\ C2—4$$

PTD 4 $CsCCpCqrCCpqCpr$

$$2\ p/Cpq,\ q/CrCpq,\ r/CCrpCrq\ *\ C4\ p/r,\ q/p,$$
$$r/q,\ s/Cpq—C1\ p/Cpq,\ q/r—5$$

PTD 5 $CCpqCCrpCrq$

$$5\ q/Cqp\ *\ C1—6$$

PTD 6 $CCrpCrCqp$

$$5\ p/CpCqr,\ q/CCpqCpr,\ r/s\ *\ C2—7$$

PTD 7 $CCsCpCqrCsCCpqCpr$

$$5\ p/CCspCsCqr,\ q/CCspCCsqCsr,\ r/CpCqr$$
$$*\ C7\ s/Csp,\ p/s—C5\ q/Cqr,\ r/s—8$$

PTD 8 $CCpCqrCCspCCsqCsr$

$$1\ p/Cpp\ *\ C3—9$$

PTD 9 $CqCpp$

$$1\ p/\ CqCpp,\ q/r\ *\ C9—10$$

PTD 10 $CrCqCpp$

By repeating the above process, we can obtain any thesis having the form:

$$I_k \qquad\qquad Cp_k\,Cp_{k-1}\dots Cp_1 p_1 \qquad\qquad\qquad (k \geqslant 1)$$

$$6\ r/p,\ p/Cqp,\ q/r\ *\ C1—11$$

PTD 11 $CpCrCqp$

$$6\ r/p,\ p/CrCqp,\ q/s\ *\ C11—12$$

PTD 12 $CpCsCrCqp$

[1] J. Łukasiewicz, op. cit.

Proceeding in an analogous way, we can prove any proposition

$$\text{II}_{k,0} \qquad CpCq_k\,Cq_{k-1}\ldots Cq_1p \qquad\qquad (k \geqslant 1)$$

and then, by means of repeated use of the syllogism PTD 5, we can prove each thesis:

$$\text{II}_{k,l} \qquad CCr_l\,Cr_{l-1}\ldots Cr_1pCr_l\,Cr_{l-1}\ldots Cr_1\,Cq_k\,Cq_{k-1}\ldots$$
$$Cq_1p \qquad\qquad (k \geqslant 1,\, l \geqslant 0)$$

as shown by the examples below.

$$5\;q/CsCrCqp,\,r/t \ast C12\!-\!13$$
$$\text{PTD 13} \qquad CCtpCtCsCrCqp$$
$$5\;p/Ctp,\,q/CtCsCrCqp,\,r/u \ast C13\!-\!14$$
$$\text{PTD 14} \qquad CCuCtpCuCtCsCrCqp$$

The proof of the third scheme containing the needed theses can be illustrated as below:

$$8\;p/CrCpq,\,q/Crp,\,r/Crq \ast C2\;p/r,\,q/p,\,r/q\!-\!15$$
$$\text{PTD 15} \qquad CCsCrCpqCCsCrpCsCrq$$
$$8\;p/CsCrCpq,\,q/CsCrp,\,r/CsCrq,\,s/t \ast C15\!-\!16$$
$$\text{PTD 16} \qquad CCtCsCrCpqCCtCsCrpCtCsCrq$$

The third scheme has the form:

$$\text{III}_k \qquad CCr_k\,Cr_{k-1}\ldots Cr_1\,CpqCCr_k\,Cr_{k-1}\ldots Cr_1pCr_k$$
$$Cr_{k-1}\ldots Cr_1q \qquad\qquad (k \geqslant 1)$$

If $k = 1$, this scheme represents PTD 2 with different variables only.

Consider now the following property Ψ: a thesis ptd is said to have the property Ψ, if and only if we can obtain a thesis PTD which is its development. Suppose that theses ptd obtained up to a given moment have the property Ψ and that ζ is the next thesis ptd subjoined to the system. In all such cases ζ has the property Ψ, as we show below.

1st case. Let ζ be a thesis obtained in virtue of the rule I. Its development then has the form

$$C\alpha_1\,C\alpha_2\ldots C\alpha_k\,\alpha_k$$

and can be obtained in the simple system by means of a substitution in the thesis I_k.

2nd case. Let ζ be a thesis which has been subjoined in virtue of the rule II. The application of that rule is possible only when some proposition β is valid in the domain of a supposition α. The development of the thesis ζ must have the form

$$(*) \qquad C\alpha_1\, C\alpha_2 \ldots C\alpha_m \beta$$

where $\alpha_1, \alpha_2, \ldots, \alpha_m$ stand for suppositions valid in the domain of α_m. The proposition β, being valid in the domain of α, must be an element (a) of the absolute domain or (b) of the domain of the supposition α_m or (c) of the domain of one of the suppositions $\alpha_1, \alpha_2, \ldots, \alpha_{m-1}$. In the case (b) this development will be identical with (*), in the case (a) it will have the form of β, and in the case (c) the form

$$C\alpha_1\, C\alpha_2 \ldots C\alpha_l \beta$$

where l satisfies: $1 \leqslant l < m$. In any case (*) follows from that development: in accordance with the thesis $\mathrm{II}_{m,0}$ in the case (a) and in accordance with the thesis $\mathrm{II}_{m-l,l}$ in the case (c).

3rd case. Let ζ be a thesis ptd obtained in virtue of the rule III. If its propositional part is β, its development must be

$$\beta$$

or

$$(**) \qquad C\alpha_1\, C\alpha_2 \ldots C\alpha_k \beta$$

If β is the development of ζ, two premisses having the form of '$C\gamma\beta$' 'γ' must be valid in the absolute domain and, being their own developments, they must be equiform with some theses PTD from which we can obtain a thesis PTD which is equiform with β and therefore a development of ζ. If now (**) is the development of ζ, β is an element of the domain of the supposition α_k and the premisses '$C\gamma\beta$' and 'γ' are valid in that domain. Thus by means of the rule II we are able to obtain theses having as developments the following propositions:

$$C\alpha_1\, C\alpha_2 \ldots C\alpha_k\, C\gamma\beta$$
$$C\alpha_1\, C\alpha_2 \ldots C\alpha_k \gamma$$

As has been shown above it is possible to obtain such developments as theses PTD. Hence (**) follows by help of the thesis III_k.

Thus we see that ζ has in all cases the property Ψ' and we can never obtain the *first* thesis ptd not having the property Ψ'. Hence follows that all the theses ptd have the property Ψ', and the theorem 4 is proved.

It remains to analyse the first of the two forms of the *reductio ad absurdum* which were mentioned in § 1. We shall consider a system in which two terms '*C*' and '*N*' are constants. Its rules are: I, II, III, and a rule IVa formulated below.

Rule IVa. *When in the domain D of a supposition α two propositions β and γ are valid and γ has the form 'Nβ', it is allowed to subjoin a proposition of the form 'Nα' to that domain in respect to which the domain D is an immediate subdomain.*

This system (it may be referred to as 'itd') corresponds to the simple system constructed by Kolmogoroff[1] for the purpose of embodying the laws of the intuitionist logic of Brouwer. As axioms we shall take: the two axioms of the positive logic which are equivalent to the four axioms of Hilbert employed by Kolmogoroff:

ITD 1   *CpCqp*

ITD 2   *CCpCqrCCpqCpr*

and the axiom subjoined by Kolmogoroff to those of the positive logic:

ITD 3   *CCpqCCpNqNp*

The system ITD is not only incomplete but does not even contain some theses belonging to the system built by Heyting for the same purpose.[2] One of Heyting's axioms,

$$CNpCpq$$

cannot be obtained as a thesis ITD, as can be shown by the Łukasiewicz–Bernays method with help of the following matrix:

C	0	1	N
0	1	1	1
1	0	1	1

---

[1] A. N. Kolmogoroff, 'O principie tertium non datur', *Matematičeski Sbornik*, 32, p. 651.

[2] A. Heyting, 'Die formalen Regeln der intuitionistischen Logik', *Sitzungsber. d. Preuss. Ak. d. Wiss. 1930. Phys.-Math. Kl.*, p. 45 ff.

THEOREM 5. *Given any thesis* ITD, *it is possible to obtain a thesis* itd *equiform with it.*

Such a theorem can be proved as was done for the theorems 1 and 3. Theses itd equiform to the axioms ITD 1 and ITD 2 can be obtained on the same way as in the system ptd. The proof of a thesis equiform to ITD 3 is given below.

itd 1	$1. SCpq$	I
itd 2	$1.1. SCpNq$	I
itd 3	$1.1.1. Sp$	I
itd 4	$1.1.1. q$	III 1, 3
itd 5	$1.1.1. Nq$	III 2, 3
itd 6	$1.1. Np$	IVa 3, 4, 5
itd 7	$1. CCpNqNp$	II 2, 6
itd 8	$CCpqCCpNqNp$	II 1, 7

THEOREM 6. *Given any thesis* itd, *it is possible to obtain a thesis* ITD *which is its development.*

The demonstration is analogous to that of theorem 4. All theses PTD used in it have their equiform theses ITD, because all axioms of the system PTD are axioms ITD. Thus the schemes of theses I_k, $II_{k,l}$ and III_k can be applied in order to show that none of the rules I, II, III can lead to the *first* among theses itd which cannot have any thesis ITD as development. It remains to prove that the same property holds for the rule IVa. Let

$$C\alpha_1 C\alpha_2 \ldots C\alpha_k \alpha_k$$

be the development of the thesis containing the supposition α_k which has led us to the contradiction 'β' and '$N\beta$'. Since those propositions are valid in the domain of α_k, we can prove, as was done in the demonstration of the theorem 4, that the following theses ITD can be obtained

(***) $$C\alpha_1 C\alpha_2 \ldots C\alpha_k \beta$$

and

(****) $$C\alpha_1 C\alpha_2 \ldots C\alpha_k N\beta$$

If $k = 1$, the thesis just now subjoined in virtue of the rule IVa has the form

$$C\alpha_1 C\alpha_2 \ldots C\alpha_{k-1} N\alpha_k$$

and its development, which has the same form, can be obtained as a thesis ITD by means of ITD 3. If $k > 1$, the development of the subjoined thesis can be obtained from (***) and (****) by means of the thesis

$$\text{IV}_k \qquad CCp_k\, Cp_{k-1} \ldots Cp_1\, qCCp_k\, Cp_{k-1} \ldots Cp_1\, NqCp_k\, Cp_{k-1} \ldots$$
$$\ldots Cp_2 Np_1$$

This scheme becomes ITD 3, when $k = 1$. As to the theses represented by this scheme for other values of k, they can be obtained from ITD 3 by use of the thesis PTD 8.

§ 4. The extended theory of deduction

The following question arises: By what means is it possible to transform the more complicated theories into systems in which our rules may hold? In those theories, beyond the rules which replace the theory of deduction, some others are required; they are those concerning the apparent variable. As to them, in the present and the next chapters we attempt to transform them into structural rules adapted to the symbolism of composite systems. For the present, we shall take as a basis the extended theory of deduction.[1] Besides all the symbols met with till now, that theory contains one more—the general quantifier 'Π' which appears in the connexions 'Πpα', 'Πqα', etc., α being a proposition which in this case may or may not contain a variable of the form 'p', 'q', etc. 'Πpα' may be read: 'α, whatever a proposition p may be' or in short: 'for every p, α (holds)', because, in the system in question, we deal with no other variables than propositional ones. We shall present examples of reasoning with the use of 'Π', leaving the symbolism of the composite systems unchanged.

$$\text{etd 1} \qquad 1.S\Pi qCqp \qquad\qquad\qquad \text{I}$$

The above supposition means: 'Cqp, whatever proposition q may be'. Hence follows '$CCppp$', because 'Cpp' is a proposition. Thus we may write

$$\text{etd 2} \qquad 1.CCppp$$

---

[1] J. Łukasiewicz, op. cit. *Elementy*, § 8, Łukasiewicz and A. Tarski, op. cit., § 5.

In fact, the rule of substitution which we shall formulate will enable us to accept the above as a thesis. In accordance to the rules of suppositions, we can deduce:

etd 3	$2.Sp$	I
etd 4	Cpp	II 3, 3
etd 5	$1.p$	III 2, 4
etd 6	$C\Pi qCqpp$	II 1, 5
etd 7	$2.1.Sq$	I
etd 8	$2.Cqp$	II 7, 3

'Cqp' is valid in a domain in which no supposition concerning 'q' is valid. In that thesis, 'q' is a quite arbitrary proposition and all the results we have obtained concerning it could be accepted for any other proposition. That is the intuitive reason for which the rule VI will allow us to write

etd 9	$2.\Pi qCqp$

Hence

etd 10	$Cp\Pi qCqp$	II 3, 9

The rule of substitution in the system etd will be formulated as follows.

Rule V. *To any domain D in which a proposition α is valid, where α is composed successively:*

(1) *of a general quantifier 'Π',*
(2) *of a variable ζ and*
(3) *of a proposition β,*

it is allowed to subjoin a proposition γ obtained from β by means of substitution for variables bound to ζ.

The above condition concerning γ must be explained by the following detailed description:

As to their form, β and γ differ only in this, that instead of all those variables of β which are bound to ζ, γ contains propositions

(1) *equiform with one another and*
(2) *having the following property: all real variables of those propositions are real variables of γ.*

We must still explain some expressions in the above by means of structural description. A variable ζ is said to be a *real variable* of a proposition α then and only then, when it does not belong to any significant part[1] of α beginning with a quantifier 'Π' and a variable equiform with ζ.

A variable η is *bound* to a variable ζ then and only then, when they are equiform with each other and they belong to a proposition composed successively:

(1) of a quantifier 'Π',
(2) of the variable ζ,
(3) of a proposition in which η is a real variable.

Rule VI. *If a proposition α is valid in the domain D, it is allowed to subjoin a proposition of the form 'Π$\zeta\alpha$' to the domain D, provided that ζ is a variable not equiform with any real variable of a supposition valid in D.*

The rules I, II, III, IV remain in force, but as to I the notion of the 'proposition' is altered. With help of all our rules we can obtain the following theses

etd 11	$2.2.SCpq$	I
etd 12	$2.2.q$	III 11, 3
etd 13	$2.CCpqq$	II 11, 12
etd 14	$CpCCpqq$	II 3, 13
etd 15	$\Pi qCpCCpqq$	VI 14
etd 16	$\Pi p\Pi qCpCCpqq$	VI 15
etd 17	$\Pi pCpp$	VI 4

Analogously we can obtain all theses of the usual theory of deduction where each is preceded by quantifiers. We shall now present a process which will illustrate how it is possible to apply definitions in the composite system.

etd 18	$3.SCu\Pi pp$	I
etd 19	$3.1.SC\Pi ppu$	I
etd 20	$3.1.1.SCtNu$	I
etd 21	$3.1.1.1.SCNut$	I

---

[1] The 'part' is understood here in such a way that it can be identical with the whole.

The above four suppositions can be considered as the definitions of two constant terms: 'u' being equivalent to 'Πpp' and 't' equivalent to 'Nu'. The domain with the prefix '3.1.1.1.' gives us the enlarged theory of deduction with those constants.

etd 22	$3.1.1.1.1.Su$	I
etd 23	$3.1.1.1.1.\Pi pp$	III 18, 22
etd 24	$3.1.1.1.1.p$	V 23
etd 25	$3.1.1.1.Cup$	II 22, 24
etd 26	$3.1.1.1.\Pi pCup$	VI 25

The false proposition 'u' implies anything.

etd 27	$3.1.1.1.2.Sp$	I
etd 28	$3.1.1.1.2.1.SNt$	I
etd 29	$3.1.1.1.2.1.1.SNu$	I
etd 30	$3.1.1.1.2.1.1.t$	III 21, 29
etd 31	$3.1.1.1.2.1.u$	IV 29, 30, 28
etd 32	$3.1.1.1.2.1.\Pi pp$	III 18, 31
etd 33	$3.1.1.1.2.1.Np$	V 32
etd 34	$3.1.1.1.2.t$	IV 28, 27, 33
etd 35	$3.1.1.1.Cpt$	II 27, 34
etd 36	$3.1.1.1.\Pi pCpt$	VI 35

The true proposition 't' is implied by anything.

etd 37	$3.1.1.CCNut\Pi pCpt$	II 21, 36
etd 38	$3.1.CCtNuCCNut\Pi pCpt$	II 20, 37
etd 39	$3.1.\Pi tCCtNuCCNut\Pi pCpt$	VI 38
etd 40	$3.1.CCNuNuCCNuNu\Pi pCpNu$	V 39
etd 41	$3.1.CNuNu$	V 17
etd 42	$3.1.CCNuNu\Pi pCpNu$	III 40, 41
etd 43	$3.1.\Pi pCpNu$	III 42, 41

The above process, leading us from etd 36 to etd 43, shows how it is possible to carry a proposition containing a defined term out of the domain of the definition by replacing at the same time the *definiendum* 't' by the *definiens* 'Nu'. With the same process, we can transport the proposition '$\Pi pCpNu$' to the absolute domain by replacing 'u' by 'Πpp':

| etd 44 | $3.CC\Pi ppu\Pi pCpNu$ | II 19, 43 |

etd 45	$CCu\Pi pp CC\Pi ppu\Pi pCpNu$	II 18, 44
etd 46	$\Pi uCCu\Pi pp CC\Pi ppu\Pi pCpNu$	VI 45
etd 47	$CC\Pi pp\Pi pp CC\Pi pp\Pi pp\Pi pCpN\Pi pp$	V 46
etd 48	$C\Pi pp\Pi pp$	V 17
etd 49	$CC\Pi pp\Pi pp\Pi pCpN\Pi pp$	III 47, 48
etd 50	$\Pi pCpN\Pi pp$	III 49, 48

The system etd corresponds to the simple system of the enlarged theory of deduction. The needed proofs are analogous to those of the theorems 1 and 2.

§ 5. Application to the calculus of functions

We shall now consider a theory analogous to the theory of apparent variables of *Principia Mathematica* and to Hilbert's calculus of functions. Contrary to the preceding chapter, we shall deal now with the individual apparent variable rather than with the propositional one. The individual variables will be symbolized by small Latin letters 'x', 'y', etc., and will be the arguments of functions having propositions as values. The symbols of those functions are the small Greek letters 'ϕ', 'ψ', etc. The arguments will follow those letters; brackets are superfluous.[1] The quantifiers 'Π' ought to be followed by an individual variable and a proposition, e.g. as in '$\Pi x\phi x$' or '$\Pi xCpq$', which are to be read: 'for every (individual) x, ϕx holds' and 'for every (individual) x, Cpq holds'.

The rules I, II, III, IV, V, and VI, when adapted to such a theory, would give a system correspondent to some simple system differing from those of *Principia Mathematica* and of Hilbert only as a consequence of the fact that the notion of significance is different. In such a system, we should be able to have a thesis of the following form:

$$C\Pi x\phi xN\Pi xN\phi x$$

The intuitive meaning of '$N\Pi xN\phi x$' is: 'for some x, ϕx holds'. The above thesis therefore means: 'If for every x, ϕx, then for some x, ϕx.' In the null field of individuals (*Individuenbereich*),

---

[1] Brackets are omitted in such functions by J. Herbrand. See his 'Recherches sur la théorie de démonstration', *Comptes rendus des séances de la Société des Sciences et des Lettres de Varsovie*, 23, 1930.

i.e. under the supposition that no individual exists in the world, this proposition is false. Thus the system states the existence of at least one individual. But whether individuals exist or not, it is better to solve this problem through other theories. We shall present therefore a system of the calculus of functions, where all the theses will be satisfied in the null field of individuals.

In that system, we must avoid any thesis which is a proposition with real individual variables, for such theses lead us to assume others requiring the existence of individuals. As to the notion of real variables in the composite system, the circumstances are quite different from those of the simple system. Symbols of variables which are not apparent variables do not merit the name of variable at all. We deal with such a term as with a given constant, though it is neither a primitive term nor a defined one. It is a constant the meaning of which, although undefined, remains unaltered through the whole process of reasoning. In practice, we often introduce such undefined constants in the course of a proof. For example we say: 'Consider an arbitrary x', and then we deduce propositions which can be said to belong to the *scope of constancy* of the symbol 'x'. This process of reasoning will be applied in our system. We shall give an example of such a reasoning in the calculus of functions.

Suppose $\Pi x \Pi y \phi x y$. *Consider* now an arbitrary individual z. According to our supposition, '$\Pi y \phi x y$' holds whatever individual x may be, hence $\Pi y \phi z y$ holds too. Repeating the same process in respect to y, we obtain $\phi z z$. We have obtained this result for an arbitrary z, therefore our result must be 'for every z, $\phi z z$ holds', i.e. $\Pi z \phi z z$. This is a consequence of the supposition '$\Pi x \Pi y \phi x y$', so we can take as a thesis

$$C \Pi x \Pi y \phi x y \Pi z \phi z z$$

We shall repeat the same proof in symbols.

cf 1   1. $S \Pi x \Pi y \phi x y$

Now we shall write the expression 'Consider an arbitrary z'

cf 2   1.1. $T z$

'T' is here a new constant analogous to the symbol of supposition 'S'.

The arbitrary constant 'z' will be called the *term* and the scope of its constancy the *domain of that term*. Further steps of our proof can be easily formalized.

cf 3 $1.1.\Pi y\phi zy$
cf 4 $1.1.\phi zz$
cf 5 $1.\Pi z\phi zz$
cf 6 $C\Pi x\Pi y\phi xy\Pi z\phi zz$

We shall make use of the abbreviated modes of speaking which were introduced in § 1 in connexion with suppositions and their domains, adapting them now to the terms and their domains. Thus in any domain D, we shall consider as *valid* any term whose prefix is equiform to an initial part of the prefix of the thesis containing the supposition or the term belonging to D. The new rules will be formulated below.

Rule Va. *It is allowed, to any domain D in which*

(1) *a term ζ and*
(2) *a proposition composed of a quantifier 'Π', of a variable η and of a proposition α*

are valid, to subjoin a proposition which differs, as to its form, from α only in this respect, that all variables bound to η are replaced by symbols equiform with ζ no one of which is an apparent variable.

Rule VIa. *If in the domain D of a term ζ a proposition α is valid, it is allowed to subjoin a proposition of the form '$\Pi\zeta a$' to that domain whereof the domain D is an immediate subdomain.*

Rule VII. *Given a domain D, it is allowed to subjoin to it any expression composed successively:*

(1) *of a number not equiform with the initial number of any element of the domain D,*
(2) *of a dot,*
(3) *of the symbol 'T' and*
(4) *of a term not equiform with any term valid in the domain D.*

The rules formulated in § 1 remain unaltered for the system cf with the exception of rule I which must be transformed into rule Ia.

Rule Ia. *Given a domain D, it is allowed to subjoin to it any expression composed successively:*

(1) *of a number not equiform with the initial number of any element of the domain D,*

(2) *of a dot,*

(3) *of the symbol 'S',*

(4) *of a proposition significant in the domain D.*

We regard a proposition α as *significant in the domain D,* if every real variable of α is equiform with some term valid in D.

We shall give some further examples of theses of the system cf.

cf 7	$1.1.1.Tv$	VII		
cf 8	$1.1.1.\phi zv$	Va	3, 7	
cf 9	$1.1.\Pi v\phi zv$	VIa	7, 8	
cf 10	$1.\Pi z\Pi v\phi zv$	VIa	2, 9	
cf 11	$1.2.Tx$	VII		
cf 12	$1.2.1.Ty$	VII		
cf 13	$1.2.1.\Pi v\phi yv$	Va	10, 12	
cf 14	$1.2.1.\phi yx$	Va	13, 11	
cf 15	$1.2.\Pi y\phi yx$	VIa	12, 14	
cf 16	$1.\Pi x\Pi y\phi yx$	VIa	11, 15	
cf 17	$C\Pi x\Pi y\phi xy\Pi x\Pi y\phi yx$	II	1, 16	
cf 18	$2.S\Pi xC\phi x\psi x$	Ia		
cf 19	$2.1.S\Pi x\phi x$	Ia		
cf 20	$2.1.1.Tx$	VII		
cf 21	$2.1.1.C\phi x\psi x$	Va	18, 20	
cf 22	$2.1.1.\phi x$	Va	19, 20	
cf 23	$2.1.1.\psi x$	III	21, 22	
cf 24	$2.1.\Pi x\psi x$	VIa	20, 23	
cf 25	$2.C\Pi x\phi x\Pi x\psi x$	II	19, 24	
cf 26	$C\Pi xC\phi x\psi xC\Pi x\phi x\Pi x\psi x$	II	18, 25	
cf 27	$3.Sp$	Ia		
cf 28	$3.1.Tx$	VII		
cf 29	$3.\Pi xp$	VIa	28, 27	
cf 30	$Cp\Pi xp$	II	27, 29	

The rules of the composite systems can be applied to different logical or mathematical systems. In such cases, it can happen that new rules may be required. For instance, if we want to build the composite system of the theory of deduction having besides '*C*' and '*N*' the new constant term of *conjunction* (logical product), it is sufficient to give three new rules. The first would permit us to subjoin to a domain a conjunction composed of propositions equiform with some propositions valid in that domain, and the others would allow us to subjoin a proposition equiform with the first and a proposition equiform with the second member of a valid conjunction.

By building composite systems containing variables of different kinds from those already taken into consideration, we must suitably adapt the rule of substitution and either the rule VI or the two rules VIa and VII to the new variables.

As to the application in mathematical theories, we can expect that the composite systems of logic will be more suited to the purposes of formalizing practical proofs, than are the simple ones. The use of the theses of the theory of deduction for that purpose is so burdensome that it is avoided even by the authors of logical systems. In more complicated theories the use of theses would be completely unproductive.

12

INVESTIGATIONS INTO THE SYSTEM OF INTUITIONIST LOGIC†

STANISŁAW JAŚKOWSKI

1. LET $A_{\mathfrak{M}}$ and $B_{\mathfrak{M}}$ be two sets without common elements, and let $X \supset_{\mathfrak{M}} Y$, $X \wedge_{\mathfrak{M}} Y$, $X \vee_{\mathfrak{M}} Y$, and $\sim_{\mathfrak{M}} X$ be four functions defined for all elements of the set $A_{\mathfrak{M}} + B_{\mathfrak{M}}$, the values of these functions belonging to $A_{\mathfrak{M}} + B_{\mathfrak{M}}$. We say then, following the terminology of Łukasiewicz and Tarski,[1] that a matrix $\mathfrak{M}$ has been defined. Let $\mathfrak{A}$ be an expression of the propositional calculus. Replace in $\mathfrak{A}$ the functional symbols by $\supset_{\mathfrak{M}}$, $\wedge_{\mathfrak{M}}$, $\vee_{\mathfrak{M}}$, $\sim_{\mathfrak{M}}$ respectively, and put, in place of the propositional variables $p, q, ...$, elements of $A_{\mathfrak{M}} + B_{\mathfrak{M}}$ denoted respectively by $h(p)$, $h(q), ...$. We obtain an expression having as its value an element of $A_{\mathfrak{M}} + B_{\mathfrak{M}}$, which element we designate by $h(\mathfrak{A})$. We call $h(\mathfrak{X})$ a value function (*Wertfunktion*); to define it, it suffices to establish its values for propositional variables $\mathfrak{X}$. An expression $\mathfrak{A}$ is satisfied by the matrix $\mathfrak{M}$, in symbols $\mathfrak{A} \in E(\mathfrak{M})$, if, with respect to the matrix $\mathfrak{M}$, the value of $\mathfrak{A}$ belongs to $B_{\mathfrak{M}}$—that is, $h(\mathfrak{A}) \in B_{\mathfrak{M}}$—for every value function h. The elements of $B_{\mathfrak{M}}$ are called designated elements.

We denote by $\mathfrak{L}_1$ the matrix whose set $A_{\mathfrak{L}_1}$ is empty, and whose set $B_{\mathfrak{L}_1}$ is composed of a single element. Every expression of the propositional calculus is satisfied by this matrix.

2. Let $\mathfrak{M}$ and $\mathfrak{N}$ be two matrices having the same element b

---

† This paper appeared originally under the title 'Recherches sur le système de la logique intuitioniste' in *Actes du Congrès International de Philosophie Scientifique* 6 (Paris, 1936), pp. 58–61. Translated by S. McCall, with one or two minor changes communicated by the author.

[1] J. Łukasiewicz and A. Tarski, 'Untersuchungen über den Aussagenkalkül', *Comptes rendus des séances de la Société des Sciences et des Lettres de Varsovie*, Cl. iii, vol. 23, 1930. [English translation in Tarski, *Logic, Semantics, Metamathematics*, Oxford, 1956—Ed.]

for their sole designated element; $B_{\mathfrak{M}} = B_{\mathfrak{N}} = \{b\}$. The set $A_{\mathfrak{N}}$ is composed of the elements of $A_{\mathfrak{M}}$ and of one additional element a; $A_{\mathfrak{N}} = A_{\mathfrak{M}} + \{a\}$. If α is a function defined by the two following conditions:

(i) $\alpha(b) = a$,
(ii) if $X \in A_{\mathfrak{M}}$, $\alpha(X) = X$,

the functions of the matrix $\mathfrak{N}$ are defined in terms of those of $\mathfrak{M}$ by means of the following tables:

	function	second argument	
	$\supset_{\mathfrak{N}}$	b	$\alpha(y)$
	b	$b \supset_{\mathfrak{M}} b$	$\alpha(b \supset_{\mathfrak{M}} y)$
first argument	$\alpha(x)$	$x \supset_{\mathfrak{M}} b$	$x \supset_{\mathfrak{M}} y$

$\wedge_{\mathfrak{N}}$	b	$\alpha(y)$
b	$b \wedge_{\mathfrak{M}} b$	$\alpha(b \wedge_{\mathfrak{M}} y)$
$\alpha(x)$	$\alpha(x \wedge_{\mathfrak{M}} b)$	$\alpha(x \wedge_{\mathfrak{M}} y)$

$\vee_{\mathfrak{N}}$	b	$\alpha(y)$
b	$b \vee_{\mathfrak{M}} b$	$b \vee_{\mathfrak{M}} y$
$\alpha(x)$	$x \vee_{\mathfrak{M}} b$	$\alpha(x \vee_{\mathfrak{M}} y)$

$\sim_{\mathfrak{N}}$	
b	$\alpha(\sim_{\mathfrak{M}} b)$
$\alpha(x)$	$\sim_{\mathfrak{M}} x$

I shall say that under these conditions $\mathfrak{N}$ is the result of the operation Γ performed upon the matrix $\mathfrak{M}$:

$$\mathfrak{N} \text{ is } \Gamma(\mathfrak{M}).$$

We have $E(\Gamma(\mathfrak{M})) \subset E(\mathfrak{M})$. On the other hand it is possible that an expression satisfied by $\mathfrak{M}$ is not satisfied by $\Gamma(\mathfrak{M})$. The operation Γ applied to $\mathfrak{L}_1$ gives the matrix of the usual two-valued logic: $\mathfrak{L}_2$ is $\Gamma(\mathfrak{L}_1)$.

3. Let $\mathfrak{M}$ and $\mathfrak{N}$ be two matrices. I shall say that the matrix $\mathfrak{P}$ is the product of $\mathfrak{M}$ and $\mathfrak{N}$,

$$\mathfrak{P} = \mathfrak{M} \times \mathfrak{N}$$

when:

(i) $A_{\mathfrak{P}} + B_{\mathfrak{P}}$ is the set of all ordered pairs $\{m, n\}$ such that

$$m \in A_{\mathfrak{M}} + B_{\mathfrak{M}}, \qquad n \in A_{\mathfrak{N}} + B_{\mathfrak{N}}$$

(ii) the functions $\supset_{\mathfrak{P}}$, $\wedge_{\mathfrak{P}}$, $\vee_{\mathfrak{P}}$, and $\sim_{\mathfrak{P}}$ are defined as follows:

$$\{m_1, n_1\} \supset_{\mathfrak{P}} \{m_2, n_2\} = \{m_1 \supset_{\mathfrak{M}} m_2, n_1 \supset_{\mathfrak{N}} n_2\}$$
$$\{m_1, n_1\} \wedge_{\mathfrak{P}} \{m_2, n_2\} = \{m_1 \wedge_{\mathfrak{M}} m_2, n_1 \wedge_{\mathfrak{N}} n_2\}$$
$$\{m_1, n_1\} \vee_{\mathfrak{P}} \{m_2, n_2\} = \{m_1 \vee_{\mathfrak{M}} m_2, n_1 \vee_{\mathfrak{N}} n_2\}$$
$$\sim_{\mathfrak{P}} \{m_1, n_1\} = \{\sim_{\mathfrak{M}} m_1, \sim_{\mathfrak{N}} n_1\}$$

and (iii) $\{m, n\} \in B_{\mathfrak{P}}$ if and only if $m \in B_{\mathfrak{M}}$ and $n \in B_{\mathfrak{N}}$.
It may be easily shown that

$$E(\mathfrak{M} \times \mathfrak{N}) = E(\mathfrak{M}) \cap E(\mathfrak{N}).$$

The system $\mathfrak{H}$ of Heyting's calculus of propositions[1] being satisfied by a matrix $\mathfrak{M}$, it is satisfied by $\Gamma(\mathfrak{M})$. Hence $\mathfrak{H}$ is satisfied by every matrix which may be represented as the result of a finite (or zero) number of multiplications and of operations Γ performed upon the matrix $\mathfrak{L}_1$. These matrices are finite. Let us denote the family of these matrices by $(\mathfrak{L}_1)_{\Pi\Gamma}$, and the set of expressions which are satisfied by them all by $\mathfrak{E}((\mathfrak{L}_1)_{\Pi\Gamma})$. We have $\mathfrak{H} \subset \mathfrak{E}((\mathfrak{L}_1)_{\Pi\Gamma})$.

4. $\mathfrak{A}$ and $\mathfrak{B}$ are said to be expressions equivalent with respect to the system $\mathfrak{H}$ if it suffices to add one of these expressions to the axioms of $\mathfrak{H}$, i.e. make it a supplementary axiom, in order to be able to deduce the other. Hence (i) $\mathfrak{A} \in \mathfrak{H}$ if and only if $\mathfrak{B} \in \mathfrak{H}$; and (ii) $\mathfrak{A} \in E(\mathfrak{M})$ if and only if $\mathfrak{B} \in E(\mathfrak{M})$, provided that $\mathfrak{H} \subset E(\mathfrak{M})$ and that $\mathfrak{M}$ is a normal matrix.

I shall say that an expression $\mathfrak{R}$ is regular if it can be put into the form

$$(\mathrm{I}) \qquad (\mathfrak{X}_1 \wedge \mathfrak{X}_2 \wedge \ldots \mathfrak{X}_n) \supset \mathfrak{Y} \quad (n = 1, \text{ or } 2, \text{ or } 3, \ldots)$$

where $\mathfrak{Y}$ is a variable, no $\mathfrak{X}_i$ contains more than two constant signs, and each $\mathfrak{X}_i$ is an implication or the negation of a single

---

[1] A. Heyting, 'Die formalen Regeln der intuitionistischen Logik', *Sitzungsberichte der Preussischen Akademie der Wissenschaften*, 1930, pp. 42–56.

variable. Let $k = d(\Re)$ be the number of these implications $\mathfrak{X}_i$ whose antecedent contains the symbol '$\supset$' or '$\sim$'. It can be shown that any arbitrary expression $\mathfrak{A}$ of the propositional calculus is equivalent with respect to $\mathfrak{H}$ either to a regular expression $\Re$ or to an expression $\Re_1 \wedge \Re_2 \wedge ... \Re_m$, composed of regular expressions joined by the symbol '$\wedge$'. I shall say that $\mathfrak{A}$ belongs to the class $\mathfrak{C}_k$, k being the maximum of $d(\Re_1)$, $d(\Re_2)$, ..., $d(\Re_m)$. We have $\mathfrak{C}_k \subset \mathfrak{C}_{k+1}$. Let us denote by $\mathfrak{J}_0, \mathfrak{J}_1, \mathfrak{J}_2,...$ certain matrices of the family $(\mathfrak{L}_1)_{\Pi\Gamma}$: $\mathfrak{J}_0 = \mathfrak{L}_2$, $\mathfrak{J}_{k+1} = \Gamma((\mathfrak{J}_k)^{k+1})$ where $(\mathfrak{J}_k)^{k+1}$ represents the product of $k+1$ matrices $\mathfrak{J}_k$.

Lemma. If $\mathfrak{A} \in \mathfrak{C}_k$, we have either $\mathfrak{A} \in \mathfrak{H}$ or $\mathfrak{A} \notin E(\mathfrak{J}_k)$.

To prove this lemma, we may confine ourselves to regular expressions $\mathfrak{A} = \Re$. If $\Re \in \mathfrak{C}_0$, then $\Re \in \mathfrak{H}$ or $\Re \notin E(\mathfrak{J}_0)$. If $\Re \in \mathfrak{C}_k$ where $k > 0$, let $\mathfrak{B}_1 \supset \mathfrak{C}_1$, $\mathfrak{B}_2 \supset \mathfrak{C}_2$,..., $\mathfrak{B}_k \supset \mathfrak{C}_k$ be those $\mathfrak{X}_i$ whose antecedents $\mathfrak{B}_1$, $\mathfrak{B}_2$,..., $\mathfrak{B}_k$ contain the symbols '$\supset$' or '$\sim$'. Consider the k expressions

$$(\text{II})_i \qquad (\mathfrak{X}_1 \wedge \mathfrak{X}_2 \wedge ... \mathfrak{X}_n) \supset \mathfrak{B}_i \qquad\qquad (i = 1, 2,..., k.)$$

It can be shown that

  (i) either $(\text{II})_i \in \mathfrak{C}_{k-1}$ or $(\text{II})_i \in \mathfrak{H}$,

  (ii) if one of the $(\text{II})_i \in \mathfrak{H}$, either $(\text{I}) \in \mathfrak{H}$ or $(\text{I}) \in \mathfrak{C}_{k-1}$,

  (iii) if, for all i, $(\text{II})_i \notin E(\mathfrak{J}_{k-1})$, we have $(\text{I}) \notin E(\mathfrak{J}_k)$.

To prove (iii) we make use of the following property of the matrices $\mathfrak{J}_k$: if $(\text{II})_i \notin E(\mathfrak{J}_{k-1})$, there exists a value function h for which $h((\text{II})_i) \in A_{\mathfrak{J}_k}$ and $h(\mathfrak{X}_l) \in B_{\mathfrak{J}_k}$, whatever l may be.

As a consequence of this lemma we obtain $\mathfrak{E}((\mathfrak{L}_1)_{\Pi\Gamma}) \subset \mathfrak{H}$.

5. Since each of the matrices $\mathfrak{J}_0, \mathfrak{J}_1, \mathfrak{J}_2,...$ is isomorphic with a part of the following matrix, we can denote the elements of $\mathfrak{J}_k$ by the signs e_i^k in such a way that each relation

$$(\text{III}) \qquad\qquad e_l^k = e_m^k \supset_{\mathfrak{J}_k} e_n^k$$
$$e_l^k = e_m^k \wedge_{\mathfrak{J}_k} e_n^k$$
$$e_l^k = e_m^k \vee_{\mathfrak{J}_k} e_n^k$$
$$e_l^k = \sim_{\mathfrak{J}_k} e_m^k$$

which holds for $k = k_1$, continues to hold for each $k = k_2$, provided $k_2 > k_1$. The designated elements have the same index; they are denoted by e_1^k.

We can construct an infinite matrix $\mathfrak{J}_\infty$ with numbers for elements, the relations

$$l = m \supset_{\mathfrak{J}_\infty} n$$
$$l = m \wedge_{\mathfrak{J}_\infty} n$$
$$l = m \vee_{\mathfrak{J}_\infty} n$$
$$l = \sim_{\mathfrak{J}_\infty} m$$

being considered as legitimate when the corresponding relations (III) are satisfied for a matrix $\mathfrak{J}_k$. The designated element of $\mathfrak{J}_\infty$ is 1 (the index of the designated elements of the $\mathfrak{J}_k$). The matrix $\mathfrak{J}_\infty$ possesses the property

$$E(\mathfrak{J}_\infty) = \mathfrak{E}((\mathfrak{L}_1)_{\mathrm{III}}) = \mathfrak{H}.$$

Ed. Note. Jaśkowski's description of the matrix $\mathfrak{N}$ on p. 260 is very condensed. To help the reader, if the two-valued matrix for implication is:

$\supset_\mathfrak{M}$	b	u
b	b	u
u	b	b

the following will be the three-valued matrix:

$\supset_\mathfrak{N}$	b	a	u
b	b	a	u
a	b	b	u
u	b	b	b

13

AXIOMATIZATION OF THE THREE-VALUED PROPOSITIONAL CALCULUS†

MORDCHAJ WAJSBERG

Introduction

THE present study contains a proof of the completeness and independence of my axiomatization of the three-valued calculus. The theorems of this three-valued logic are those implicative-negative expressions of the theory of deduction which satisfy the following table, whose designated value is 1.

C	0	1	2	N
0	1	1	1	1
1	0	1	2	0
2	2	1	1	2

The author of the three-valued calculus, and more generally of the n-valued calculus, for every natural number n, as well as of the calculus of denumerably many values, is Professor Jan Łukasiewicz.[1]

In the present paper I prove that every theorem of the three-valued calculus is a consequence of the following set of theorems:

1. $CqCpq$
2. $CCpqCCqrCpr$
3. $CCCpNppp$
4. $CCNqNpCpq$

† This paper originally appeared under the title 'Aksjomatyzacja trójwarto-ściowego rachunku zdań' in *Comptes rendus des séances de la Société des Sciences et des Lettres de Varsovie*, Cl. iii, 24 (1931), pp. 126–45. Translated by B. Gruchman and S. McCall.

[1] See J. Łukasiewicz and A. Tarski, 'Untersuchungen über den Aussagen-kalkül', *Comptes rendus des séances de la Société des Sciences et des Lettres de Varsovie*, Cl. iii, 23 (1930). See also J. Łukasiewicz, 'Philosophische Bemerkungen zu mehrwertigen Systemen des Aussagenkalküls', ibid. [Paper 3 in this volume—Ed.]

Rules of inference are those accepted in the theory of deduction; the rule of substitution and the rule of detachment.

The consequences of the set of axioms will be called theses. Implicative-negative expressions will be denoted by 'wd'. I shall not show, for any explicitly mentioned wd, that it is a theorem of the three-valued calculus, because this can easily be done by means of the table.

Section I of this paper contains the deduction of certain theses from the above-mentioned axioms, using the stated rules. Section II contains a proof of the completeness of this axiomatization, and Section III a proof of its independence.

In order to show the completeness of the axiomatization I shall prove successively that:

(a) Every wd α is a thesis, if for a certain wd β the following wds are theses:

$$C\beta C\beta\alpha$$
$$CN\beta CN\beta\alpha$$
$$CC\beta N\beta CCN\beta\beta\alpha.$$

(b) Denoting by $\alpha\beta/\alpha'$, for any wds α and α' and for any variable β, the wd which results from α by substituting the wd α' for β, we have that, if a certain wd α contains two non-equiform variables β and γ, then

(1) $C\beta C\beta\alpha$ is a thesis, if $\alpha\beta/C\gamma\gamma$ is a thesis,
(2) $CN\beta CN\beta\alpha$ is a thesis, if $\alpha\beta/NC\gamma\gamma$ is a thesis, and
(3) $CC\beta N\beta CCN\beta\beta\alpha$ is a thesis, if the following are theses:

$$\alpha\gamma/C\beta\beta$$
$$\alpha\gamma/NC\beta\beta$$
$$\alpha\beta/\gamma.$$

(c) In the light of (a) and (b) every theorem α of the three-valued calculus which contains two non-equiform variables is a thesis if certain substitutions of it which include fewer non-equiform variables are theses. It is, consequently, a thesis if all theorems of the three-valued calculus which do not contain two non-equiform variables are theses.

(d) If a certain wd α is a theorem of the three-valued calculus, and if every variable contained in α is equiform with the variable β, then the following wds are theses:

$$\alpha\beta/C\beta\beta$$
$$\alpha\beta/NC\beta\beta$$
$$CC\beta N\beta CCN\beta\beta\alpha$$

In consequence of this, the following wds are theses:

$$C\beta C\beta\alpha$$
$$CN\beta CN\beta\alpha$$
$$CC\beta N\beta CCN\beta\beta\alpha.$$

Hence α is also a thesis, according to (a).

(e) From (c) and (d), every theorem of the three-valued calculus is a thesis.

I consider it my pleasant duty to thank most cordially Professor Jan Łukasiewicz, whose many kindly offered suggestions contributed greatly to the improvement of this paper, both as to its form and to its content.

Section I

Proofs of theses I shall write in the same way as they are given in Łukasiewicz's *Elements of Mathematical Logic*.[1] The proof of thesis 6 by means of thesis 5, as well as the proof of thesis 15 by means of theses 6 and 14, come from this work. The following is an almost literal quotation from it (pp. 66–67).

'Every proven thesis is provided with a number and preceded by a derivational line. For instance theorem 5 is preceded by the following derivational line:

$$2p/Cpq, \ q/CCqrCpr, \ r/s * C2\text{---}5$$

The derivational line always consists of two parts which are divided by an asterisk. The first part in our example indicates that the rule of substitution is to be applied to thesis 2. The part following the asterisk indicates that the substitution of thesis 2 obtained in this way has the form of a conditional, the

---

[1] J. Łukasiewicz, *Elementy logiki matematycznej*, authorized mimeographed script prepared by M. Presburger, Warsaw, 1929. [Translated as *Elements of Mathematical Logic*, Oxford and Warsaw, 1963—Ed.]

antecedent of which is equiform with thesis 2 and the consequent of which is equiform with theorem 5. We can now assert theorem 5, applying the rule of detachment.'

Inscriptions of the type '*rs/α*' replace '*r/α, s/α*'.

1. *CqCpq*
2. *CCpqCCqrCpr*
3. *CCCpNppp*
4. *CCNqNpCpq*
   *
   2*p/Cpq, q/CCqrCpr, r/s* * C2—5
5. *CCCCqrCprsCCpqs*
   5*q/Cqr, r/Csr, s/CCsqCpCsr* * C5*p/s, s/CpCsr*—6
6. *CCpCqrCCsqCpCsr*
   5*s/CCCprsCCqrs* * C2*p/Cqr, q/Cpr, r/s*—7
7. *CCpqCCCprsCCqrs*
   2*p/q, q/Cpq* * C1—8
8. *CCCpqrCqr*
   8*q/Cqr, r/CCsqCpCsr* * C6—9
9. *CCqrCCsqCpCsr*
   9*q/CCrNrr, s/q* * C3*p/r*—10
10. *CCqCCrNrrCpCqr*
    10*q/CqCCrNrr, r/Cqr* * C10*p/CCqrNCqr*—11
11. *CpCCqCCrNrrCqr*
    11*p/CqCpq* * C1—12
12. *CCqCCrNrrCqr*
    5*p/CrNr, s/CCqrr* * C12*q/Cqr*—13
13. *CCCrNrqCCqrr*
    8*p/CrNr, r/CCqrr* * C13—14
14. *CqCCqrr*
    6*p/q, q/Cqr, s/p* * C14—15
15. *CCpCqrCqCpr*
    15*p/Cpq, q/Cqr, r/Cpr* * C2—16
16. *CCqrCCpqCpr*
    15*p/Cpq, q/CCprs, r/CCqrs* * C7—17
17. *CCCprsCCpqCCqrs*
    15*r/p* * C1*q/p, p/q*—18

18. $CqCpp$
    $18q/CqCpp * C18$—19
19. Cpp
    $8p/Nq, q/Np, r/Cpq * C4$—20
20. $CNpCpq$
    $15p/CNqNp, q/p, r/q * C4$—21
21. $CpCCNqNpq$
    $21p/Cpp * C19$—22
22. $CCNqNCppq$
    $8p/Nq, q/NCpp, r/q * C22$—23
23. $CNCppq$
    $2p/Np, q/Cpq * C20$—24
24. $CCCpqrCNpr$
    $24p/Np, q/NCpp, r/p * C22q/p$—25
25. $CNNpp$
    $4q/NNp * C25p/Np$—26
26. $CpNNp$
    $7p/NNq * C25p/q$—27
27. $CCCNNqrsCCqrs$
    $27r/NCpp, s/Nq * C22q/Nq$—28
28. $CCqNCppNq$
    $27r/Np, s/CpNq * C4q/Nq$—29
29. $CCqNpCpNq$
    $2p/Cqr, q/CCpqCpr, r/s * C16$—30
30. $CCCCpqCprsCCqrs$
    $30s/CCCprsCCpqs * C2p/Cpq, q/Cpr, r/s$—31
31. $CCqrCCCprsCCpqs$
    $31r/NNq * C26p/q$—32
32. $CCCpNNqsCCpqs$
    $32s/CNqNp * C29q/p, p/Nq$—33
33. $CCpqCNqNp$
    $15p/Cpq, q/Nq, r/Np * C33$—34
34. $CNqCCpqNp$
    $16q/Np, r/Cpq * C20$—35
35. $CCpNpCpCpq$
    $13r/p, q/CpCpq * C35$—36

36. $CCCpCpqpp$
   $17p/CpCpq,\ rs/p,\ q/r * C36$—37
37. $CCCpCpqrCCrpp$
   $6p/Cqr,\ q/Cpq,\ r/Cpr * C16$—38
38. $CCsCpqCCqrCsCpr$
   $38s/Cqr,\ p/CCprs,\ q/CCpqs,\ r/t * C31$—39
39. $CCCCpqstCCqrCCCprst$
   $39q/Cpq,\ s/q,\ t/CCqpp,\ r/q * C37r/q$—40
40. $CCCpqqCCCpqqCCqpp$
   $40p/CpCpq,\ q/p * C36$—$C36$—41
41. $CCpCpCpqCpCpq$
   $33p/q,\ q/Cpq * C1$—42
42. $CNCpqNq$
   $30s/CCspCCpqCsr * C6p/Cpq,\ q/p$—43
43. $CCqrCCspCCpqCsr$
   $17p/q,\ s/CCspCCpqCsr,\ q/t * C43$—44
44. $CCqtCCtrCCspCCpqCsr$
   $44q/NCpq,\ t/Nq,\ p/t * C42$—45
45. $CCNqrCCstCCtNCpqCsr$
   $16q/CqCCrNrr,\ r/Cqr * C12$—46
46. $CCpCqCCrNrrCpCqr$
   $46p/CNqq,\ q/CpCpq,\ r/Cpq * C45r/q,\ s/p,\ t/Cpq$—47
47. $CCNqqCCpCpqCpq$
   $6p/CNqq,\ q/CpCpq,\ r/Cpq * C47$—48
48. $CCsCpCpqCCNqqCsCpq$
   $48s/CpNp * C35$—49
49. $CCNqqCCpNpCpq$

Section II

In definitions the sign '$/\!/$' separates the *definiendum* from the *definiens*. In proofs of theorems I do not mention definitions.

Df. 1. $\alpha \rightarrow \beta \;/\!/$ If α is a thesis, then β is a thesis.

Note. No further reference will be made to the following statements:

(a) $\alpha \rightarrow \alpha$.
(b) If $\alpha \rightarrow \beta$ and $\beta \rightarrow \gamma$, then $\alpha \rightarrow \gamma$.

(c) If α is a thesis and $\alpha \to \beta$, then β is a thesis.

(d) If $C\alpha\beta$ is a thesis, then $\alpha \to \beta$.

Df. 2. $C^0\alpha\beta \parallel \beta$, and, if k is a natural number, $C^k\alpha\beta \parallel C\alpha C^{k-1}\alpha\beta$. (For example, thesis 41 takes the form CC^3pqC^2pq when this definition is applied.)

Theorem T1. *If m is a natural number and α, β, α_1, α_2,..., α_m are wds, then the following is a thesis*:

$$CC\alpha\beta CC\alpha_1 C\alpha_2 \dots C\alpha_m \alpha C\alpha_1 C\alpha_2 \dots C\alpha_m\beta.$$

This theorem we prove by means of thesis 16, using induction on m.

Theorem T2. *If m and n are natural numbers, $m < n$, and α, α_1, α_2,..., α_n are wds, then*

$$C\alpha_1 C\alpha_2 \dots C\alpha_m \alpha \to C\alpha_1 C\alpha_2 \dots C\alpha_n \alpha.$$

Proof by means of thesis 1 and T1.

Theorem T3. *If k and m are natural numbers, and α, α_1, α_2,..., α_m are wds, then*

$$C\alpha_1 C\alpha_2 \dots C^{k+2}\alpha_m \alpha \to C\alpha_1 C\alpha_2 \dots C^2\alpha_m \alpha.$$

Proof by means of thesis 41 and T1.

Theorem T4. *If α is a wd, m a natural number such that $m > 1$, and the sequences of wds α_1, α_2,..., α_m and β_1, β_2,..., β_m differ only in the order of their terms, then*

$$C\alpha_1 C\alpha_2 \dots C\alpha_m \alpha \to C\beta_1 C\beta_2 \dots C\beta_m \alpha.$$

Proof by means of thesis 15 and T1.

Theorem T5. *If α and β are wds and β results from α by replacing, in a proper part[1] of α, wd γ by wd δ, then we have as theses either*:

$$\left.\begin{array}{l} CC\gamma\delta C\alpha\beta \\ C\alpha CC\gamma\delta\beta \\ CC\delta\gamma C\beta\alpha \\ C\beta CC\delta\gamma\alpha \end{array}\right\} \text{ (group A)}$$

and

or:

$$\left.\begin{array}{l} CC\delta\gamma C\alpha\beta \\ C\alpha CC\delta\gamma\beta \\ CC\gamma\delta C\beta\alpha \\ C\beta CC\gamma\delta\alpha \end{array}\right\} \text{ (group B)}$$

and

[1] A proper part = a part which is different from the whole.

Note 1. In proving this theorem and theorems T6 and T8 I use only theses 2, 16, 33, and 34 and their consequences.

Note 2. When proving the theorems of this paper which take the form of conditionals, I shall denote, following Leśniewski, the antecedents of the theorems by 'Hp' and the consequents by 'Tz'.

Proof. Let us denote Tz by $T(\alpha, \beta, \gamma, \delta)$.

(*a*) If α', β', γ', δ', and ϵ are wds, and α' is equiform with one of the three wds $C\epsilon\gamma'$, $C\gamma'\epsilon$, and $N\gamma'$, and β' is equiform with one of the three corresponding wds $C\epsilon\delta'$, $C\delta'\epsilon$, and $N\delta'$, then $T(\alpha', \beta', \gamma', \delta')$.

Indeed if γ', δ', and ϵ are wds, the following wds, as well as those formed from them by exchanging γ' for δ' and vice versa, are theses:

$$CC\gamma'\delta'CC\epsilon\gamma'C\epsilon\delta' \quad (16), \qquad CC\epsilon\gamma'CC\gamma'\delta'C\epsilon\delta' \quad (2),$$

$$CC\delta'\gamma'CC\gamma'\epsilon C\delta'\epsilon \quad (2), \qquad CC\gamma'\epsilon CC\delta'\gamma'C\delta'\epsilon \quad (16),$$

$$CC\delta'\gamma'CN\gamma'N\delta' \quad (33), \qquad CN\gamma'CC\delta'\gamma'N\delta' \quad (34).$$

It follows from this that, when we assign to the wds α' and β', in one of the three possible combinations, a shape in accordance with our assumption, we get in each case that $T(\alpha', \beta', \gamma', \delta')$.

If Hp, then

(*b*) There exist three sequences of wds—α_1, α_2,..., α_n, β_1, β_2, ..., β_n, and ϵ_1, ϵ_2,..., ϵ_{n-1}—such that $\alpha_1 = \gamma$, $\alpha_n = \alpha$, $\beta_1 = \delta$, $\beta_n = \beta$; and for $k = 1, 2,..., n-1$ α_{k+1} is equiform with one of the three wds $C\epsilon_k\alpha_k$, $C\alpha_k\epsilon_k$, and $N\alpha_k$, while β_{k+1} is equiform with the corresponding wds $C\epsilon_k\beta_k$, $C\beta_k\epsilon_k$, or $N\beta_k$.

In the light of (*a*) it follows that

(*c*) $T(\alpha_2, \beta_2, \alpha_1, \beta_1)$,

and (*d*) For $l = 1, 2,..., n-1$, $T(\alpha_{l+1}, \beta_{l+1}, \alpha_l, \beta_l)$.

It is easy to demonstrate that

(*e*) If $k = 1, 2,..., n-2$ and $T(\alpha_{k+1}, \beta_{k+1}, \alpha_k, \beta_k)$, then

$$T(\alpha_{k+2}, \beta_{k+2}, \alpha_k, \beta_k).$$

Indeed if we denote by A_1, A_2, B_1, B_2 the following pairs of wds:

$$CC\alpha_k\beta_k\,C\alpha_{k+1}\beta_{k+1} \quad \text{and} \quad C\alpha_{k+1}CC\alpha_k\beta_k\beta_{k+1} \qquad (A_1)$$

$$CC\beta_k\alpha_k\,C\alpha_{k+1}\beta_{k+1} \quad \text{and} \quad C\alpha_{k+1}CC\beta_k\alpha_k\beta_{k+1} \qquad (A_2)$$

$$CC\alpha_{k+1}\beta_{k+1}\,C\alpha_{k+2}\beta_{k+2} \quad \text{and} \quad C\alpha_{k+2}CC\alpha_{k+1}\beta_{k+1}\beta_{k+2} \qquad (B_1)$$

$$CC\beta_{k+1}\alpha_{k+1}\,C\alpha_{k+2}\beta_{k+2} \quad \text{and} \quad C\alpha_{k+2}CC\beta_{k+1}\alpha_{k+1}\beta_{k+2} \qquad (B_2)$$

and by A_1', A_2', B_1', B_2' those pairs of wds which are derived from the above by replacing 'α' by 'β' and vice versa; then from (d) it follows that we obtain as theses either the pairs A_1, B_1, A_1', and B_1', or A_2, B_1, A_2', and B_1', or A_1, B_2, A_1', and B_2', or A_2, B_2, A_2', and B_2'. In examining each of these possibilities, we easily reach the conclusion that $\mathrm{T}(\alpha_{k+2},\beta_{k+2},\alpha_k,\beta_k)$. For instance, in the first case we use the following two substitutions:

$$2p/C\alpha_k\beta_k,\ q/C\alpha_{k+1}\beta_{k+1},\ r/C\alpha_{k+2}\beta_{k+2},$$
$$6p/\alpha_{k+2},\ q/C\alpha_{k+1}\beta_{k+1},\ r/\beta_{k+2},\ s/C\alpha_k\beta_k.$$

Applying the rule of detachment twice to each of these theses we get

$$CC\alpha_k\beta_k\,C\alpha_{k+2}\beta_{k+2} \quad \text{and} \quad C\alpha_{k+2}CC\alpha_k\beta_k\beta_{k+2}.$$

Now we can easily prove

(f) $\mathrm{T}(\alpha_n,\beta_n,\alpha_1,\beta_1)$ (c, e);

$\mathrm{T}z$ (f).

Note. It can be easily shown that, if in the sequence α_1, α_2,..., α_n of wds referred to in (b) the case in which α_k ($k = 2, 3,..., n$) has the form $C\epsilon_{k-1}\alpha_{k-1}$ occurs exactly m times ($m = 0$ not excluded) then the expressions of group A are theses when $n-m$ is an odd number—whereas the opposite holds for the wds of group B.

As examples of this we may take theses 2, 6, 7, 16, 17, 31, 33, 34, 38, and 39, as well as the group of theses under theorem T1.

THEOREM T6. *If α and β are wds, if n is a natural number, and if there exists a certain sequence of wds α_0, α_1,..., α_n such that $\alpha_0 = \alpha$, $\alpha_n = \beta$ and for $l = 1, 2,..., n$ α_l comes from α_{l-1} by* *replacing in* α_{l-1} *a proper part* γ_l *by* δ_l *(the latter being wds), then certain wds of the form*

$$C\phi_1\,C\phi_2\dots C\phi_n\,C\alpha\beta \quad \text{and} \quad C\alpha C\phi_1\,C\phi_2\dots C\phi_n\beta$$

are theses, where the expressions ϕ_k *(k = 1, 2,..., n) are equiform with* $C\gamma_k\delta_k$ *or* $C\delta_k\gamma_k$.

Proof. Let us denote Tz by $T(\beta, n)$. If Hp then

(a) By the previous theorem certain wds of the form $C\phi C\alpha\alpha_1$ and $C\alpha C\phi\alpha_1$ are theses, where ϕ is of the form $C\gamma_1\delta_1$ or $C\delta_1\gamma_1$. Consequently $T(\alpha_1, 1)$.

(b) For $k = 1, 2,..., n-1$, if $T(\alpha_k, k)$, then certain wds of the form $C\phi_1\,C\phi_2\dots C\phi_k\,C\alpha\alpha_k$ and $C\alpha C\phi_1\,C\phi_2\dots C\phi_k\,\alpha_k$ are theses, where for $l = 1, 2,..., k$ ϕ_l is of the form $C\gamma_l\delta_l$ or $C\delta_l\gamma_l$. In addition, owing to the fact that α_{k+1} is derived from α_k, and consequently $C\alpha\alpha_{k+1}$ from $C\alpha\alpha_k$, by replacing the proper part γ_{k+1} of the latter by δ_{k+1}, according to the previous theorem we shall have as theses certain wds of the form $C\alpha_k\,C\phi_{k+1}\,\alpha_{k+1}$ and $CC\alpha\alpha_k\,C\phi'_{k+1}\,C\alpha\alpha_{k+1}$, where ϕ_{k+1} and ϕ'_{k+1} are of the form $C\gamma_{k+1}\delta_{k+1}$ or $C\delta_{k+1}\gamma_{k+1}$. Now, applying T1, we can get from the theses $CC\alpha\alpha_k\,C\phi_{k+1}\,C\alpha\alpha_{k+1}$† and $C\phi_1\,C\phi_2\dots C\phi_k\,C\alpha\alpha_k$ the thesis $C\phi_1\,C\phi_2\dots C\phi_{k+1}\,C\alpha\alpha_{k+1}$, and from the theses $C\alpha_k\,C\phi_{k+1}\,\alpha_{k+1}$ and $C\alpha C\phi_1\,C\phi_2\dots C\phi_k\,\alpha_k$ the thesis $C\alpha C\phi_1\,C\phi_2\dots C\phi_{k+1}\,\alpha_{k+1}$.

From (a) and (b) we conclude that $T(\alpha_n, n)$, i.e. $T(\beta, n)$, i.e. Tz.

Note 1. The sequence described in Hp exists for wds α and β in all cases in which wd α contains n proper parts $\gamma_1, \gamma_2,..., \gamma_n$, none of which is a proper part of any other, and β is derived from α by replacing these parts by the corresponding wds $\delta_1, \delta_2,..., \delta_n$. For example, β may be derived from α by the substitution γ_1/δ_1, where α contains just n parts equiform with the variable γ_1.

Note 2. In Tz we can replace 'α' by 'β' and vice versa, since one of the consequences of Hp is the sentence obtained from Hp by interchanging 'α' and 'β', 'γ' and 'δ'.

Df. 3. $\alpha \equiv \beta \parallel C\alpha\beta$ and $C\beta\alpha$ are theses.

Note. '$\alpha \equiv \beta$' is read 'α is equivalent to β'.

† [The original text has $C\alpha C\alpha_k\,C\phi_{k+1}\,C\alpha\alpha_{k+1}$ here—Ed.]

THEOREM T7. The following theorems are correct upon the assumption that the variables they contain stand for expressions of the theory of deduction.

(a) If $\alpha \equiv \beta$, then $\beta \equiv \alpha$.

(b) If $\alpha \equiv \beta$ and $\beta \equiv \gamma$, then $\alpha \equiv \gamma$. $\qquad$ (2)

(c) If α is a thesis and $\alpha \equiv \beta$, then β is a thesis.

(d) If α and β are theses, then $\alpha \equiv \beta$. $\qquad$ (1)

(e) $NN\alpha \equiv \alpha$. $\qquad$ (25, 26)

(f) If α is a thesis, then $C\alpha\beta \equiv \beta$. $\qquad$ (14, 1)

(g) $CC\beta\beta\alpha \equiv \alpha$. $\qquad$ (f, 19)

(h) $CC\gamma C\beta\beta\alpha \equiv \alpha$. $\qquad$ (f, 18)

(i) $CCNC\gamma\gamma\beta\alpha \equiv \alpha$. $\qquad$ (f, 23)

(j) $C\beta C\alpha\alpha \equiv C\gamma\gamma$. $\qquad$ (d, 18, 19)

(k) $CNC\beta\beta\alpha \equiv C\gamma\gamma$. $\qquad$ (d, 23, 19)

(l) $C\beta NC\alpha\alpha \equiv N\beta$. $\qquad$ (28, 20)

THEOREM T8. *If in the hypothesis of theorem* T6 $\gamma_l \equiv \delta_l$, *for* $l = 1, 2, ..., n$, *then* $\alpha \equiv \beta$.

Proof. In this case, according to T6 and the note 2 after T6, certain wds of the forms $C\phi_1 C\phi_2 ... C\phi_n C\alpha\beta$ and $C\psi_1 C\psi_2 ... C\psi_n C\beta\alpha$ will be theses, where the wds ϕ_k and ψ_k ($k = 1, 2, ..., n$) are of the form $C\gamma_k \delta_k$ or $C\delta_k \gamma_k$. But then the latter will be theses, since $\gamma_k \equiv \delta_k$. It follows that $C\alpha\beta$ and $C\beta\alpha$ are theses, hence $\alpha \equiv \beta$.

Note. It follows that, upon replacing some of the parts of a certain wd α by equivalent expressions, we obtain a wd equivalent to α.

THEOREM T9. *If α and β are wds and β is derived from α by replacing certain of its proper parts equiform with the wd γ by wds equiform with δ, then $C^2C\gamma\delta C^2C\delta\gamma C\alpha\beta$ and $C\alpha C^2C\gamma\delta C^2C\delta\gamma\beta$ are theses.*

Proof by means of T6, T4, T2, and T3.

Df. 4. α_1 and $\alpha_2 ...$ and $\alpha_n \rightarrow \beta \,/\!/.$ If $\alpha_1, \alpha_2, ...,$ and α_n are theses, then β is a thesis.

THEOREM T10. $C^2\alpha\beta$ *and* $C^2N\alpha\beta$ *and* $CC\alpha N\alpha CCN\alpha\alpha\beta \rightarrow \beta$.

Note. Instead of this theorem, one could prove that

$$CC^2pqCC^2NpqCCCpNpCCNppqq$$

is a thesis. Similarly one could replace all theorems which appear in the following proof with theses.

Proof.

(a) $CC\alpha N\alpha\beta \rightarrow CC\beta\alpha\alpha$ (13)

(b) $CC\beta\alpha\alpha \rightarrow CC\alpha\beta\beta$ (40)

(c) $CC\alpha N\alpha\beta \rightarrow CC\alpha\beta\beta$ (a, b)

(d) $CC\alpha N\alpha\beta$ and $C\alpha\beta \rightarrow \beta$ (c)

(e) $CCN\alpha NN\alpha\beta \rightarrow CCN\alpha\alpha\beta$ (T8, T7 part *e*)

(f) $CCN\alpha\alpha\beta$ and $CN\alpha\beta \rightarrow \beta$ $(d\ \alpha/N\alpha, e)$

(g) $CC\alpha N\alpha\beta \rightarrow CC\alpha N\alpha C\alpha\beta$ (T2)

(h) $C\alpha C\alpha\beta$ and $CC\alpha N\alpha C\alpha\beta \rightarrow C\alpha\beta$ $(d\ \beta/C\alpha\beta)$

(i) $CC\alpha N\alpha\beta$ and $C^2\alpha\beta \rightarrow C\alpha\beta$ (g, h)

(j) $CC\alpha N\alpha\beta$ and $C^2\alpha\beta \rightarrow \beta$ (d, i)

(k) $CCN\alpha\alpha\beta$ and $C^2N\alpha\beta \rightarrow \beta$ $(j\ \alpha/N\alpha, e)$

(l) $C^2\alpha\beta \rightarrow C^2\alpha CCN\alpha\alpha\beta$ (T2)

(m) $CC\alpha N\alpha CCN\alpha\alpha\beta$ and $C^2\alpha CCN\alpha\alpha\beta \rightarrow CCN\alpha\alpha\beta$

                                    $(j\ \beta/CCN\alpha\alpha\beta)$

(n) $CC\alpha N\alpha CCN\alpha\alpha\beta$ and $C^2\alpha\beta \rightarrow CCN\alpha\alpha\beta$ (l, m)

Tz (k, n)

THEOREM T11. *The following are theses:*

α. $CC^2CpNpqCCpNpq$

α'. $CC^2CNppqCCNppq$

Proof.

(a) $CCpNp\alpha$ $(14q/CpNp, r/CCpNpq)$

(b) $C^2p\alpha$ (35, T2, T4)

(c) α   (T10 part *j*, *a*, *b*)

(d) α' $(c\ p/Np,$ T7 part *e*, T8)

THEOREM T12. *If the antecedent of theorem* T9 *is satisfied, then* $C\alpha CC\gamma N\gamma CCN\gamma\gamma CC\delta N\delta CCN\delta\delta\beta$ *is a thesis.*

Proof. If Hp, then by T9 $C\alpha C^2C\gamma\delta C^2C\delta\gamma\beta$ is a thesis. By 49 $CCN\delta\delta CC\gamma N\gamma C\gamma\delta$ and $CCN\gamma\gamma CC\delta N\delta C\delta\gamma$ are theses.

Applying T1 and T4 it is easy to show that as a consequence the following wd will be a thesis:

$$C\alpha C^2 C\gamma N\gamma C^2 CN\gamma\gamma C^2 C\delta N\delta C^2 CN\delta\delta\beta.$$

Using T4 and T11 we may now prove Tz.

THEOREM T13. *If α is a wd containing the variables 'p' and 'q', then $\alpha p/Cqq$ and $\alpha p/NCqq$ and $\alpha q/Cpp$ and $\alpha q/NCpp$ and $\alpha p/q \to \alpha$.*

Proof. If Hp, then

(a) $\alpha p/Cqq \to C^2 CCqqpC^2 CpCqq\alpha$ (T9 γ/Cqq, δ/p)

(b) $\alpha p/Cqq \to C^2 p\alpha$ (T8, a, T7 parts g and h)

(c) $\alpha p/NCqq \to C^2 CNCqqpC^2 CpNCqq\alpha$ (T9 $\gamma/NCqq$, δ/p)

(d) $\alpha p/NCqq \to C^2 Np\alpha$ (T8, c, T7 parts i and l)

Similarly we obtain

(e) $\alpha q/Cpp \to C^2 q\alpha$ and $\alpha q/NCpp \to C^2 Nq\alpha$

Let us now denote the wd $CCpNpCCNpp\alpha$ by 'β'.

(f) $C^2 p\alpha$ and $C^2 Np\alpha$ and $\beta \to \alpha$ (T10)

(g) $C^2 q\alpha \to C^2 q\beta$ and $C^2 Nq\alpha \to C^2 Nq\beta$ (T2)

(h) $\alpha p/q \to CCqNqCCNqqCCpNpCCNpp\alpha$ (T12)

(i) $\alpha p/q \to CCqNqCCNqq\beta$ (h)

(j) $C^2 q\alpha$ and $C^2 Nq\alpha$ and $\alpha p/q \to \beta$ (T10, g, i)

(k) $C^2 p\alpha$ and $C^2 Np\alpha$ and $C^2 q\alpha$ and $C^2 Nq\alpha$ and $\alpha p/q \to \alpha$ (f, j)

Tz   (b, d, e)

THEOREM T14. *If every theorem of the three-valued calculus which does not contain two non-equiform variables is a thesis, then, if α is a theorem of the three-valued calculus, α is a thesis.*

We obtain a proof from T13, taking into account that if any theorem α of the three-valued calculus contains two non-equiform variables, for instance 'p' and 'q', then the expressions

$$\alpha p/Cqq, \ \alpha p/NCqq, \ \alpha q/Cpp, \ \alpha q/NCpp, \text{ and } \alpha p/q$$

are also theorems of the three-valued calculus, being substitutions of α. Moreover, they have fewer variables of different form than α.

Theorem T15. *If α is a wd containing the variable 'p' as a proper part, then*

$$\alpha p / Cpp \quad and \quad \alpha p / NCpp \quad and \quad CCpNpCCNpp\alpha \to \alpha.$$

Proof. If Hp, then

(a) $\alpha p / Cpp \to C^2 p\alpha$ and $\alpha p / NCpp \to C^2 Np\alpha$ (cf. the proof of T13).

Tz   (T10, a)

Theorem T16. *If α is a wd, and if every variable contained in α is equiform with 'p', then, if α' has the form $\alpha p / Cpp$ or $\alpha p / NCpp$, $\alpha' \equiv Cpp$ or $\alpha' \equiv NCpp$.*

Proof. Let us denote this theorem by $T(\alpha)$.

(a) If α is a variable, then $T(\alpha)$.

Indeed, α' then takes the form Cpp or $NCpp$, hence $\alpha' \equiv Cpp$ or $\alpha' \equiv NCpp$.

(b) If a certain wd α has the form $C\alpha_1\alpha_2$, and $T(\alpha_1)$ and $T(\alpha_2)$, then $T(\alpha)$. Indeed, if α is a wd and every variable it contains is of the form 'p', then, since α takes the form $C\alpha_1\alpha_2$, α_1 and α_2 are also wds which do not contain variables distinct from 'p'. Consequently if $T(\alpha_1)$ and $T(\alpha_2)$, then each of the wds $\alpha_1 p / Cpp$, $\alpha_2 p / Cpp$, $\alpha_1 p / NCpp$, $\alpha_2 p / NCpp$ is equivalent to one of the wds Cpp and $NCpp$. In the light of T8 we conclude from this that α' is equivalent to one of the four wds

$$CCppCpp, \qquad\qquad CCppNCpp,$$
$$CNCppCpp, \qquad\qquad CNCppNCpp.$$

From T7 parts (g) and (k) it follows that each of these four expressions is equivalent with Cpp or $NCpp$, and hence α' is too.

(c) If a certain wd α has the form $N\alpha_1$ and $T(\alpha_1)$, then $T(\alpha)$.

Proof using T8 and T7 part (e).

Tz   (a, b, c)

Theorem T17. *If α is a theorem of the three-valued calculus containing no variables distinct from 'p', then if α' has the form $\alpha p / Cpp$ or $\alpha p / NCpp$, α' is a thesis.*

Proof. If Hp, then α' is a substitution of a theorem of the three-valued calculus, hence α' is also a theorem of this calculus, hence by T16 $\alpha' \equiv Cpp$, since no theorem of the three-valued calculus is equivalent to $NCpp$. Therefore by 19 α' is a thesis. Q.e.d.

Df. 5. If α and β are wds, then 'α is shorter than β' means 'α has fewer signs of the form "C" or "N" than β'.

For example, '$NCpq$' is shorter than '$NNCpq$', but not shorter than '$CpCpq$'.

Note. I shall use expressions of the type 'α is the shortest wd with the property F' to mean 'α is a wd with the property F and there is no shorter wd with the property F'.

THEOREM T18. *If a certain wd α has the property F and if every wd with the property F, not being the shortest wd with that property, is equivalent to a shorter one possessing it, then α is equivalent to a certain wd which is the shortest wd with the property F.*

THEOREM T19. *If α is a wd and has a proper part of the form 'Cpp', then α is of the form '$NCpp$' or is equivalent to a certain wd which is shorter than α.*

Proof. If α is not of the form '$NCpp$', then by Hp α contains a part γ which is either equiform with '$NNCpp$' or equiform for a certain ϕ with one of the wds $CCpp\phi$, $C\phi Cpp$, $CNCpp\phi$, and $C\phi NCpp$. The wd γ is equivalent to a certain wd δ, i.e. in each case to one of the following wds: Cpp, ϕ, Cpp, Cpp, and $N\phi$ (T7 parts e, g, j, k, l). Hence by T8 α is equivalent to the wd β which one gets from α by replacing γ by a certain wd equiform with δ. β is shorter than α, because δ is shorter than γ.

Note. Every variable which appears in β is equiform with some variable contained in α.

THEOREM T20. *If α is a wd containing only variables equiform with 'p', then α has the property either of being equiform with 'p', or of having a part of the form 'Cpp' or 'Np'.*

Proof. Let us denote the above property by 'F'. Then

(a) 'p' has the property F.

(*b*) If certain wds ϕ and ψ have the property F, then $C\phi\psi$ also has the property F, since either (i) at least one of the wds ϕ and ψ has a part of the form 'Cpp' or 'Np', in which case $C\phi\psi$ has this part too, or (ii) ϕ and ψ are of the form 'p', in which case $C\phi\psi$ is of the form 'Cpp'.

(*c*) If a certain wd ϕ has the property F, then the wd $N\phi$ also has this property (proof as in (*b*)).

Tz $\hspace{4cm}$ (*a, b, c,* Hp)

THEOREM T21. *If a certain theorem β of the three-valued cal-culus has the form '$CCpNpCCNpp\alpha$', where α contains no variables not equiform with 'p', then β either is of the form $CCpNpCCNppCpp$ or is equivalent to some shorter theorem of the three-valued calculus of the form '$CCpNpCCNpp\alpha'$', where α' too contains no variables not equiform with 'p'.*

Proof. If Hp, then

(*a*) α is not of the form 'p' or 'Np' because, as one can easily check, neither the wd $CCpNpCCNppp$ nor the wd $CCpNpCCNppNp$ is a theorem of the three-valued calculus.

(*b*) Either α is of the form 'Cpp', and hence β is of the form '$CCpNpCCNppCpp$', or α contains some proper part of the form 'Cpp' or 'Np' (T20, *a*).

(*c*) α is not of the form '$NCpp$' (similar to *a*).

(*d*) If α contains a proper part of the form 'Cpp', then, by T19 with the note and (*c*), α is equivalent to a certain shorter wd α' also containing variables of the form 'p' only. Conse-quently by T8 the wd $CCpNpCCNpp\alpha$, i.e. β, is equivalent to $CCpNpCCNpp\alpha'$, an expression which is shorter than β.

(*e*) If α contains a proper part γ of the form 'Np' then, denoting by α' a wd which is derived from α by replacing γ by 'p', and denoting by β' a wd of the form $CCpNpCCNpp\alpha'$, we have that β' is also derived from β by means of this replace-ment, and β from β' by means of a converse replacement. Hence, by T9, $C\beta C^2CNppC^2CpNp\beta'$ and $C\beta'C^2CpNpC^2CNpp\beta$ are theses. Using T4 and T11 one can easily show that in the light of the foregoing $C\beta\beta'$ and $C\beta'\beta$ are theses; hence that

$\beta \equiv \beta'$. Now β' is shorter than β because α' is shorter than α, and furthermore α' contains variables only of the form 'p'.

Tz (b, c, d, e)

THEOREM T22. *If Hp of theorem T21, then β is a thesis.*

Proof. If Hp, then by T18 and T21 β is either of the form $CCpNpCCNppCpp$ or is the equivalent to that expression which is the shortest theorem of the three-valued calculus of the form $CCpNpCCNpp\alpha'$, where α' contains variables of the form 'p' only. From this, taking into account the fact that the wd $CCpNpCCNppCpp$ is a consequence of theses 1 and 18, we have Tz.

THEOREM T23. *If α is a theorem of the three-valued calculus containing variables of the form 'p' only, then α is a thesis.*

Proof. If Hp, then

(a) α contains 'p' as a proper part.

(b) $\alpha p/Cpp$ and $\alpha p/NCpp$ and $CCpNpCCNpp\alpha \rightarrow \alpha$ (T15)

(c) $\alpha p/Cpp$, $\alpha p/NCpp$, and $CCpNpCCNpp\alpha$ are theorems of the three-valued calculus, in the light of the rule of substitution and thesis 1.

(d) $\alpha p/Cpp$, $\alpha p/NCpp$, and $CCpNpCCNpp\alpha$ are theses.

                                                    (T17, T22)

Tz (b, d)

THEOREM T24. *If α is a theorem of the three-valued calculus, then α is a thesis.*   (T14, T23)

Section III

In this section I shall deal with the question of the independence of each of the theses 1, 2, 3, and 4 from the other three.

For the theses 1, 3, and 4 I shall construct an independence proof using the tabular method—the following tables were kindly given to me by Łukasiewicz. The designated value in them is '1'.

For 1:

C	0	1	N
0	1	0	0
1	0	1	1

$C0C00 = 0$

For 3:

C	0	1	2	3	N
0	1	1	1	1	1
1	0	1	2	3	0
2	3	1	1	1	3
3	2	1	3	1	2

$CCC3N333 = 3$

For 4:

C	0	1	N
0	1	1	0
1	0	1	1

$CCN0N1C10 = 0$

For the thesis 2 I should like to give, while I am about it, a more general proof that it is not a consequence of any set of expressions of the theory of deduction, the elements of which contain at most two non-equiform variables, and which does not have every wd as a consequence.

One can easily see that the set of theses 1, 3, and 4 has the required properties.

We argue along these lines. Let us denote by '$T(\xi, \eta, \alpha, \beta)$' the following sentence: 'For a certain wd ϕ, $C\xi\eta$ is of the form $\phi p/\alpha q/\beta$.'[1] We easily see that

(a) for any wds ξ, η, α, and β, $T(\xi, \eta, \alpha, \beta)$ if and only if for certain wds ϕ_1, ϕ_2, ϕ_3, and ϕ_4 at least one of the following holds:

(1) ξ is of the form $\phi_1 p/\alpha$ or $\phi_2 p/\alpha q/\beta$, while η is of the form $\phi_3 q/\beta$ or $\phi_4 p/\alpha q/\beta$.

(2) Like (1) but interchanging ξ and η.

Let us further assume for certain α and β that

(b) (1) α and β are wds,

(2) α is not equiform with β, and

(3) if, for all wds α' and β', $T(\alpha, \beta, \alpha', \beta')$, then α and β are equiform respectively with α' and β', or with β' and α'.

---

[1] Wherever in the argument I use the expression '$\phi p/\xi q/\eta$' (or '$\phi p/\xi$') I omit the very important reservation that ϕ contains no variables other than 'p' and 'q' (or 'p').

For example, $C\alpha\beta$ can take the forms Cpq, $CpCqr$, $CCpqCqr$, but not Cpp, $CpCpq$, $CCpqCqp$.

We state without proof that for such wds α and β we have

(c) For arbitrary wds ϕ, ϕ_1, ξ, η, and ϕ_2, if ϕ has the form $\phi_1 p/\xi q/\eta$ and simultaneously the form $\phi_2 p/\alpha q/\beta$, then $T(\xi, \eta, \alpha, \beta)$.

$$(a, b)$$

The foregoing is obvious in the case where α and β are variables. From (c) follows

(d) For arbitrary wds ξ and η: if for certain wds ϕ, ϕ_1, and ϕ_2, ϕ is of the form $\phi_1 p/\xi q/\eta$ and simultaneously of the form $\phi_2 p/\alpha q/\beta$, then each wd ψ, which for a certain wd ψ_1 is of the form $\psi_1 p/\xi q/\eta$, is at the same time for a certain wd ψ_2 of the form $\psi_2 p/\alpha q/\beta$.

$$(c, a)$$

Further we have

(e) For arbitrary wds ϕ, ϕ_1, ξ, and ϕ_2: if ϕ is of the forms $\phi_1 p/\xi$ and $\phi_2 p/\alpha q/\beta$, then, for a certain wd ϕ_3, ξ is of the form $\phi_3 p/\alpha q/\beta$. Indeed, under the given assumption either ϕ takes one of the forms 'ξ', '$N\xi$', '$NN...N\xi$', in which case the conclusion is obvious, or ϕ is for a certain wd ϕ' of the form $\phi' p/\xi q/\xi$, aud then, according to (c), $T(\xi, \xi, \alpha, \beta)$.

In the light of (a) it follows from the foregoing that either

(1) for a certain wd ϕ_3 ξ is of the form $\phi_3 p/\alpha q/\beta$, or

(2) for certain wds ψ_1 and ψ_2 ξ is of the form $\psi_1 p/\alpha$ and $\psi_2 q/\beta$.

However, only the first case is possible, because it follows from the second that either, for a certain wd α_1, α is of the form $\alpha_1 p/\beta$, or, for a certain wd β_1, β is of the form $\beta_1 p/\alpha$, which is contrary to (a). Consequently our conclusion is correct.

Let us assume further that $C\alpha\beta$ contains just n variables of different form, for instance 'p_1', 'p_2',..., 'p_n'; and let 'q_1', 'q_2',..., 'q_n' be a sequence of variables.

Now let us, for every wd ϕ, denote by 'ϕ^z' a certain wd ϕ' such that either

(1) ϕ' is derived from ϕ by replacing in ϕ every part ϕ_1, which for a certain wd ϕ_2 is of the form $\phi_2 p/\alpha q/\beta$, by $\phi_1 p_1/q_1$, p_2/q_2,..., p_n/q_n, or

(2) $\phi' = \phi$, if ϕ does not contain such a part ϕ_1.

Examples: Let us take $C\alpha\beta$, ϕ, and ϕ^z respectively as of the form:

1. Cp_1p_2, $CCp_1qCCqp_2Cp_1p_2$, $CCp_1qCCqp_2Cq_1q_2$
2. $Cp_1Cp_2p_3$, $CCp_2Cp_1p_3Cp_1Cp_2p_3$, $CCp_2Cp_1p_3Cq_1Cq_2q_3$
3. Cp_1p_2, $Cp_1Cp_2p_3$, $Cp_1Cp_2p_3$.

Accepting the above definition (of ϕ^z) we may prove

(f) If some wd ϕ is a consequence of a certain set U of expressions of the theory of deduction which contain at most two distinct variables, then ϕ^z is also a consequence of U.

To show this we demonstrate in turn (for arbitrary wds ϕ, ϕ_1, ξ, and η):

(α) If ϕ is of the form $\phi_1p/\xi q/\eta$, then ϕ^z is of the form $\phi p_1/q_1$, $p_2/q_2,..., p_n/q_n$, or of the form $\phi p/\xi^z$, q/η^z.

Indeed, if under the above assumption ϕ contains a certain part ψ, which for a certain wd ψ_1 is of the form $\psi_1p/\alpha q/\beta$, then we have at least one of the three following cases:

(1) for a certain wd ψ_2, ψ is of the form $\psi_2p/\xi q/\eta$;
(2) for a certain wd ψ_2, ψ is of the forms ψ_2p/ξ or ψ_2p/η;
(3) ψ is a part of a certain part of ϕ, which latter part is equiform with ξ or with η.

In the first case, by (d), for a certain wd ϕ', ϕ is of the form $\phi'p/\alpha q/\beta$, hence ϕ^z is of the form $\phi p_1/q_1$, $p_2/q_2,..., p_n/q_n$. In the second case, by (e), for a certain wd ϕ_3, ξ or η is of the form $\phi_3p/\alpha q/\beta$. Consequently in the last two cases when replacing in ϕ every part equiform with ξ or η by ξ^z or η^z respectively, we replace at the same time ψ by $\psi p_1/q_1$, $p_2/q_2,..., p_n/q_n$. Therefore if ϕ^z is not of the form $\phi p_1/q_1$, $p_2/q_2,..., p_n/q_n$, then ϕ^z is of the form $\phi p/\xi^z q/\eta^z$. Q.e.d.

(β) If ϕ does not contain the sign 'C', then $\phi^z = \phi$.

(γ) If ϕ is a substitution of a certain wd of the set U, then ϕ^z is a substitution of the same wd.   (α, β)

(δ) If ϕ is of the form $C\xi\eta$, and ϕ, ϕ^z, ξ, and ξ^z are consequences of U, then η^z is also a consequence of U.

Indeed, under the given assumption either $\eta^z = \eta$ or η contains for a certain wd η_1 a part of the form $\eta_1p/\alpha q/\beta$. In that case ϕ^z has the form $C\xi'\eta^z$, where ξ' is equiform with ξ^z or is a substitution of ξ (hence ξ' is a consequence of U). In both cases η^z is a consequence of U.

It is easy now to demonstrate (f) by means of (γ) and (δ) if one takes into account the following known theorem:

If some wd ϕ is a consequence of a certain set of expressions of the theory of deduction, then it will remain a consequence of this set if we restrict the rule of substitution in such a way that substitution is allowed only in the expressions of this set.

Let us now come back to example 1 following the definition of 'ϕ^z'. It is evident that for a wd ϕ which is a substitution of thesis 2, ϕ^z has the form $CCp_1qCCqp_2Cq_1q_2$. Now every wd is a consequence of ϕ^z; hence in the light of (f) if ϕ is a consequence of the set U with properties mentioned under (f), then every wd is a consequence of U. From this it follows that, if not every wd is a consequence of U, the thesis 2 is not a consequence of U. Q.e.d.

In a similar fashion one can prove the following theorem. If some set U of expressions of the theory of deduction satisfies a certain table T with a finite number of values, and if for a certain natural number k the wd C^kpp satisfies the table T, then either 'p' is a consequence of U, or some wd which satisfies T is not a consequence of U, or at least one expression of the set U contains more than two variables.

In conclusion it is worth while adding that, as was remarked by Łukasiewicz, the set of theses 1, 2, 3, and 4 can be replaced by the set 1, 2, 36, and 4.

14

CONTRIBUTIONS TO METALOGIC†

MORDCHAJ WAJSBERG

Introduction

THIS paper contains various contributions to the metacalculus of propositions.

§ 1 contains three- and four-axiom systems for implicational logic, and the proof that in the Tarski–Bernays axiom system consisting of the following three propositions:

1. $CpCqp$
2. $CCCpqpp$
3. $CCpqCCqrCpr$

the second proposition can be replaced by the rule: from $CCpqp$ infer p.

In § 2 certain logical matrices are presented.

§ 3 contains theorems on the transformation of logical functions, including the theorem that if in a set X of propositions the connective Np (or any other connective with the exception of Cpq) is replaced by a logically equivalent proposition $\phi(p)$, the set of all consequences of X which do not contain the sign N remains unchanged. Besides p, $\phi(p)$ may contain no other variable occurring in the propositions of X.

§ 4 contains a new plan for a proof of the completeness of the implicational calculus.

§ 5 contains the proof that if we add to any axiom system of implicational logic (i.e. that containing the sign C only) the proposition $C0p$ (where 0 denotes the False), we obtain an axiom system for C and 0.

† This paper originally appeared under the title 'Metalogische Beiträge' in *Wiadomości Matematyczne* 43 (1937), pp. 1–38. Translated by S. McCall and P. Woodruff, with certain small changes in the numbering of the formulae of §§ 1 and 4.

Finally §§ 6 and 7 contain, respectively, completeness proofs of the axiom systems of the implicational and equivalential logics given in my paper 'Ein neues Axiom des Aussagenkalküls in der Symbolik von Sheffer' (A new axiom of the propositional calculus in the symbolism of Sheffer), *Monatshefte für Math. u. Phys.*, vol. 39, no. 2. The proofs are accomplished through the deduction of the above-mentioned three axioms of Bernays and Tarski, and the following axiom system for equivalential logic due to Leśniewski:

1. $EEEprEqpErq$
2. $EEpEqrEEpqr$.

The completeness proof for this system is to be found in the latter's 'Grundzüge eines neuen Systems der Grundlagen der Mathematik', *Fundamenta Mathematicae*, Vol 14,Warsaw, 1929.[1]

Explanation of terminology

Logical propositions are written (in the style of Łukasiewicz) without brackets or dots; Cpq denotes implication, Np negation, Epq equivalence. Propositions which contain no other functional symbol besides C are called *C-propositions*, the terms *C-N-proposition* and *C-0-proposition* being similarly employed.

Propositions which satisfy the following matrix of the propositional calculus, whose designated value is 1,

C	0	1	N
0	1	1	1
1	0	1	0

(or which satisfy the corresponding matrices for other functions) will be said to be *valid*. The class of all valid C-propositions forms the *C-calculus*; similarly we speak of the *C-N-calculus* and the *C-0-calculus*. Every set of valid C-propositions, from which the whole C-calculus is deducible, is an *axiom system* of the C-calculus (and similarly for the C-N-calculus, etc.).

The rule of substitution and the rule of detachment (i.e. the

---

[1] My first axiom systems for the C-calculus and the E-calculus were constructed in 1925 without the knowledge of similar work ; they were at that time communicated to Professor Leśniewski in writing.

inference-schema: from $C\alpha\beta$ and α infer β) are used for deducing propositions; in the equivalential calculus the latter rule reads: from $E\alpha\beta$ and α infer β.

The expression $\alpha p/\beta$, $q/\gamma,...$ denotes the proposition resulting from the substitution of β, $\gamma,...$ for p, $q,...$ respectively in α (not all the variables p, $q,...$ need occur in α). From every proposition $C\alpha\beta$ may be derived a rule of inference: if α, then β (in symbols, $\alpha \to \beta$). In general from every proposition of the type $C\alpha_1 C\alpha_2 ... C\alpha_k \beta$ may be derived the rule α_1 & α_2 ... & $\alpha_k \to \beta$, with the meaning that if $\alpha_1, \alpha_2,..., \alpha_k$ (are deducible), then (so is) β. If we denote the former proposition by α, the rule will be named α^k. (For example, if α is $CpCCpqq$, the rule p & $Cpq \to q$ is α^2.)

In order to draw consequences from rules of the type α_1 & α_2 ... & $\alpha_k \to \beta$ (where α_1, $\alpha_2,..., \alpha_k$, β are propositions), it is necessary to use the rule of substitution. If we write $\vdash\alpha_1$ & $\vdash\alpha_2$ ... & $\vdash\alpha_k \to \vdash\beta$, we mean that if the propositions (or rules) α_1, $\alpha_2,..., \alpha_k$ are deducible from whatever propositions (and rules) are in question at the time, then so is β.

$C^k\alpha\beta$ denotes, for $k = 0$, the proposition β, and for $k = 1, 2,...$ the proposition $C\alpha_1 C\alpha_2 ... C\alpha_k \beta$, where $\alpha_i = \alpha$.

§ 1. Axiom systems of the C-calculus

THEOREM 1. *The propositions $CCpqCCqrCpr$, $CCCpqpp$, and $CpCCpqq$ form an axiom system of the C-calculus.*

Note. We shall in future designate these propositions by *Syl*, *Peir*, and *Pon* respectively (*Syl* from syllogism, *Peir* from Peirce, the discoverer of this proposition, and *Pon* from *modus ponens*).[1]

Proof. It suffices to deduce the proposition $CpCqp$ from these propositions (cf. the Tarski–Bernays system).

A. Consequences of Syl

    1. $CCpqCCqrCpr$ (= Syl)
        $1p/Cpq$, $q/CCqrCpr$, $r/s * C1—2$
    2. $CCCCqrCprsCCpqs$
        $2q/Cqr$, r/Csr, $s/CCsqCpCsr * C2p/s$, $s/CpCsr—3$

[1] *Modus ponens* can also be expressed by the proposition $CCpqCpq$.

3. $CCpCqrCCsqCpCsr$
   $2s/CCCprsCCqrs * C1p/Cqr, q/Cpr, r/s$—4
4. $CCpqCCCprsCCqrs$
   $1q/Cpr * 5$
5. $CCpCprCCCprrCpr$
   $4p/Cpq, r/p, s/p * 6$
6. $CCCpqqC$ Peir $CCqpp$
   $4s/q * 7$
7. $CCpqCCCprqCCqrq$
$7^2.$ Cpq & $CCprq \rightarrow CCqrq.$

B. Consequences of Syl and Peir

8. $CCCpqpp$ (= Peir)
   $1p/CpCpr, q/CCCprrCpr, r/Cpr * C5$—$C8p/Cpr, q/r$—9
9. $CCpCprCpr$
10. $CCpqq \rightarrow CCqpp$ (6, 8)
11. Cpq & $CCprq \rightarrow q.$ $(7^2, 8p/q, q/r).$

C. Consequences of Syl and Pon

12. $CpCCpqq$ (= Pon)
   $3q/Cpq, r/q * C12$—13
13. $CCsCpqCpCsq$ (= Comm)[1].

D. Consequences of Syl, Peir, and Pon

   $13s/CCpqq, p/$Peir$, q/CCqpp * C6$—C Peir—14
14. $CCCpqqCCqpp$
   $1q/CCpqq, r/CCqpp * C12$—$C14$—15
15. $CpCCqpp$
   $3q/CCqpp, r/p, s/q * C15q/Cqp$—$C12p/q, q/p$—16
16. $CpCqp$ (= Simpl).[1]   Q.e.d.

THEOREM 2. *The propositions* Syl, Peir, $CpC\alpha p$, *and* $CC\alpha qq$ *form an axiom system of the C-calculus—provided that the last two expressions represent valid C-propositions.*

Proof. Because of theorem 1 it is sufficient to show that
$\vdash$ Syl & $\vdash CpC\alpha p$ & $\vdash CC\alpha qq \rightarrow \vdash$ Pon.

17. $CpC\alpha p$

---

[1] *Comm* from 'commutation', *Simpl* from 'simplification'—these notations from *Principia Mathematica*.

18. $CC\alpha qq$

    $1q/C\alpha p,\ r/CCpqC\alpha q * C17—C1p/\alpha,\ q/p,\ r/q—19$

19. $CpCCpqC\alpha q$

$19^1.\ p \to CCpqC\alpha q$

    $1^2p/Cpq,\ q/C\alpha q,\ r/q * 20$

20. $CCpqC\alpha q\ \&\ CC\alpha qq \to CCpqq$

21. $p \to CCpqq$ $(19^1,\ 20,\ 18)$

    $21p/Cqr,\ q/Cpr * 22$

22. $Cqr \to CCCqrCprCpr$

    $2^1s/Cpr * 23$

23. $CCCqrCprCpr \to CCpqCpr$

24. $Cqr \to CCpqCpr$ $(22,\ 23)$

    $24q/Cpq,\ r/Cpr,\ p/s * 25$

25. $CCpqCpr \to CCsCpqCsCpr$

26. $Cqr \to CCsCpqCsCpr$ $(24,\ 25)$

    $26q/C\alpha q,\ r/q,\ s/p,\ p/Cpq * C18 \to C19—\text{Pon.}$

This proves both Th. 2 and also (as is evident from lines 21–26) the following:

THEOREM 3. $\vdash \text{Syl} \&\ \vdash (p \to CCpqq) \to \vdash (Cqr \to CCs_1 Cs_2 \ldots Cs_n qCs_1 Cs_2 \ldots Cs_n r)\ (n = 1, 2, \ldots)$.

If we take $\alpha = Cqp$ in Th. 2 we obtain $CC\alpha qq = CCCqpqq$, the latter being a substitution of Peir, hence:

THEOREM 4. *The propositions* Syl, Peir, *and* $CpCCqpp$ *form an axiom system of the C-calculus.*

The following two theorems are obtained by putting, in Th. 2, $\alpha = Cpp$ and Cqq respectively.

THEOREM 5. *The propositions* Syl, Peir, $CpCCppp$, *and* $CCCppqq$ *form an axiom system of the C-calculus.*

THEOREM 6. *The propositions* Syl, Peir, $CpCCqqp$, *and* $CCCqqqq$ *form an axiom system of the C-calculus.*

THEOREM 7. $\vdash (Cpq \to CCqrCpr)\ \&\ \vdash (CCpqp \to p) \to \vdash (CpCpr \to Cpr)$.

To prove this make the substitutions q/Cpr in the first of these formulae and $p/Cpr,\ q/r$ in the second.

We shall denote by $C^*\alpha\beta$ any proposition of the type $C^k\alpha\beta$

$(k = 1, 2, ...)$—it is evident that the rule $CpCpr \to Cpr$ entails the stronger rule $C^*pr \to Cpr$. Accordingly there follows, from the previous theorem,

THEOREM 8. $\vdash (Cpq \to CCqrCpr)$ & $\vdash (CCpqp \to p) \to$
$$\vdash (C^*pr \to Cpr).$$

If we now notice that the following formulae hold:
$$\vdash C^*CpqCCqrCpr \to \vdash (Cpq \to CCqrCpr)$$
and $\qquad \vdash C^*CCpqpp \to \vdash (CCpqp \to p),$
we obtain

THEOREM 9. $\vdash C^*CpqCCqrCpr$ & $\vdash C^*CCpqpp \to$
$$\vdash (C^*pr \to Cpr).$$

If we now apply the rule $C^*pr \to Cpr$ to the first two formulae of the theorem proved above, we obtain

THEOREM 10. $\vdash C^*CpqCCqrCpr$ & $\vdash C^*CCpqpp \to$
$$\vdash \text{Syl} \ \& \ \vdash \text{Peir}.$$

THEOREM 11. $\vdash \text{Syl}$ & $\vdash \text{Peir}$ & $\vdash C^*\alpha C^*\beta\gamma \to \vdash C\alpha C\beta\gamma$.

Proof.

27. $\vdash \text{Syl}$ & $\vdash \text{Peir} \to \vdash CCpCprCpr$ $\qquad$ (see 9 above)
28. $\vdash CCpCprCpr$ & $\vdash C^k\alpha C^l\beta\gamma \to \vdash C\alpha C^l\beta\gamma$
29. $\vdash \text{Syl}$ & $\vdash CCpCprCpr \to \vdash CC^l\beta\gamma C\beta\gamma$
30. $\vdash \text{Syl}$ & $\vdash C\alpha C^l\beta\gamma$ & $\vdash CC^l\beta\gamma C\beta\gamma \to \vdash C\alpha C\beta\gamma$
31. $\vdash \text{Syl}$ & $\vdash \text{Peir}$ & $\vdash C^*\alpha C^*\beta\gamma \to \vdash C\alpha C\beta\gamma$. $\quad$ Q.e.d. (27–30)

The last two theorems yield together

THEOREM 12. $\vdash C^*CpqCCqrCpr$ & $\vdash C^*CCpqpp$ & $\vdash C^*pC^*qp \to \vdash CpCqp$ (= Simpl)

and thence, in conjunction with Th. 10,

THEOREM 13. *The propositions $C^*CpqCCqrCpr$, $C^*CCpqpp$, and C^*pC^*qp form an axiom system of the C-calculus.* (Cf. the Tarski–Bernays axiom system.)

Similarly we obtain, by means of Ths. 1 and 4 respectively, the following two theorems:

THEOREM 14. *The propositions $C^*CpqCCqrCpr$, $C^*CCpqpp$, and C^*pC^*Cpqq form an axiom system of the C-calculus.*

THEOREM 15. *The propositions* $C^*CpqCCqrCpr$, $C^*CCpqpp$, *and* C^*pC^*Cqpp *form an axiom system of the C-calculus.*

THEOREM 16. $\vdash (Cpq \to CCqrCpr)$ & $\vdash (CCpqp \to p) \to$
$$\vdash (Cpq\ \&\ CCprq \to q).$$

Proof. Let us denote these formulae in order by A, B, C; we then get, from A and A p/Cqr, q/Cpr, r/q ($= CCqrCpr \to CCCprq$ $CCqrq$) the formula $Cpq \to CCCprqCCqrq$, from which the rule Cpq & $CCprq \to CCqrq$ follows immediately, and, with the help of B p/q, q/r ($= CCqrq \to q$), we then obtain the formula C.

THEOREM 17. *The formulae* $C^*CpqCCqrCpr$, $CCpqp \to p$, *and* C^*pCqp *form an axiom system of the C-calculus.*

Proof. In consideration of Th. 14 it suffices to show that the propositions Peir and Pon are deducible from the above formulae.

32. $C^*pr \to Cpr$	(Th. 8)
33. Syl & Simpl	(32)
34. Simpl & Syl q/p, $p/q \to CpCCprCqr$	(Syl²)
35. $CpCCprCqr$	(33, 34)
36. $CpCCprCCprr$	(35q/Cpr)
37. $p \to CCprCCprr$	(36¹)
38. $p \to CCprr$	(37, 32 p/Cpr)
39. $CCCpCqpqq$	(38$p/CpCqp$, r/q; 33)
40. $CpCCpqCCpCqpq$	(35r/q, $q/CpCqp$)
41. Cqr & $CsCpq \to CsCpr$	(38r/q; Th. 3)
42. $CpCCpqq\ (= \text{Pon})$	(41$q/CCpCqpq$, r/q, s/p, p/Cpq; 39, 40)
43. Cpq & $CCprq \to q$	(Th. 16)
44. $CpCCCpqpp$	(Simpl $q/CCpqp$)
45. $CCpqCCCpqpp$	(42p/Cpq, q/p)
46. $CCCpqpp\ (= \text{Peir})$.	(43q/Peir, r/q; 44, 45)

As an immediate consequence we obtain

THEOREM 18. Syl, Simpl, *and the rule* $CCpqp \to p$ *form an axiom system of the C-calculus.*

THEOREM 19. $\vdash C^*CpqCCqrCpr$ & $\vdash (CCpqp \to p)$ & $\vdash C^*p$ $CCpqp \to \vdash$ Pon.

Proof. From the first three formulae there follows

47. $C^*pr \rightarrow Cpr$ (Th. 8)
48. Syl & $CpCCpqp$ (47)
49. $CpCCpqp$ & Syl $p/Cpq,\ q/p \rightarrow CpCCprCCpqr$ (Syl²)
50. $CpCCprCCpqr$ (48, 49)
51. $CpCCprCCprr$ (50q/r)
52. $p \rightarrow CCprCCprr$ (51¹)
53. $p \rightarrow CCprr$ (52, 47p/Cpr)
54. $CCCpCCpqpqq$ (53p/CpCCpqp, r/q; 48)
55. $CpCCpqCCpCCpqpq$ (50r/q, q/CCpqp)
56. Cqr & $CsCpq \rightarrow CsCpr$ (53r/q; Th. 3)
57. Pon. (56q/CCpCCpqpq, r/q, s/p, p/Cpq; 54, 55)

If we now observe that

$$\vdash C^*CCpqpp \rightarrow \vdash (CCpqp \rightarrow p)$$

and that

$$\vdash C^*CpqCCqrCpr\ \&\ \vdash C^*CCpqpp\ \&\ \vdash C^*pC^*Cpqp \rightarrow$$
$$\rightarrow\ \vdash CpCCpqp \quad \text{(Ths. 10 and 11)}$$

THEOREM 20. $\vdash C^*CpqCCqrCpr$ & $\vdash C^*CCpqpp$ & $\vdash C^*pC^*$
$Cpqp \rightarrow \vdash$ Pon,

and then, by means of Th. 14,

THEOREM 21. $C^*CpqCCqrCpr$ & $C^*CCpqpp$ & C^*pC^*Cpqp
form an axiom system of the C-calculus.

We prove further

THEOREM 22. $C^*CpqCCqrCpr,\ CCpqp \rightarrow p,\ C^*pCCpqp$ and
$C^*pCCCpqp\alpha$ *form an axiom system of the C-calculus—provided
that the last of these expressions represents a valid C-proposition.*

Proof. We denote these propositions in order by A, B, C, D
and deduce from them the propositions Syl, Peir, and Pon (see
below, 63, 70, and 60; cf. Th. 1).

58. Cpq & $CCprq \rightarrow q$ (Th. 16; A, B)
59. Cp Peir & $CCpq$ Peir $\rightarrow$ Peir (58q/Peir, r/q)
60. Pon (Th. 19; A, B, C)
61. $CCpq$ Peir (Pon p/Cpq, q/p)
62. $CpCCCpqp\alpha$ (Th. 8; A, B, D)

63. Syl   (Th. 8; A, B)
64. Comm   (60, 63—cf. the proof of 13)
65. $CCCpqpCp\alpha$ (62, 64)
66. $CpCqr$ & $Csq \to CpCsr$ (Syl—cf. the proof of 3)
67. $CpCCp\alpha p$ & $CCCpqpCp\alpha \to Cp$ Peir
                                 $(66 \ q/Cp\alpha, \ r/p, \ s/CCpqp)$
68. $CpCCp\alpha p$ (Th. 8; A, B, C q/α)
69. Cp Peir   (67, 68, 65)
70. Peir.   (59, 69, 61)

If in the above theorem we put $\alpha = p$, so that

$$D = C^*pCCCpqpp,$$

both D and C become substitutions of $C^*pCCqrp$. Hence we have

THEOREM 23. $C^*CpqCCqrCpr$, $CCpqp \to p$ and $C^*pCCqrp$ form an axiom system of the C-calculus.

If we denote by Syl* the proposition $CCqrCCpqCpr$, we obtain

THEOREM 24. Syl*, Peir, and Pon form an axiom system of the C-calculus.

Proof. Because of Th. 1 it will suffice to deduce the proposition Syl from Syl* and Pon.

71. $CCsCqrCsCCpqCpr$ (Syl* $q/Cqr, \ r/CCpqCpr, \ p/s$)
72. $CpCCsCpqCsq$ (= Comm*) $(71s/p, \ q/Cpq, \ r/q, \ p/s$; Pon)
73. $CCsCpqCpCsq$ (= Comm)
         (Comm* $p/$Comm*, $s/CsCpq, \ q/CpCsq = C$ Comm*—
                      C Comm* $p/CsCpq, \ s/p, \ q/Csq$—Comm)
74. Syl.   (Syl*, Comm)

Note. Since Comm* is also deducible from Comm (Comm $s/CsCpq, \ q/Csq = C$ Comm—Comm*) these two propositions are logically equivalent.

Appendix. Consequences of Syl and $CCCppqq$. Note particularly 87, 88, and 89.

75. $CCpqCCqrCpr$ (= Syl)
76. $CCCppqq$

77. $CCpqCCCpqrCCqqr$ $\qquad$ $(4r/q,\,s/r)$

78. $Cpq \rightarrow CCCpqrCCqqr$ $\qquad$ (77^1)

79. $CCCCpqrCCqqr—C76p/q,\,q/r—CCCpqrr$

$\qquad\qquad\qquad\qquad (75p/CCpqr,\,q/CCqqr)$

80. $Cpq \rightarrow CCCpqrr$ $\qquad$ $(78;\,79,\,75,\,76)$

81. $Cpq \rightarrow CCCpqCrqCrq$ $\qquad$ $(80\,r/Crq)$

82. $CCCpqCrqCrq \rightarrow CCrpCrq$ $\qquad$ $(2q/p,\,r/q,\,p/r,\,s/Crq)$

83. $Cpq \rightarrow CCrpCrq$ $\qquad$ $(81,\,82)$

84. $CCrpCrq \rightarrow CCsCrpCsCrq$ $\qquad$ $(83\,p/Crp,\,q/Crq,\,r/s)$

85. $Cpq \rightarrow CCsCrpCsCrq$ $\qquad$ $(83,\,84)$

86. $76p/q,\,q/r \rightarrow C77—87$

$\qquad\qquad\qquad\qquad (85p/CCqqr,\,q/r,\,s/Cpq,\,r/CCpqr)$

87. $CCpqCCCpqrr$ $\qquad$ $(86,\,76,\,77)$

88. $CCsCCpqrCCpqCsr$ $\qquad$ $(3p/Cpq,\,q/CCpqr;\,87)$

89. Syl*. $\qquad$ $(88s/Cpq,\,p/q,\,q/r,\,r/Cpr;\,75)$

§ 2. Proofs of independence

In what follows I construct four matrices $A–D$ and specify which of the propositions mentioned below, and which of the rules of the type $\alpha \rightarrow \beta$ derived from them, satisfy or fail to satisfy these matrices. To say that a rule of the type α_1 & $\alpha_2 \ldots \rightarrow \beta$ satisfies a given matrix (or is satisfied by it) will mean that every substitution which simultaneously transforms all the propositions α_1, α_2,... into propositions which satisfy the matrix in question also transforms β into a proposition which satisfies the matrix.[1]

Each of the matrices $A–D$ contains the matrix of the ordinary propositional calculus:

C	0	1	N
0	1	1	1
1	0	1	0

whose designated value is 1. For each of these matrices 1 is the sole designated value, and they all possess the property of allowing propositions which are not variables to take only the

---

[1] In accordance with this definition a matrix is utilizable for independence proofs only when it satisfies the rule p & $Cpq \rightarrow q$.

values 0 and 1.† Hence if we replace the variables in a valid proposition by non-variables,[1] the proposition is satisfied by these matrices.

The propositions in question are:

1. $CCpqCCqrCpr$ (Syl)
2. $CCqrCCpqCpr$ (Syl*)
3. $CpCCpqq$ (Pon)
4. $CCCpqpp$ (Peir)
5. $CpCqp$ (Simpl)
6. $CqCpp$
7. $CCpCqrCqCpr$ (Comm)
8. $CCCpqqCCqpp$
9. $CCCpqCqpCqp$
10. $CCCppqq$

11. $CCpCpqCpq$
12. Cpp
13. $CpNNp$
14. $CNNpp$
15. $CCpqCNqNp$
16. $CCNqNpCpq$
17. $CpCNpq$
18. $CNpCpq$
19. $CCNppp$

Matrix A^2

C	0	1	2	3	N
0	1	1	1	1	1
1	0	1	0	0	0
2	0	1	1	0	3
3	0	1	0	1	2

Matrix A satisfies propositions 1, 2, 6, 10–16, 19, and rules 3^1–5^1, 9^1, 17^1, 18^1.

The rules 3^1, 5^1, and 17^1 are satisfied because their right-hand sides always take the value 1 for $p = 1$.

Rule 4^1, i.e. $CCpqp \rightarrow p$, is satisfied because if the proposition $CC\alpha\beta\alpha$ is satisfied α must always, for the following reason, assume the value 1. If $CC\alpha\beta\alpha = 1$, α cannot take the value 0 (since $CC0p0 = 0$ always) and is hence not a variable; but from this it follows that α can never assume the value 2, and thus must always be equal to 1.

---

† [*Ed. note*. As was pointed out by C. H. Langford in his review of Wajsberg's paper in *The Journal of Symbolic Logic* 2 (1937), p. 93, this is not true of matrix A, though it is of B–D, and consequently the proofs that rules 4^1 and 9^1 are satisfied by matrix A must be slightly modified.]

[1] i.e. propositions of the type $C\alpha\beta$ or $N\alpha$.

[2] I constructed this matrix in 1926 for the purpose of showing that C. I. Lewis's system of strict implication (in his work *A Survey of Symbolic Logic*) differs from the classical propositional calculus (C, N stand here for $\prec$, $\sim$).

Rule 9^1, $CCpqCqp \rightarrow Cqp$, is satisfied because only in two cases (as may be easily verified) do we have $CCpqCqp = 1$ and $Cqp = 0$, namely when $p = 2$ and $q = 3$ or vice versa. But if we note that only variables can take these values, and that $CCpqCqp$ is not satisfied, we see that whenever $CC\alpha\beta C\beta\alpha$ is satisfied, $C\beta\alpha$ must also be satisfied.

Rule 18^1, $Np \rightarrow Cpq$, is satisfied because if $Np = 1$ then $p = 0$; and $C0q$ is satisfied.

The remaining propositions are not satisfied, as we see from the following substitutions of them, each equal to 0: $C2CC222$ (3 and 5), $CCC2022$ (4), $CC1C22C2C12$ (7), $CCC200CC022$ (8), $CCC23C32C32$ (9), $C2CN20$ (17), and $CN2C20$ (18).

Rule 7^1, $CpCqr \rightarrow CqCpr$, is not satisfied, since $CqCpp$ is satisfied but $CpCqp$ is not.

Rule 8^1, $CCpqq \rightarrow CCqpp$, is not satisfied, since $CCpCpqCpq$ is satisfied but $CCCpqpp$ is not.

Matrix B

C	0	1	2	N
0	1	1	1	1
1	0	1	1	0
2	0	0	1	0

This matrix satisfies the rule p & $Cpq \rightarrow q$ because, if the propositions α and $C\alpha\beta$ are both satisfied, β cannot take the value 0 (since $C10 = 0$) and is thus not a variable; it follows that β cannot take the value 2 either and so must always assume the value 1. Propositions 1, 2, 4, 8–12, 14, 15, 18, 19 and rules 3^1, 6^1, 13^1, 17^1 are satisfied. The following conclusions retain their validity if 2 is added as a new designated value.

Rules 3^1, 13^1, and 17^1 are satisfied, because their right sides always yield 1 for $p = 1$. Rule 6^1 is satisfied because Cpp is satisfied. Propositions 3, 5, 6, 13, and 17 are not satisfied since every proposition of the type $Cp\alpha$—with the sole exception of Cpp—takes for $p = 2$ the value 0. That the propositions 7 and 16 are not satisfied is shown by the fact that the following substitutions of them are equal to 0:

$$CC0C20C2C00, \quad CCN1N2C21.$$

Rule 5^1, $p \to Cqp$, is not satisfied, for Cpp is satisfied but not $CqCpp$. On the other hand the rules $p \to CCqrp$ and $p \to CNqp$ are satisfied, since their right-hand sides yield 1 for $p = 1$. Rule 7^1, $CpCqr \to CqCpr$, is not satisfied, since $CCppCpp$ is satisfied while $CpCCppp$ is not. Rule 16^1, $CNqNp \to Cpq$, is not satisfied since $CNCppNp$ is satisfied while $CpCpp$ is not.

If in matrix B we put $C22 = 0$, formulae 9, 12, and 6^1 are no longer satisfied, but the remaining results continue to hold.[1]

Matrix B satisfies every valid one-dimensional proposition† of the type $C\alpha p$, for it is equal to 1 when $p = 2$. But if we put $C22 = 0$, we must exclude $\alpha = p$.

Matrix C

C	0	1	2	N
0	1	1	0	1
1	0	1	0	0
2	1	1	0	1

Matrix C satisfies 1, 2, 5, 13, and 15; 12^1, 14^1, and 19^1. 14^1 is satisfied because $NNp = 1$ only if $p = 1$. 19^1 is satisfied because $CNpp = 1$ only if $p = 1$. The rules 3^1, 4^1, 6^1–11^1, and 16^1–18^1 are not satisfied, as we see from the following:

(3^1) $CC\alpha22 = 0$ for $\alpha = 1$.

(4^1) For $\alpha = CNCCppCpppp = C0p$, $CC\alpha p\alpha$ is satisfied, but α itself is not satisfied ($C02 = 0$).

(6^1) Cpp is not satisfied (take $p = 2$).

(7^1) $CCpqCpq$ is satisfied, but not $CpCCpqq$.

(8^1) $CCp\alpha\alpha$ is satisfied for every satisfied proposition α (since $Cq1 = 1$), but $CC\alpha22$ is not.

(9^1 and 10^1) $CCppCpp$ is satisfied, but not Cpp (for $p = 2$).

(11^1) $CN\alpha CN\alpha p$ is satisfied for every satisfied proposition α (since $C0C0p = 1$), but $CN\alpha2 = 0$.

(16^1) $CNpNp$ is satisfied, but not Cpp.

(17^1) For $\alpha = 1$, $CN\alpha2 = 0$. (The propositions $CpCNpCqr$ and $CpCNpNq$ are satisfied.)

---

[1] In the proofs that 5^1 and 16^1 are not satisfied Cpp must be replaced by $CCppCpp$.

† [*Ed. note.* A 'one-dimensional' proposition is a proposition containing only one variable.]

(18^1) For $\alpha = 1$, $NN\alpha = 1$ but $CN\alpha2 = 0$. (The propositions $CNpCpCqr$ and $CNpCpNq$ are satisfied.)

If we put $C22 = 1$ in matrix C, the propositions 6, 9, 12 (as well as the rules corresponding to them) and the rule 10^1 are satisfied. The other results above retain their validity except that the proof that 16^1 is not satisfied must read '$CNpNNCpp$ is satisfied, but not $CNCppp$'.

The matrix C satisfies no proposition of the type $C\alpha p$, for $Cq2 = 0$, but on the other hand it satisfies every valid one-dimensional proposition of the type $Cp\alpha$ (with the exception of Cpp), for $C2\alpha = 1$ if α is not a variable and hence cannot take the value 2.

If we put $C22 = 1$, matrix C satisfies no proposition of the type $C\alpha p$ except Cpp; but it satisfies every valid one-dimensional proposition of the type $Cp\alpha$.

Matrix D

C	0	1	2	N
0	1	1	0	1
1	0	1	0	0
2	0	0	0	0

This matrix satisfies only the propositions 1, 2, 15, and the rules 12^1–14^1, 19^1. 13^1 is satisfied, because $NN1 = 1$. 14^1 is satisfied, for $NN\alpha$ can be satisfied only when α never takes the value 0, and is hence not a variable; the latter can therefore never take the value 2, i.e. $\alpha = 1$ always. 19^1 is satisfied because $CNpp = 1$ only if $p = 1$.

The remaining propositions and rules are not satisfied, as

(i) is evident for 12 from the equation $C22 = 0$,
(ii) may be shown for 3^1, 4^1, 6^1, 8^1–11^1, 17^1, and 18^1 in the same way as in the case of the preceding matrix, and
(iii) is seen for 5^1, 7^1, and 16^1 from the following:

(5^1) $C21 = 0$.
(7^1) $CCppCpp$ is satisfied, but not $CpCCppp$ (for $p = 2$).
(16^1) $CNpNp$ is satisfied, but not Cpp.

The above matrix satisfies no proposition of either of the types $Cp\alpha$ or $C\alpha p$ (since $C2q = Cq2 = 0$).

If we put $C22 = 1$, Cpp is satisfied but the remaining results hold, except that the proof that 16^1 is not satisfied should now read, '$CNCppNp$ is satisfied, but not $CpCpp$'.

Matrix D possesses the further property that it satisfies every valid proposition of the type $CCpq\alpha$, where α contains no variable other than p and q; for a valid proposition can take the value 0 only when the value 2 is substituted for certain of its variables, and if in Cpq we give to either p or q the value 2, we have $Cpq = 0$ and hence $CCpq\alpha = 1$ (for in no valid proposition of the type $CCpq\alpha$ can α be a variable, and so take the value 2).

Similarly it can be shown that every valid proposition of the type $CCpp\alpha$, containing the variable p only, satisfies the matrices C and D.

§ 3. On transformations of propositional functions

The following results concern certain properties of propositions which remain constant when a particular propositional connective (distinct from Cpq) undergoes a single-valued transformation. The contents of this section apply to any proposition at all, unless it is explicitly stated otherwise.

Definition. Let $\phi\ (q_1, q_2, ..., q_r)$ denote in what follows a proposition (held constant in the sequel) beginning with the function sign M (of k arguments) and containing the variables q_1, $q_2, ..., q_r$ and no others. We then introduce the two propositional transformations α^ϕ and α^N by means of the following conditions:

(a) $\alpha^\phi = \alpha$ for every variable α.

(b) $(F\alpha_1\alpha_2...\alpha_m)^\phi = F\alpha_1^\phi\alpha_2^\phi...\alpha_m^\phi$, for every m-place function-sign F distinct from (the negation-function) N.

(c) $(N\alpha)^\phi = \phi(\alpha^\phi, q_2, ..., q_r)$.

(a_1) $\alpha^N = \alpha$ for every variable α.

(b_1) $(F\alpha_1\alpha_2...\alpha_m)^N = F\alpha_1^N\alpha_2^N...\alpha_m^N$ for every m-place function-sign F distinct from M.

(c_1) $\phi(\alpha_1, \alpha_2,..., \alpha_r)^N = N\alpha_1^N.$

(d_1) If for no propositions β_1, β_2,..., β_r is $M\alpha_1\alpha_2...\alpha_k$ of the form $\phi(\beta_1, \beta_2,..., \beta_r)$, then $M(\alpha_1\alpha_2...\alpha_k)^N = M\alpha_1^N\alpha_2^N...\alpha_k^N.$[1]

It may easily be seen that for every proposition α the expressions α^ϕ and α^N are uniquely specified propositions, for in the case that α is a variable, $\alpha^\phi = \alpha^N = \alpha$, and, for each more complicated proposition α, α^ϕ and α^N are uniquely specified provided that the same is true of every proper part of α.

Further, it is seen without difficulty that α^ϕ and α^N possess the following properties A and B.

A. For every proposition α, α^ϕ is a proposition formed from α by replacing in it every part of the type $N\beta$ by $\phi(\beta, q_2,..., q_r)$.

B. For every proposition α, α^N is a proposition formed from α by replacing in it every part of the type $\phi(\alpha_1, \alpha_2,..., \alpha_r)$ by $N\alpha$, where of two parts, one of which includes the other, the larger is always replaced first.

Note. In general it is not the case either that $(\alpha^\phi)^N = \alpha$ or that $(\alpha^N)^\phi = \alpha$, as we see from the following examples. If $\phi(p) = MMp$ and $\alpha = MMp$, then $\alpha^\phi = \alpha$, hence

$$(\alpha^\phi)^N = \alpha^N = Np \neq \alpha;$$

if we take $\alpha = Np$, then $\alpha^N = Np$, and $(\alpha^N)^\phi = MMp \neq \alpha$. On the other hand it may be shown that for every M-free proposition α the relation $(\alpha^\phi)^N = \alpha$ holds (cf. theorem $S(\alpha)$ from the proof of Lemma 2 of this section, with $\beta = \alpha^\phi$) and that in the case where $\phi(q_1, q_2,..., q_r)$ contains only one variable (i.e. where $r = 1$) and α is N-free, $(\alpha^N)^\phi = \alpha$. As a step toward the proof of Th. 1, we now demonstrate

LEMMA 1. *If* $\beta = \alpha p_1/\gamma_1, p_2/\gamma_2,..., p_n/\gamma_n,$ *then* $\beta^\phi = \alpha^\phi p_1/\gamma_1^\phi, p_2/\gamma_2^\phi,..., p_n/\gamma_n^\phi.$

Proof. Let us denote this proposition by $S(\alpha)$. $S(\alpha)$ holds at least when α is a variable, for in this case $\alpha^\phi = \alpha = p_1$, therefore $\beta = \gamma_1$ and $\beta^\phi = \alpha^\phi p_1/\gamma_1^\phi$. It remains to be shown, therefore, that when $\alpha = F\alpha_1\alpha_2...\alpha_m$, where F is any m-place function,

---

[1] As long as it is not explicitly stated otherwise, we must take $M \neq C$ in what follows. We can, on the other hand, have $M = N$, but for the sake of clarity we shall assume in our proof that the sign N does not occur in $\phi(q_1, q_2, ..., q_r)$. (We may for this purpose replace it by, e.g., N'.)

and when all the formulae $S(\alpha_1)$, $S(\alpha_2)$,..., $S(\alpha_m)$ hold, then so does $S(\alpha)$. To prove this we notice that β in this case is of the form $F\beta_1\beta_2...\beta_m$, where the propositions $\beta_1, \beta_2,..., \beta_m$ are formed from the propositions $\alpha_1, \alpha_2,..., \alpha_m$ respectively through the substitutions $p_1/\gamma_1, p_2/\gamma_2,..., p_n/\gamma_n$. If we now take $F = N$, we have $\alpha = N\alpha_1$, $\beta = N\beta_1$, $\alpha^\phi = \phi(\alpha_1^\phi, q_2,..., q_r)$, and $\beta^\phi = \phi\ (\beta_1^\phi,$ $q_2,..., q_r)$; from the assumption that $\beta = \alpha p_1/\gamma_1, p_2/\gamma_2,..., p_n/\gamma_n$ it follows that $\beta_1 = \alpha_1 p_1/\gamma_1, p_2/\gamma_2,..., p_n/\gamma_n$, and hence in virtue of the assumption $S(\alpha_1)$, that $\beta_1^\phi = \alpha_1^\phi p_1/\gamma_1^\phi, p_2/\gamma_2^\phi,..., p_n/\gamma_n^\phi$. If we now observe that in this case $\alpha^\phi = \phi(\alpha_1^\phi, q_2,..., q_r)$ and $\beta^\phi = \phi$ $(\beta_1^\phi, q_2,..., q_r)$, we see that β^ϕ is derived from α^ϕ by the same substitution as was β_1 from α_1, i.e. $\beta^\phi = \alpha^\phi p_1/\gamma_1^\phi, p_2/\gamma_2^\phi,..., p_n/\gamma_n^\phi$, as was to be shown.

If we now assume $F \neq N$, we have $\alpha^\phi = F\alpha_1^\phi \alpha_2^\phi ... \alpha_m^\phi$ and $\beta^\phi = F\beta_1^\phi \beta_2^\phi ... \beta_m^\phi$. From the fact that $S(\alpha_1)$, $S(\alpha_2)$,..., $S(\alpha_m)$ hold we conclude that the propositions $\beta_1^\phi, \beta_2^\phi,..., \beta_m^\phi$ are formed from the propositions $\alpha_1^\phi, \alpha_2^\phi,..., \alpha_m^\phi$ respectively through the substitutions $p_1/\gamma_1^\phi, p_2/\gamma_2^\phi,..., p_n/\gamma_n^\phi$. Consequently $F\beta_1^\phi \beta_2^\phi ... \beta_m^\phi$ $(= \beta^\phi)$ follows from $F\alpha_1^\phi \alpha_2^\phi ... \alpha_m^\phi$ $(= \alpha^\phi)$ by the same substitution, which completes our proof.

We are now able to demonstrate

THEOREM 1. *Let X be a set of propositions in whose elements the variables $q_2, q_3,..., q_r$ do not appear, and let X^ϕ be the set of all propositions α^ϕ, where α is an element of X. If Y is the set of all N-free consequences of X and Y^ϕ the set of all N-free consequences of X^ϕ, then $Y \subset Y^\phi$.*

Proof. It suffices to show that for every element α of Y the proposition α^ϕ belongs to Y^ϕ, since for every N-free proposition it is plain that $\alpha^\phi = \alpha$.

We demonstrate this by means of an induction on the rules of inference (whereby it should be observed that one need use the rule of substitution on the elements only of a set X in order to obtain, with the help of the rule of detachment, all the consequences of X) and show that

1. If β is a substitution of a proposition α, in which the

variables q_2, q_3,..., q_r do not occur, then β^ϕ is a substitution of α^ϕ.

2. If $(C\beta\gamma)^\phi$ and β^ϕ are consequences of certain propositions (e.g. of X^ϕ), then the same is true of γ^ϕ.

The first of these assertions is a direct consequence of the lemma just proved, and the second follows from the observation that $(C\beta\gamma)^\phi = C\beta^\phi\gamma^\phi$. This completes the proof of our theorem.

Note. The theorem just proved remains valid if $M = C$.

Example. If we replace, in any axiom system X of the C–N-calculus the expression Np by a certain proposition $\phi(p)$, e.g. NNp, $CNpp$, or $CpNp$, we obtain a set of propositions X^ϕ from which all valid C-propositions are deducible.

Moreover, if we take for $\phi(p)$ a proposition equivalent to Np, e.g. $NNNp$ or $CpNp$, X^ϕ will consist of valid propositions and consequently only valid C-propositions will be deducible from it. If, for example, we choose for $\phi(p)$ the proposition $CpNp$ and for X the following three-proposition axiom system of the C-N-calculus of Łukasiewicz,

1. $CCNppp$
2. $CpCNpq$
3. $CCpqCCqrCpr$,

then X^ϕ will consist of the following three propositions:

1′. $CCCpNppp$
2′. $CpCCpNpq$
3′. $CCpqCCqrCpr$,

and we then obtain the result that all (and only) valid C-propositions are deducible from the latter three propositions. May it be parenthetically noted that 1′ is a substitution of Peir ($= CCCpqpp$), from which we conclude that all valid C-propositions are consequences of Peir, Syl, and 2′.

From the preceding we infer without difficulty

THEOREM 1′. *If in a set X of valid propositions not containing the variables q_2, q_3,..., q_r the function Np is replaced by any proposition $\phi(p, q_2,..., q_r)$ which is logically equivalent to it, the set of all N-free consequences of X remains unaltered.*

We show now as a preliminary to Th. 2.

LEMMA 2. *If α is an M-free proposition and β a substitution of α^ϕ, β^N is a substitution of α.*

Proof. We shall prove the following stronger theorem $S(\alpha)$: If the sequence p_1, p_2,..., p_n represents the variables contained in a certain M-free proposition α and if β is a substitution of α^ϕ in which the propositions β_1, β_2,..., β_n replace the variables p_1, p_2,..., p_n respectively, then $\beta^N = \alpha p_1/\beta_1^N$, p_2/β_2^N,..., p_n/β_n^N. It is plain that $S(\alpha)$ holds at least when α is a variable, for then $\alpha = p_1$, and $\beta^N = \beta_1^N = \alpha p_1/\beta_1^N$. Consequently it remains to show (since α is M-free) that, for every m-place function F distinct from M, the validity of $S(F\alpha_1 \alpha_2 ... \alpha_m)$ follows from that of $S(\alpha_1)$, $S(\alpha_2)$,..., and $S(\alpha_m)$. For this purpose let $\alpha = F\alpha_1 \alpha_2 ... \alpha_m$ and let β be a substitution of α^ϕ in which the propositions β_1, β_2,..., β_n replace, respectively, the variables p_1, p_2,..., p_n found in α. There are two cases; either $F = N$ or not. In the first case $\alpha = N\alpha_1$, and $\alpha^\phi = \phi(\alpha_1^\phi, q_2,..., q_r)$; β is then of the type $\phi(\gamma, \delta_2,..., \delta_r)$, where γ is formed from α_1^ϕ by the same substitution as was β from α^ϕ. From the validity of $S(\alpha_1)$ we conclude immediately that $\gamma^N = \alpha_1 p_1/\beta_1^N$, p_2/β_2^N,..., p_n/β_n^N, and from this it follows, since in this case $\beta^N = N\gamma^N$ and $\alpha = N\alpha_1$, that $\beta^N = \alpha p_1/\beta_1^N$, p_2/β_2^N,..., p_n/β_n^N, q.e.d.

In the second case $\alpha^\phi = F\alpha_1^\phi \alpha_2^\phi ... \alpha_m^\phi (F \neq N)$, hence β is of the type $F\gamma_1 \gamma_2 ... \gamma_m$, where the propositions γ_1, γ_2,..., γ_m are respectively substitutions of α_1^ϕ, α_2^ϕ,..., α_m^ϕ; in each case the propositions β_1, β_2,..., β_n being substituted for the variables p_1, p_2, ..., p_n respectively. If we assume the validity of the formulae $S(\alpha_1)$, $S(\alpha_2)$,..., $S(\alpha_m)$, then the γ_i^N are respectively of the form $\alpha_i p_1/\beta_1^N$, p_2/β_2^N,..., p_n/β_n^N. If we now note that when $F \neq M$, $\beta^N = F\gamma_1^N \gamma_2^N ... \gamma_m^N$, and that $\alpha = F\alpha_1 \alpha_2 ... \alpha_m$, then it follows that $\beta^N = \alpha p_1/\beta_1^N$, p_2/β_2^N,..., p_n/β_n^N. This completes our proof.

We now demonstrate

THEOREM 2. *Let X be a set of M-free propositions and X^ϕ the set of all propositions α^ϕ, where α is an element of X. Then every M-free consequence of X^ϕ is at the same time a consequence of X.*

Proof. It suffices to show that for every consequence α of X^ϕ the proposition α^N belongs to [*Ed.* the set of consequences of] X, since for every M-free proposition α, $\alpha^N = \alpha$. Every consequence of X^ϕ can be obtained from substitution-instances of elements of X^ϕ by the sole application of the rule of detachment. For this reason the demonstration of the following two assertions suffices for the proof of our theorem:

1. If β is a substitution-instance of an element of X^ϕ then β^N is a consequence of X, and

2. If $(C\alpha\beta)^N$ and α^N are consequences of X, so is β^N.

To establish 1, we note that if β is a substitution-instance of an element of X^ϕ then this element is of the form α^ϕ, where α is a member of X—and then it follows from the just-proven lemma that β^N is a substitution-instance of α and thereby a consequence of X.

The correctness of 2 follows from the remark that

$$(C\alpha\beta)^N = C\alpha^N\beta^N,$$

and herewith our theorem is fully demonstrated. (The latter equation, and hence the whole theorem, loses its validity if we take $M = C$.)

If we replace M by N in the propositions of X^ϕ, then the theorem demonstrated above takes the following more convenient form:

THEOREM 2′. *Let X be a proposition-set and X^ϕ the set of all propositions α^ϕ, where α is an element of X. Then every N-free consequence of X^ϕ is also a consequence of X.*

An immediate consequence of Th. 2′ is

THEOREM 3. *If a proposition-set X is consistent,[1] then the same holds for the set X^ϕ of all propositions α^ϕ such that α is an element of X.*

Proof. By the preceding theorem, every N-free consequence of X^ϕ is at the same time a consequence of X. Hence no variable can be derived from X^ϕ, for otherwise this variable—as an N-

---

[1] A proposition-set is commonly called inconsistent (consistent) if every proposition is derivable from it (not every proposition is derivable from it).

free proposition—would be a consequence of X, contrary to our assumption that X is consistent.

Example. If we set $\phi(p) = NCpp$, we obtain (writing N for M) the theorem: The (previously cited) axiom-system for the C-N-calculus, due to Łukasiewicz, which consists of the following three propositions:

1. $CCNppp$
2. $CpCNpq$
3. $CCpqCCqrCpr$

remains consistent when Np is replaced by $NCpp$, i.e. the following proposition-triple is consistent:

1'. $CCNCppp p$
2'. $CpCNCppq$
3'. $CCpqCCqrCpr.$

Theorems 1 and 2' together yield:

THEOREM 4. *Let X be a proposition-set, in whose elements the variables $q_2, ..., q_r$ do not appear, and X^ϕ the set of all propositions α^ϕ such that α is an element of X. Then the sets of all N-free consequences of X and X^ϕ are identical.* (It is essential here that one take $M \neq C$.)

Note. The above results, as can easily be seen, admit of the generalization that in them one may replace Np by an arbitrary primitive connective, say $Gp_1 p_2 ... p_l$, different from Cpq (e.g. disjunction, conjunction, and the like) on the condition that the number r of arguments of the expression $\phi(q_1, q_2, ..., q_r)$ cannot be assumed less than l.

If Cpq is not a primitive connective, but defined by means of other connectives (e.g. by disjunction and negation), then G cannot be taken to be any of the latter connectives.

In this fashion one obtains, from Th. 3 for example, a more general theorem which one can assert for all of logic (and mathematics) roughly as follows:

If in a set X of formulae one replaces a particular primitive connective $Gx_1 x_2 ... x_l$, other than Cpq, with another expression beginning with G (of the same semantic type) $\phi(x_1, x_2, ..., x_m)$

$(m \geqslant l)$, where the variables $x_{l+1}, \ldots, x_m$ do not occur in X, then the set of all consequences of X in which the connective G does not appear remains unchanged. Here either Cpq is assumed as a primitive connective, or the condition is satisfied that the above replacement does not affect the form of the expression which defines implication.

§ 4. General schema of a completeness-proof for the C-calculus

Let a set M of C-propositions be given, for which it is to be proved that all valid C-propositions are derivable from it.

We divide the proof into two parts:

A. The demonstration that from M all valid 1-dimensional (i.e. containing only one variable) C-propositions are derivable.

B. The proof that if for a given k all valid k-dimensional C-propositions are derivable from M, then the same is true of all $k+1$-dimensional C-propositions.

To establish A, we first derive from M the propositions Syl and Syl* (i.e. $CCpqCCqrCpr$ and $CCqrCCpqCpr$) and then show

LEMMA 1. Cpq & $Cqp \to C\phi(p)\phi(q)$ where $\phi(q)$ comes from $\phi(p)$ by (one or more) substitutions of q for p.[1]

We next derive from M (e.g. by means of Cpp, $CpCqp$, and $CpCCpqq$) the propositions $C\alpha\beta$ and $C\beta\alpha$ for the following three pairs of propositions α and β:

1. $\alpha = CpCpp,\ \beta = Cpp$
2. $\alpha = CCppCpp,\ \beta = Cpp$
3. $\alpha = CCppp,\ \beta = p$

and conclude from Lemma 1 that, for each of these pairs, the propositions $C\phi(\alpha)\phi(\beta)$ and $C\phi(\beta)\phi(\alpha)$ are derivable from M. If we now write 0 for p and 1 for $C00$, then it follows that the following propositions 4–6 and their converses 7–9 are derivable from M for every C-proposition $\phi(p)$:

4. $C\phi(1)\phi(C01)$	7. $C\phi(C01)\phi(1)$
5. $C\phi(1)\phi(C11)$	8. $C\phi(C11)\phi(1)$
6. $C\phi(0)\phi(C10)$	9. $C\phi(C10)\phi(0)$.

[1] See theorem 5 of my paper 'Beiträge zum Metaaussagenkalkül', *Monatsh. f. Math. u. Phys.*, 1935, no. 2.

If we now note that every valid C-proposition must satisfy the following matrix, whose designated value is 1:

$$
\begin{array}{c|cc}
C & 0 & 1 \\
\hline
0 & 1 & 1 \\
1 & 0 & 1
\end{array}
$$

then it follows from this that when, in a valid C-proposition ϕ written only with the variable p, one replaces the variable p by 0 and writes 1 for $C00$, one obtains from ϕ the numeral 1 by repeated replacement of $C01$ and $C11$ by 1, and of $C10$ by 0.

It follows that from 1 (i.e. $C00$), by the converse replacement of 1 by $C01$ or $C11$ and of 0 by $C10$, one can obtain every expression ϕ which comes from a valid proposition, written with the variable p alone, by the substitution of 0 for p. If we now note, as is obvious, that the given replacement can be carried out (in the sense that $CC00\phi$ is derivable) by repeated applications of formulae 4–6 and Syl; we obtain (writing p for 0 again) the result that from Cpp, by means of formulae 4–6 and Syl, every valid proposition written solely with the variable p is derivable. Upon the derivation of Cpp from M, the truth of A is proven.

We now go on to the proof of B.

The following holds.[1]

LEMMA 2. *If the C-proposition $\phi(\beta)$ comes from $\phi(\alpha)$ by (one or more) replacements of α by β, then for some k, l $(= 0, 1, ...)$ the proposition $C\phi(\alpha)C^kC\alpha\beta C^lC\beta\alpha\phi(\beta)$ is derivable from Syl and Syl*.*

If one now derives from M the propositions $CpCqp$ and $CCpCpqCpq$, then it can be shown with the help of Syl* that one can assume the indices k and l to be equal to 1. In other words, the following holds:

LEMMA 3. *If the C-proposition $\phi(\beta)$ comes from $\phi(\alpha)$ by replacement of α by β, then the proposition $C\phi(\alpha)CC\alpha\beta CC\beta\alpha\phi(\beta)$ is derivable from M.*

Furthermore, one may derive from M the rule

($\mathfrak{A}$) Cpr & $CCpqr \rightarrow r$

---

[1] See the proof of theorem 9 of my paper cited above.

(see the proof of 11, § 1) and then the formula

(𝔅) Cpr & Cqr & $CCpqCCqpr \to r$

which can be obtained in the following way by means of 𝔄:

10.	$CrCCqpr$	(Simpl)
11.	$Cpr \to CpCCqpr$	(Syl, 10)
12.	$CpCCqpr$ & $CCpqCCqpr \to CCqpr$	($\mathfrak{A}\,r/CCqpr$)
13.	Cqr & $CCqpr \to r$	($\mathfrak{A}\,p/q, q/p$)
14.	𝔅.	(11, 12, 13)

Now let α be a $k+1$-dimensional proposition ($k = 1, 2,...$) in which the variables p and q occur, and let us assume that the (clearly k-dimensional) propositions $\alpha q/Cpp$, $\alpha p/Cqq$, and $\alpha q/p$ are consequences of M. In accordance with lemma 3 the following three propositions are then derivable from M:

15. $C\alpha q/CppCCCppqCCqCpp\alpha$
16. $C\alpha p/CqqCCCqqpCCpCqq\alpha$
17. $C\alpha q/pCCpqCCqp\alpha$.

If we now derive from M (via $CqCCppq$, $CqCpp$, Syl, and Pon) the propositions $CCCCppqrCqr$ and $CCCqCpprr$, we obtain with the aid of Syl from 15 and 16 respectively:

18. $C\alpha q/CppCq\alpha$
19. $C\alpha p/CqqCp\alpha$

and then from 18, 19, and 17 (since we are assuming the derivability of their antecedents) the propositions $Cp\alpha$, $Cq\alpha$, and $CCpqCCqp\alpha$, from which as a consequence of 𝔅 the proposition α is derivable. With this B is established and our task is finished.

If one now asks the question, whether the set of valid C-propositions is complete in the sense that the addition of any non-valid C-proposition makes the set inconsistent, the problem can be answered affirmatively as follows:

If α is not valid, then its variables can be replaced by 0 and 1 in such a manner that α receives the value 0 according to the above-mentioned matrix. If we now write p for 0 and Cpp for 1 it follows that by replacing the variables of α by p and Cpp one can obtain a proposition β such that, by means of the formulae 7–9 (writing p for 0 and Cpp for 1) and Syl, the proposition $C\beta p$ can be derived (as was $CCpp\phi$ via 4–6 above).

But from this it follows immediately that the addition of β (and hence also the addition of α) to the set of all valid C-propositions (more precisely, to the propositions Syl, Syl*, and the above-introduced proposition-pairs $C\alpha\beta$, $C\beta\alpha$) makes the set inconsistent. Q.e.d.

§ 5. An axiom-system for the C-0-calculus

If we let 0 designate 'the False', then the proposition $C0p$ is valid. The set of all propositions formed from C and 0 we shall call the C-0-calculus.

We shall now show:

THEOREM 1. *If the proposition $C0p$ is added to an arbitrary axiom-system of the C-0-calculus, an axiom system of the C-0-calculus is obtained.*

Proof. If in Łukasiewicz's axiom-system of the C-N-calculus, already mentioned several times,

1. $CCNppp$
2. $CpCNpq$
3. $CpqCCqrCpr$,

we replace the proposition Np by the logically equivalent proposition $Cp0$, we obtain the following valid C-0-propositions:

1′. $CCCp0pp$
2′. $CpCCp0q$
3′. $CpqCCqrCpr$

If herein we again abbreviate $Cp0$ by Np, it is evident that all valid C-N-propositions are derivable from 1′–3′ (since the same is true of 1–3). Hence it follows immediately that from 1′–3′ all valid propositions constructed out of the connectives Cpq and $NCpp$ are derivable. If we now add to 1′–3′ the propositions

4′. $CCCpp00$ $(= CNCpp0)$ and
5′. $C0CCpp0$ $(= C0NCpp)$,

then from 1′–5′ every proposition of the type $C\phi(NCpp)\phi(0)$ is derivable, where $\phi(p)$ is a C-proposition. This follows from the fact that Syl and Syl* are derivable from $1'-3'$ (cf. lemma 1 of the previous section). If we now note the logical equivalence of $NCpp$ and 0, then the set of all valid C-0-propositions is obtained from the set of all valid propositions formed with Cpq and $NCpp$, through replacement of $NCpp$ by 0. That is, the system $1'-5'$ is an axiom system of the C-0-calculus. It thus remains to show that propositions $1'-5'$ are derivable from valid C-propositions and the proposition $C0p$. This is proved for each case as follows:

(a) $1' = CCCp0pp$ is a substitution-instance of $CCCpqpp$ $(= \text{Peir})$.

(b) If in the valid C-proposition $CCrqCpCCprq$ we replace r by 0, we obtain $CC0q2'$ and then $2'$ by means of $C0p$.

(c) $3'$ is a C-proposition.

(d) If in $2'$ we put Cpp for p, $4'$ is obtained by means of Cpp.

(e) $5'$ is a substitution-instance of $C0p$.

The C-0-calculus is also complete in the sense that the addition of a non-valid C-0-proposition renders it inconsistent. For in such a proposition, say ϕ, one can replace the variables by 0 and 1 (i.e. $C00$) in such a way that it receives the value 0 according to the matrix of the propositional calculus. But then one can derive in the C-calculus the proposition $C\phi0$ (analogously to the derivation of $C\beta p$ in the previous section), and from $C\phi0$ and $C0p$ the variable p is obtained.

§ 6. Axiom-systems of the C-calculus[1]

I.

A_1. $CCsrCCCpqrCCpsr$
A_2. $CpCqCrp$.

$\quad A_1s/Cpq,\ r/CrCsCpq * C^2A_2p/Cpq,\ q/r,\ r/s$—1
1. $CCpCpqCrCsCpq$
$\quad 1\ p/q,\ q/Crq,\ r,s/CqCqCrq * C^3A_2p/q$—2

---

[1] In the following proofs the Tarski–Bernays axioms (possibly with some changes of variable) are indicated by asterisks.

*2. $CqCrq$

    $1\ p,q/r,\ r,s/CrCrr * C^3\ 2\ q/r$—3

3. Crr

    $A_1 s/p,\ r/CqCrp,\ p/q,\ q/Crp * CA_2$—$C3\ r/CqCrp$—4

4. $CCqpCqCrp$

    $A_1 s/Cqp,\ r/CqCrp,\ p/r,\ q/p * C4$—$C2\ q/Crp,\ r/q$—5

5. $CCrCqpCqCrp$

    $5\ r/Cqp,\ p/Crp * C4$—6

6. $CqCCqpCrp$

    $A_1 s/q,\ r/CCqpCrp,\ p/r,\ q/p * C6$—$C2\ q/Crp,\ r/Cqp$—7

*7. $CCrqCCqpCrp$

    $A_1 s/r * C3$—8

8. $CCCpqrCCprr$

    $5\ r/CCpqp,\ q/Cpp,\ r/p * C3\ r/p$—9

*9. $CCCpqpp.$

II.

$A.$ $CCCpqCCrstCCuCCrstCCpuCst.$

    $A\ p/Cpq,\ q/CCrst,\ r/u,\ s/CCrst,\ t/CCpuCst,$

      $u/CCpqCCrst * C^2 A$—1

1. $CCCpqCCpqCCrstCCCrstCst$

    $1\ p/CCpqCCrst,\ q/CCuCCrstCCpuCst,\ r/Cpq,$

    $s/CCpqCCrst,\ t/CCCrstCCpuCst,\ u/CCuCCrst$

    $CCpuCst * CAp/Cpq,\ q/CCrst,\ r/u,\ s/CCrst,$

    $t/CCpuCst,\ u/CCpqCCrst$—$C1$—CA—2

2. $CCCpqCCrstCCCrstCCpuCst$

    $A\ p/Cpq,\ q/CCrst,\ r/Crs,\ s/t,\ t/CCpuCst,$

      $u/CCpqCCrst * C^2\ 2$—3

3. $CCCpqCCpqCCrstCtCCpuCst$

    $3\ p/Cpq,\ q/CCrst,\ r/p,\ s/Cpq,\ t/Cst * CAu/Cpq$—4

4. $CCstCCCCpquCCpqCst$

    $A\ p/s,\ q/t,\ r/Cpq,\ s/u,\ t/CCpqCst,\ u/Cst * C^2\ 4$—5

5. $CCsCstCuCCpqCst$

    $3\ p/s,\ q/Cst,\ r/p,\ s/q,\ t/Cst * C5\ u/CsCst$—6

6. $CCstCCsuCqCst$

    $5\ s/Cst,\ t/CqCst,\ u/6,\ p/Cst,\ q/CCsu\ CqCst * C6\ u/t$—$C^2\ 6$

    —7

7. $CCstCqCst$

    1 p/Crs, $q/t * C7$ s/Crs, $q/CCrst$—8

8. $CCCrstCCCrsuCst$

    8 r/s, s/t, $t,u/CCsuCqCst * C^2 6$—9

9. $CtCCsuCqCst$

    A p/q, q/Cst, r/s, s/u, $t/CqCst$, $u/t * C7$ s/q, t/Cst, q/Csu

    —$C9$—10

10. $CCqtCuCqCst$

    A p/s, q/t, r/s, s/u, $t/CqCst$, $u/Cqt * C6$—$C10$ u/Csu—11

11. $CCsCqtCuCqCst$

    11 $s/CsCqt$, q/u, $t/CqCst$, $u/11 * C^2 11$—12

12. $CuCCsCqtCqCst$

    12 $u/12 * C12$—13

13. $CCsCqtCqCst$

    13 s/Cqt, q/u, $t/CqCst * C10$—14

14. $CuCCqtCqCst$

    14 $u/14 * C14$—15

15. $CCqtCqCst$

    13 s/Cqt, $t/Cst * C15$—16

16. $CqCCqtCst$

    15 $t/CCqtCst$, $s/Csu * C16$—17

17. $CqCCsuCCqtCst$

    A p/s, q/t, r/s, s/u, $t/CCqtCst$, $u/q * C6$ q/Cqt—$C17$—18

18. $CCsqCuCCqtCst$

    13 s/Csq, $u/13$, $t/CCqtCst * C18$ $u/13$—$C13$—19

*19. $CCsqCCqtCst$

    13 $s,t/Cst$, $q/7 * C7$ $q/7$—C 7—20

20. $CCstCst$

    8 r/s, s/t, $t,u/Cst * C^2 20$—21

*21. $CtCst$

    13 s/Cst, $q/s * C20$—22

22. $CsCCstt$

    15 q/s, $t/CCstt$, $s/CrCpCqp * C22$—23

23. $CsCCrCpCqpCCstt$

    23 s/Cpq, t/r, p/t, $q/s * 24$

24. $CCpqCCrCtCstCCCpqrr$

    15 q/t, t/Cst, $s/q * C21$—25

25. $CtCqCst$

    25 t/r, $q/CrCtCst$, $s/CCpqr * 26$

26. $CrCCrCtCstCCCpqrr$

    $A\ s/CtCst$, $t/CCCpqrr$, $u/r * C24$—$C26$—27

27. $CCprCCtCstCCCpqrr$

    13 s/Cpr, $q/CtCst$, $t/CCCpqrr * C27$—$C21$—28

28. $CCprCCCpqrr$

    13 s/t, $q/21 * C21\ s/21$—$C21$—29

29. Ctt

    28 $r/p * C29\ t/p$—30

*30. $CCCpqpp.$

III.

*A_1. $CCCpqpp$

A_2. $CCCCpqrsCCqrCps.$

    $A_2\ r,s/p * CA_1$—1

1. $CCqpCpp$

    1 $q/CCpqp * CA_1$—2

2. Cpp

    $A_2\ p/Cpq$, q/r, r/s, $s/CCqrCps * CA_2$—3

3. $CCrsCCpqCCqrCps$

    3 $s/r * C2$—4

*4. $CCpqCCqrCpr$

    4 p/Cpq, $q/CCqrCpr$, $r/s * C4$—5

5. $CCCCqrCprsCCpqs$

    $A_2\ s/CCpqr * C2\ p/CCpqr$—6

6. $CCqrCpCCpqr$

    6 q/r, $p/q * C2\ p/r$—7

7. $CqCCqrr$

    4 p/q, $q/CCqrr$, $r/Cpr * C7$—8

8. $CCCCqrrCprCqCpr$

    5 q/Cqr, $s/CqCpr * C8$—9

9. $CCpCqrCqCpr$

    8 $r/p * C1\ q/Cqp$—10

10. $CqCpp$

    9 p/q, $q,r/p * C10$—11

*11. $CpCqp.$

§ 7. Axiom-systems of the E-calculus[1]

I.

*A_1. $EEpEqrEEpqr$
A_2. $EEpqEqp$.

    $A_2\ p/EpEqr,\ q/EEpqr * EA_1$—1
1. $EEEpqrEpEqr$
    $1\ p/Epq,\ q/r,\ r/EpEqr * E1$—2
2. $EEpqErEpEqr$
    $2\ p/EpEqr,\ q/EEpqr,\ r/s * EA_1$—3
3. $EsEEpEqrEEEpqrs$
    $A_1\ p/s,\ q/EpEqr,\ r/EEEpqrs * E3$—4
4. $EEsEpEqrEEEpqrs$
    $4\ s/Epq,\ p/r,\ q/p,\ r/Eqr * E2$—5
*5. $EEErpEqrEpq$.

II.

A_1. $EEpEqrErEqp$
A_2. $EEEpppp$.

    $A_1\ p/EpEqr,\ q/r,\ r/Eqp * EA_1$—1
1. $EEqpErEpEqr$
    $1\ q/EEppp,\ r/p * EA_2$—2
2. $EpEpEEEpppp$
    $A_1\ q/p,\ r/EEEpppp * E2$—$EA_2$—3
3. Epp
    $1\ q/p * E3$—4
4. $ErEpEpr$
    $4\ r/Epp,\ p/Eqq * E3$—$E3\ p/q$—5
5. $EEqqEpp$
    $1\ q,r/Eqp,\ p/ErEpEqr * E1$—6
6. $EEqpEErEpEqrEEqpErEpEqr$
    $A_1\ p/Eqp,\ q/ErEpEqr,\ r/EEqpErEpEqr * E6$—7
7. $EEEqpErEpEqrEEErEpEqrEqp * E1$—8
8. $EErEpEqrEqp$
    $1\ q/ErEpEqr,\ p/Eqp,\ r/EEqrEpr * E8$—9

---

[1] In the following, asterisks are used to distinguish (i) the axioms of Leśniewski, or (ii) the axioms of system I of this section—with occasional changes of variable.

9. $EEEqrEprEEqpEErEpEqrEEqrEpr$
   $A_1 p/EEqrEpr, q/Eqp, r/EErEpEqrEEqrEpr * E9$
   $—EA_1 p/r, q/p, r/Eqr—10$

10. $EEqpEEqrEpr$
    $A_1 p/Eqp, q/Eqr, r/Epr * E10—11$

11. $EEprEEqrEqp$
    $10 q/Eqq, p/Epp * E5—12$

12. $EEEqqrEEppr$
    $10 q/EEEpqEpqEpq, p/Epq, r/EEqqEpq *$
    $EA_2 p/Epq—E12 q,r/Epq, p/q—13$

13. $EEpqEEqqEpq$
    $11 p/Eqp, r/EEqqEpq, q/Epq * E10 r/q—E13—14$

*14. $EEpqEqp$
    $10 q/Epq, p/Eqp * E14—15$

15. $EEEpqrEEqpr$
    $11 p/ErEqp, r/EEqpr, q/EEpqr * E14 p/r, q/Eqp—E15$
    $—16$

16. $EEEpqrErEqp$
    $11 p/EEpqr, r/ErEqp, q/EpEqr * E16—EA_1—17$

*17. $EEpEqrEEpqr.$

III.

A. $EEEEpqrsEsEpEqr.$
   $A p/Epq, q/r, r/s, s/EsEpEqr * E A—1$

1. $EEsEpEqrEEpqErs$
   $1 s/EsEpEqr, p/Epq, q/r, r/s * E1—2$

2. $EEEpqrEsEsEpEqr$
   $1 s/EEpqr, p, q/s, r/EpEqr * E2—3$

3. $EEssEEpEqrEEpqr$
   $3 s/EEppEpEpp * E1 s/Epp, q, r/p—4$

*4. $EEpEqrEEpqr$
   $4 p/EpEqr, q/Epq * E4—5$

5. $EEEpEqrEpqr$
   $A q/Eqr, r/Epq, s/r * E5—6$

6. $ErEpEEqrEpq$
   $4 p/r, q/p, r/EEqrEpq * E6—7$

7. $EErpEEqrEpq$
   4 $p/Erp, q/Eqr, r/Epq * E7$—8
*8. $EEErpEqrEpq.$

IV.

A. $EEEpEqrEErssEpq.$

A $p/EEstEtr, q/Ers, r/s, s/t * EA p/Est, q/t$—1
1. $EEEstEtrErs$
   1 $s/EEErssEqr, t/EErss, r/q * EA p/EErss$—2
2. $EqEEErssEqr$
   1 $s/q, t/EErss, r/EEqEErssr * E2 q/EqEErss$—3
3. $EEEqEErssrq$
   3 $q/EErsEsEErss, r/EErss * EA p/Ers, q/s, r/EErss$—4
4. $EErsEsEErss$
   1 $s/r, t/s, r/EErss * E4$—5
5. $EEErssr$
   5 $r/EErsEsr, s/EErss * EA p/Ers, q/s$—6
*6. $EErsEsr$
   6 $r/EEstEtr, s/Ers * E1$—7
7. $EErsEEstEtr$
   7 $r/EErss, s/r, t/EpEqr * E5$—8
8. $EErEpEqrEEpEqrEErss$
   7 $r/ErEpEqr, s/EEpEqrEErss, t/Epq * E8$—$EA$—9
9. $EEpqErEpEqr$
   7 $r/EEEpqrr, s/Epq, t/ErEpEqr * E5 r/Epq, s/r$—$E9$—
   10
10. $EErEpEqrEEEpqrr$
    6 $r/ErEpEqr, s/EEEpqrr * E10$—11
11. $EEEEpqrrErEpEqr$
    1 $s/EEpqr, t/r, r/EpEqr * E11$—12
*12. $EEpEqrEEpqr.$

Added in proof. In 2.2 (p. 4) of my paper 'Untersuchungen über Unabhängigkeitsbeweise nach der Matrizenmethode' [Investigations concerning Independence-proofs by the matrix-method]† the following concept $E'(\mathfrak{M})$ was used by an oversight

† [This paper appeared in *Wiadomości Matematyczne* 41 (1936), pp. 33–70. —*Ed.*]

instead of the usual one $E(\mathfrak{M})$—this was most kindly pointed out to me by Dr. J. C. C. McKinsey (U.S.A.):

$\alpha \in E'(\mathfrak{M}) = \alpha \in E(\mathfrak{M})$ and α does not assume two different values.

The condition for the satisfaction of a proposition given in 2.2 is thus (for the conventional concept $E(\mathfrak{M})$) sufficient but not necessary. The remaining results given in this paper on congruence matrices still retain their validity, for they concern only matrices whose $B(\mathfrak{M})$ consists of a single element and hence for which the concepts $E(\mathfrak{M})$ and $E'(\mathfrak{M})$ collapse into one another; only in the case of 5 does it turn out that the above condition is merely sufficient.

Note. I proved in 1927 that for every finite (non-empty) domain there exist two different functions, such that by means of each of them all functions of the domain can be constructed (such functions I call *generators* of the domain). These functions can be defined in the following way, if one designates the elements of the domain by 0, 1, 2,..., $n-1$ and the smallest non-negative root of x modulo n by $[x]_n$:

1. $\max\left([-x]_n, [y+1]_n\right)$
2. $\min\left([-x]_n, [y+1]_n\right)$

These results were then communicated in writing to Łukasiewicz and later presented in his seminar.

I have developed these results considerably by demonstrating, among others, the following theorems I–IV; in them I designate the function $[x+1]_m$ by $z(x)$, the ith iteration of $f(x)$ $(i=1, 2,...)$ by $f^i(x)$, the function $f^{n-1}(x)\ddagger$ by $f^{-1}(x)$, and finally by $e_i(x)$ the function which for $x = n-1-i$ equals 1 and otherwise equals 0.

I. The following functions are generators of the domain 0, 1, ..., $n-1$:

1. $\max\big(z^l(x), z^m(y)\big)$ $\qquad\Big\{$ (l, m, n have no
2. $\min\big(z^i(x), z^m(y)\big)$ $\qquad\Big\{$ common divisor)

3. $z(x)+z^m(y)$ $\qquad\Big\{$ m, n have no common divisor; $a+b$ has
4. $e^m(x)+z^m(y)$ $\qquad\quad$ the ordinary sense when $a+b \leqslant n-1$,
$\qquad\qquad\qquad\qquad\quad$ otherwise $a+b \leqslant n-1$)

$\ddagger$ For infinite domains the inverse of $f(x)$.

II. If $\phi(x,y)$ is a generator of a certain (possibly infinite) domain B and $f(x)$ an arbitrary function which permutes the elements of B, then the function $\psi(x,y) = f^{-1}\big(\phi(f(x),f(y))\big)$ is again a generator of B. To prove this it suffices to note that from $\phi(x,y) = z$ it always follows that $\psi(f^{-1}(x),f^{-1}(y)) = f^{-1}(z)$, and hence that one obtains $\phi(x,y)$ from $\psi(x,y)$ by a renaming of the elements of B.

III. If a function $\phi(x,y)$ suffices for the construction of all one-place functions of a finite domain B, then it is a generator of B.

IV. If B is a finite domain, then for every permutation $f(x)$, for which $f(x) \neq x$ always, there exists at least one generator of B, $\phi(x,y)$, with the property that $\phi(x,x) = f(x)$.

The proofs of the above I shall publish shortly.

In connexion with the above I would like to add that Dr. Donald L. Webb (U.S.A.), without knowledge of my results, has recently constructed the following two generators of the domain $0, 1,..., n-1$:

1. $\phi(x,y)$ with the property that $\phi(x,x) = z(x)$, $\phi(x,y) = 0$ for $x \neq y$.

2. $z(\min(x,y))$

Cf. his publications:

(1) 'Generation of any n-valued logic by one binary operation', *Proceedings of the National Academy of Sciences* 21 (1935), pp. 252–4.

(2) 'Definition of Post's generalized negative and maximum in terms of one binary operation', *American Journal of Mathematics* 58 (1936).

See also the paper of E. L. Post, *Amer. Journ. of Math.* 43 (1921), pp. 163–85, where he claims to have constructed a generator for finite domains without presenting it explicitly.

Łomża (Poland) 2. XI. 1935

15

CONTRIBUTIONS TO METALOGIC—II†

MORDCHAJ WAJSBERG

Introduction

THIS work is a sequel to my paper 'Metalogische Beiträge' (this journal, volume 43),‡ which in future will be designated MB. It contains axiom systems for the signs C, C and N, C and 0 (C = implication, N = negation, 0 = the False). In this paper it is proved that in the Tarski–Bernays axiom system for implicational logic:

1. $CpCqp$ $p \supset (q \supset p)$
2. $CCCpqpp$ $((p \supset q) \supset p) \supset p$
3. $CCpqCCqrCpr$ $(p \supset q) \supset ((q \supset r) \supset (p \supset r))$

the propositions 1 and 2 can be replaced by the proposition $CrCCCpqpp$. $r \supset (((p \supset q) \supset p) \supset p)$

Explanation of terminology

1. The expression α_1 & α_2 & ... $\to \beta_1$ & β_2... denotes the rule, if α_1, α_2, etc., then β_1, β_2, etc. If such a rule is a consequence of a certain set M of formulae (i.e. propositions and rules) then the following theorem holds: Whenever (given a certain simultaneous substitution for their variables) the propositions $\alpha_1, \alpha_2 ...$ are all consequences of M, then (for the same substitution of their variables) the propositions $\beta_1, \beta_2...$ are also all consequences of M.

For example, the rule $p \to Cqp$ $p \vdash q \supset p$ means that whenever any proposition α is a consequence of certain formulae in question at any given time, then $C\beta\alpha$ is a consequence of the same formulae.

† This paper appeared originally under the title 'Metalogische Beiträge II' in *Wiadomości Matematyczne* 47 (Warsaw, 1939), pp. 119–39. Translated by S. McCall.   ‡ [Paper 14 above—Ed.]

If a rule of the above form is provable as a theorem, it means that the formulae β_1, β_2... are deducible from the formulae α_1, α_2.... (In the first part of this work the sign $\vdash$ was used for this purpose.)

2. If $\phi = C\alpha_1 C\alpha_2 ... C\alpha_k \beta$, then $\phi^k = (\alpha_1 \,\&\, \alpha_2 \,\&\, ... \alpha_k \to \beta)$. If, for example, $\phi = C\alpha\beta$, ϕ^1 is the rule $\alpha \to \beta$.

3. $Ax_C (\alpha_1, \alpha_2...)$ means that the class of formulae α_1, α_2... forms a (complete) axiom system of the C-calculus (of implicational logic). The symbols Ax_{CN} and Ax_{CO} are to be understood similarly.

4. $C^k \alpha\beta$ denotes for $k = 0$ the proposition β and for $k = 1, 2...$ the proposition $C\alpha_1 C\alpha_2 ... C\alpha_k \beta$, where $\alpha_i = \alpha$ $(i = 1, 2,... k)$. $C\alpha_1 ... \alpha_k \beta$ serves as an abbreviation for $C\alpha_1 C\alpha_2 ... C\alpha_k \beta$.

5. Provable formulae of the customary propositional calculus are designated as valid.

6. The rule of substitution and the rule of detachment $p \,\&\, Cpq \to q$ are used for the deduction of consequences.

7. For brevity's sake the following propositions will be denoted by the names on the left (which derive in part from *Principia Mathematica*).

Comm	$CCpCqrCqCpr$	$(P \supset (Q \supset R)) \supset (Q \supset (P \supset R))$
Comm*	$CqCCpCqrCpr$	$(Q \supset ((P \supset (Q \supset R)) \supset (P \supset R))$
Fals	COp	$\sim A \supset (A \supset P)$
Freg	$CCpCqrCCpqCpr$	$(P \supset (Q \supset R)) \supset ((P \supset Q) \supset (P \supset R))$
Freg*	$CCpqCCpCqrCpr$	
Id	Cpp	
Peir	$CCCpqpp$	
Pon	$CpCCpqq$	
Red	$CCpCpqCpq$	
Simpl	$CpCqp$	
Simpl*	$CqCpp$	
Syl	$CCpqCCqrCpr$	
Syl*	$CCpqCCrpCrq$	
Transp	$CCNpNqCqp$	
Transp*	$CqCCNpNqp$	
Transp$_1$	$CCpqCNqNp$	$(P \supset Q) \supset (\sim Q \supset \sim P)$

Correspondingly, Comm^1 denotes the rule $CpCqr \to CqCpr$.

§ 1. Axiom systems of the C-calculus

THEOREM 1. Ax_C (Syl, Simpl, Peir) (Tarski–Bernays).

Theorem 2 reproduces the most important results of MB, § 1.

THEOREM 2.

(a) Ax_C (Syl, Simpl, Peir¹)   (Th. 18 of MB)
(b) Ax_C (Syl, Pon, Peir)   (Th. 1)
(c) Ax_C (Syl, Peir, $CpCCqpp$)   (Th. 4)
(d) Ax_C (Syl, Peir¹, $CpCCqrp$)   (Th. 23)
(e) Ax_C (Syl*, Pon, Peir)   (Th. 24)
(f) Ax_C (Syl, Peir, $CpC\alpha p$, $CC\alpha qq$), provided the last two expressions represent valid C-propositions.   (Th. 2)
(g) Ax_C (Syl, Peir¹, $CpCCPqp$, $CpCCCPqp\alpha$), provided the last expression represents a valid C-proposition.
                                                       (Th. 12)
(h) Ax_C (Syl, Peir, $CpCCppp$, $CCCppqq$)   (Th. 13)
(i) Ax_C (Syl, Peir, $CpCCqqp$)   (Th. 6, Peir, p/q)
(j) Syl¹ & Peir¹ → Red¹   (Th. 7)
(k) Syl¹ & Peir¹ → (Cpq & $CCprq$ → q)   (Th. 16)
(l) Syl & Peir¹ & $CpCCpqp$ → Pon   (Th. 19)
(m) Syl → $CCpCqrCCsqCpCsr$ (Proof of Th. 1)
(n) Syl → $CCpqCCCprsCCqrs$ (as above)
(o) Syl & Peir → Red   (as above)

THEOREM 3. Syl & $C\alpha CC\alpha\beta\beta → CC\gamma C\alpha\beta C\alpha C\gamma\beta$

To prove this make the substitution p/α, $q/C\alpha\beta$, r/β, s/γ in the consequence $CCpCqrCCsqCpCsr$ of Syl.

THEOREM 4. Syl & $CCpCpqCpq$ (A) & $C\alpha CC\alpha\beta\alpha$ (B) → $C\alpha CC\alpha\beta\beta$.

Proof.

Syl p/α, $q/CC\alpha\beta\alpha$, $r/CC\alpha\beta CC\alpha\beta\beta = CB$—$C$ Syl $p/C\alpha\beta$, q/α, r/β—a

(a) $C\alpha CC\alpha\beta CC\alpha\beta\beta$
    Syl p/α, $q/CC\alpha\beta CC\alpha\beta\beta$, $r/CC\alpha\beta\beta = Ca$—$C$ A $p/C\alpha\beta$, q/β—b

(b) $C\alpha CC\alpha\beta\beta$

THEOREM 5. Syl & $CCpCpqCpq$ & $C\alpha CC\alpha\beta\alpha \rightarrow CC\gamma C\alpha\beta C\alpha$ $C\gamma\beta$.   (Th. 3, Th. 4)

THEOREM 6. Syl & Peir & $C\alpha Cq\alpha \rightarrow CC\gamma C\alpha\beta C\alpha C\gamma\beta$, provided the variable q does not occur in α.   (Th. 5, Th.2o)

THEOREM 7. Syl & Peir & $CCp\alpha Cq Cp\alpha$ (A) $\rightarrow C\alpha Cq\alpha$, provided p and q do not occur in α.

Proof.

(a) $CCqCCp\alpha\alpha CCp\alpha Cq\alpha$	(Th. 6, $\alpha/Cp\alpha$, γ/q, β/α)
(b) $CpCCC\alpha q\alpha\alpha$	($Ap/CC\alpha q\alpha$, q/p; Peir p/α)
(c) $CCC\alpha q\alpha Cp\alpha$	(*a*. q/p, $p/C\alpha q$, *b*)
(d) $CCC\alpha q\alpha C\alpha Cp\alpha$	(Syl^2 $p/CC\alpha q\alpha$, $q/Cp\alpha$, $r/C\alpha Cp\alpha$;
	c, Aq/α)
(e) $CCC\alpha Cq\alpha\alpha C\alpha Cq\alpha$	(*d*. $q/Cq\alpha$, p/q)
(f) $C\alpha Cq\alpha$	(Peir $p/C\alpha Cq\alpha$, q/α; *e*)

THEOREM 7a.† Ax_C (Syl, Peir, $CCprCqCpr$)

Proof. If we put $\alpha = r$ in the previous theorem, we see that Simpl is deducible from the above propositions—but we have Ax_C (Syl, Peir, Simpl).

THEOREM 8. Ax_C (Syl, Peir, $CCp_1...p_n rCqCp_1...p_n r$).

Proof by induction with the help of Th. 7.

THEOREM 9. Syl & Peir & $CqC\alpha\alpha(A) \rightarrow C\alpha Cq\alpha$, provided q does not occur in α.

Proof. (a) $CCp\alpha CqCp\alpha$ ($CCpCqrCCsqCpCsr -$
                                                              p/q, $q,r/\alpha$, s/p; A)

If we now assume that p does not occur in α, we obtain

        (b) $C\alpha Cq\alpha$ (Th. 7, *a*)

From this theorem we obtain, putting $\alpha = p$:

THEOREM 10. Syl & Peir & $CqCpp \rightarrow CpCqp$, whence

THEOREM 11. Ax_C (Syl, Peir, $CqCpp$).

In general we obtain by means of Ths. 8 and 9:

THEOREM 12. Ax_C (Syl, Peir, $CqCCp_1...p_n rCp_1...p_n r$).

THEOREM 13. Ax_C (Syl, Peir, $CqCCCpppp$ (A)).

† [*Ed. note.* Two theorems numbered 7 appear in the original.]

Proof. Because of Th. 11 it suffices to derive the proposition $CqCpp$ from the above propositions.

(a) $CCpqCCCprsCCqrs$ (Syl)

(b) $CCpCqrCCsqCpCsr$ (Syl)

(c) $CCCpppCCCCppppCCppp$ ($a.\ p/Cpp,\ q,r,s/p$)

(d) $CCCpppCqCCppp$ ($b.\ p,r/CCppp,q/CCCpppp,s/q;c,A$)

(e) $CCCpppCqp$ (Th. 6, $\alpha/CCppp,\ \beta/p,\ \gamma/q;\ d,\ A$)

(f) $CCCppCppCpp$ ($a.\ p/CCppp,q,r,s/Cpp;e.\ q/p;$ Peir $p/Cpp,q/p$)

(g) $CqCpp$ ($e.\ p/Cpp;f$)

If we now notice that from CrPeir the propositions Peir and $CqCCCpppp$ are deducible, we obtain

THEOREM 14. Ax_C (Syl, CrPeir).

THEOREM 15. Ax_C (Syl, Peir, $CqCCCpppCCppp$ (A)).

Proof. Because of Th. 13 it suffices to derive $CqCCCpppp$ from these propositions.

(a) $CqCCCCppppCCCppppp$ (Syl² $p/q,q/CCCpppCCppp$, $r/CCCCppppCCCppppp;\ A;$ Syl $p,q/CCppp,r/p$)

If we denote $CCCpppp$ by α, then (a) is identical with

(b) $CqC\alpha\alpha$

(c) $C\alpha Cq\alpha$ (Th. 9, b)

(d) α (Peir q/p)

(e) $Cq\alpha = CqCCCpppp$ ($c,\ d$)

THEOREM 16. Ax_C (Syl, Peir, $CqCCppCpp$ (A)).

Proof. On the basis of Th. 15 it suffices to show that from the above axioms the proposition $CqCCCpppCCppp$ is derivable.

(a) $CCCppCppCCCpppCCppp$ (Syl $p,q/Cpp,r/p$)

(b) $CqCCCpppCCppp$ (Syl $p/q,q/CCppCpp$, $r/CCCpppCCppp;\ A,a$).

THEOREM 17. Ax_C (Syl, Peir, $CqCCppCCppCpp$ (A)).

Proof. We reduce this theorem to the previous one by deriving the proposition $CqCCppCpp$ from the above axiom-triple.

(a) $CCpCpqCpq$ (Syl, Peir)

(b) $CqCCppCpp$ (Syl $p/q,q/CCppCCppCpp,r/CCppCpp;$ $A;a.\ p,q/Cpp$).

If we now note that from CsSyl the propositions Syl and $CqCCppCCppCpp$ are derivable, we obtain

THEOREM 18. Ax_C (CsSyl, Peir).

THEOREM 19. Ax_C (Syl, Peir, Simpl[1]).

Proof by either Th. 14 or 18.

THEOREM 20. Ax_C (Syl, $CCCpqpCrp$ (A)).

Proof. By Th. 11 it is sufficient to deduce $CqCpp$ and Peir from the above formulae.

(a) $CCCpppCpp$	$(A\ q,r/p)$
(b) $CqCpp$	$(A\ p/Cpp,\ q/p,\ r/q\ ;\ a)$
(c) $CCpCpqCCCpqqCpq$	(Syl $q/Cpq,\ r/q$)
(d) $CCCpqqCpq \rightarrow Cpq$	$(A\ p/Cpq,\ r/A)$
(e) $CpCpq \rightarrow Cpq$	$(c,\ d)$
(f) $CCCpqpp =$ Peir	$(e.\ p/CCpqp,\ q/p\ ;\ A\ r/CCpqp)$

THEOREM 21. Ax_C ($CCpqCsCCqrCpr$ (A), Peir).

Proof. By Th. 18 it suffices to deduce CsSyl from the above propositions.

(a) $CCpCprCsCCCprrCpr$	$(A\ q/Cpr)$
(b) $CCpqCCpqCCqrCpr$	$(A\ s/Cpq)$
(c) $Cpq\ \&\ Cqr \rightarrow Cpr$	(b)
(d) $CpCpr \rightarrow CsCCCprrCpr$	(a^1)
(e) $CCCCprrCprCpr$	(Peir $p/Cpr,\ q/r$)
(f) $CpCpr \rightarrow CsCpr$	$(c.\ p/s,\ q/CCCprrCpr,\ r/Cpr\ ;\ d,\ e)$
(g) Cs Syl	$(f.\ p/Cpq,\ r/CCqrCpr\ ;\ b)$

THEOREM 22. $Ax_C(CpCqCrp(A), CCpqr\ \&\ Csr \rightarrow CCpsr(B))$.

As a proof we deduce from these formulae the formulae Simpl, Syl, and Peir[1] (see Th. 2a).

(a) $CpCpq \rightarrow Cpq$	$(B\ r/CrCsCpq,\ s/Cpq\ ;\ A\ p/Cpq,\ q/r,\ r/s)$
(b) $CpCqp$ (Simpl)	$(a.\ q/Cqp\ ;\ A\ q/p,\ r/q)$
(c) Cpp	$(a.\ q/p\ ;\ b.\ q/p)$
(d) $CCqrCqCpr$	$(B\ p/q,\ q/Cpr,\ r/CqCpr,\ s/r\ ;\ c.\ p/CqCpr\ ;$
	$A\ p/r,\ r/p)$
(e) $CCpCqrCqCpr$	$(B\ q/r,\ r/CqCpr,\ s/Cqr\ ;\ b.\ p/Cpr\ ;\ d)$
(f) $CqCCqrCpr$	$(e.\ p/Cqr,\ r/Cpr\ ;\ d)$

(g) $CCpqCCqrCpr = \mathrm{Syl}$ (B q/r, $r/CCqrCpr$, s/q;
                                      b. p/Cpr, q/Cqr; f)

(h) $CCpqp \rightarrow p$ (Peir[1])   (B $r,s/p$; c)

THEOREM 23. Ax_C (Simpl, Pon, Syl[1], Syl*[1], Peir[1]).

To prove this we derive the two formulae of the preceding theorem.

(a) $CCrpCrCqp$ (Simpl, Syl*[1])
(b) $CpCqCrp = A$ (a, Simpl)
(c) $Cpq \rightarrow CCCprsCCqrs$ (Syl[1])
(d) $Cpq \ \& \ CCprq \rightarrow CCqrq$ (c. s/q)
(e) $Cpq \ \& \ CCprq \rightarrow q$ (d. Peir[1] p/q, q/r)
(f) $CCpqr \rightarrow CCpqCCprr$ (a. r/Cpq, p/r, q/Cpr)
(g) $CpCCprr \ \& \ CCpqCCprr \rightarrow CCprr$ (e. $q/CCprr$, r/q)
(h) $CCpqr \rightarrow CCprr$ (g, i. Pon q/r; f)
(i) $Csr \rightarrow CCpsCpr$ (Syl*[1] p/s, q/r, r/p)
(j) $CCpsCpr \rightarrow CCCprrCCpsr$ (Syl[1] p/Cps, q/Cpr)
(k) $Csr \rightarrow CCCprrCCpsr$ (i, j)
(l) $Csr \ \& \ CCprr \rightarrow CCpsr$ (k)
(m) $CCpqr \ \& \ Csr \rightarrow CCpsr = B$ (h, l)

THEOREM 24. Pon & Syl[2] & Comm[1] $\rightarrow$ Syl[1]

Proof.

(a) $Cpq \rightarrow CpCCqrr$ (Syl[2], Pon p/q, q/r)
(b) $Cpq \rightarrow CCqrCpr$ (a. Comm[1] q/Cqr)

THEOREM 25. Id & Comm[1] $\rightarrow$ Pon.

Proof.

(a) $CCpqCpq$ (Id p/Cpq)
(b) $CpCCpqq = \mathrm{Pon}$ (a, Comm[1])

THEOREM 26. Id & Syl[2] & Comm[1] $\rightarrow$ Syl[1] (Th. 24, Th. 25).

THEOREM 27. Simpl & Comm[1] $\rightarrow$ Id.

Proof.

(a) $CqCpp$ (Simpl, Comm[1])
(b) Cpp (a)

THEOREM 28. Simpl & Syl[2] & Comm[1] $\rightarrow$ Syl[1] & Pon (Ths. 25–27).

THEOREM 29. Ax_C (Simpl, Syl*[1], Peir[1], Comm[1]) (Ths. 23, 28).

THEOREM 30. Syl & Pon & Simpl[1] & Peir[1] → Comm & Red & Id p/Cpq & Simpl p/Cpr & Peir p/Cpr.

Proof.

(a) $CpCpq \to Cpq$		(Syl q/Cpq; Peir[1] p/Cpq)
(b) Comm		(Syl, Pon)
(c) $CCpqCpq = \text{Id } p/Cpq$		(Comm, Pon)
(d) $CpCpCCpCpqq$		(c. q/Cpq; Comm, Syl)
(e) $CCpCpqCpq = \text{Red}$		(a. $q/CpCpq$; d; Comm)
(f) $CrCCpqCpq$		(c, Simpl[1])
(g) $CCprCqCpr = \text{Simpl } p/Cpr$		(Comm; f. r/q, q/r)
(h) Cpr & $CCpqr \to r$		(Syl, Peir[1]; see MB, Th. 16)
(i) $CCprCCCCprqCprCpr$		(g. $q/CCCprqCpr$)
(j) $CCCprqCCCCprqCprCpr$		(Pon $p/CCprq$, q/Cpr)
(k) $CCCCprqCprCpr = \text{Peir } p/Cpr$		
		(h. p/Cpr, $r/CCCCprqCprCpr$; i, j)

If one adds to the above propositions the proposition Id, one obtains (by Simpl[1]) $CqCpp$, and then (by Comm) $CpCqp = \text{Simpl}$. If one notes (MB, Th. 18) that Ax_C (Syl, Simpl, Peir[1]), one obtains

THEOREM 31. Ax_C (Syl, Pon, Id, Simpl[1], Peir[1]).

THEOREM 32. Simpl & Freg[1] → Syl[1] & Syl*[1] & Comm[1] & Id & Pon.

Proof.

(a) $Cqr \to CpCqr$		(Simpl p/Cqr, q/p)
(b) $Cqr \to CCpqCpr$		(a, Freg[1])
(c) Syl[1]		(b. q/p, r/q, p/r)
(d) $CCpqCpr$ & $CqCpq \to CqCpr$		(b. q/Cpq, r/Cpr, p/q)
(e) $CCpqCpr \to CqCpr$		(d, Simpl p/q, q/p)
(f) $CpCqr \to CqCpr$ (Comm[1])		(Freg[1], e)
(g) $CqCpp$		(Comm[1], Simpl)
(h) Cpp (Id)		(g)
(i) $CpCCpqq$ (Pon)		(h. p/Cpq, Comm[1])
(j) $CCrpCrCCpqq$		(Syl*[1], i)
(k) $Cpq \to CpCCqrr$		(j. r/p, p/q, q/r)

(*l*) Syl[1]   (*k*, Comm[1])

(*m*) $Cpq \rightarrow CCCprsCCqrs$ (*l*)

It follows that Simpl & Freg[1] → Syl*[1] & Comm[1]. With the help of Th. 29 we obtain

THEOREM 33. Ax_C (Simpl, Freg[1], Peir[1]).

THEOREM 34. Ax_C (Syl, Peir, $CpCCppp$ (*A*)).

Proof. It suffices to deduce $CpCCqpp$ from these propositions (Th. 2c).

(*a*) $CCpCqrCCsqCpCsr$		(Syl)
(*b*) $CCqpCCsqCCprCsr$	(Syl $p/q, q/p$; *a.* p/Cpr; Syl[2])	
(*c*) $CCqpCCpqCCppCpp$	(*b.* $s, r/p$)	
(*d*) $CCpCpqCpq$	(Syl, Peir)	
(*e*) $CpCqr \rightarrow CCpqCpCpr$	(a[1] s/p)	
(*f*) $CCpqCpCpr \rightarrow CCpqCpr$	(*d.* q/r; Syl[2])	
(*g*) $CpCqr \rightarrow CCpqCpr =$ Freg[1]	(*e, f*)	
(*h*) $CCCpqCqrCCpqCpr$	(*g.* $p/Cpq, q/Cqr, r/Cpr$; Syl)	
(*i*) Cpp	(*A*, Peir q/p; Syl[2])	
(*j*) $CCCppCppCpp$	(*A* p/Cpp; *i*)	
(*k*) $CCpCCpqpCCppCpp$	(*c.* $q/CCpqp$; Peir)	
(*l*) $CCpCCpqpCpp$	(*k, j*; Syl[2])	
(*m*) $CCqCCppCppCCCppCppCqCpp$	(*b.* $q/CCppCpp, p,$ $r/Cpp, s/q$; *j*)	
(*n*) $CCqCCppCppCqCqCpp$	(*h.* $p/q, q/CCppCpp, r/CqCpp$; *m*)	
(*o*) $CCqCCppCppCqCpp$	(*n*; *d.* $p/q, q/Cpp$; Syl[2])	
(*p*) $CCqpCCpqCpp$	(*c, o.* q/Cpq; Syl[2])	
(*q*) $CCsCppCpCsp$	(*a.* $q/Cpp, r/p$; *A*)	
(*r*) $CCCpqCppCpp$	(*q.* s/Cpq; *l*; Syl[2])	
(*s*) $CCqpCpp$	(*p, r*; Syl[2])	
(*t*) $CpCCqpp$	(*q. s/Cqp*; *s*)	

THEOREM 35. Syl & $CpC\alpha p$ & $\alpha \rightarrow CpCCppp$.

Proof.

LEMMA 1. Syl → $CCppC\alpha\alpha$, *if α contains only the variable p, but is distinct from p.*

Proof. We denote this statement by $S(\alpha)$. Then $S(p)$ is trivially true, and it suffices to show that $S(C\alpha\beta)$ follows from $S(\alpha)$.

Suppose that $S(\alpha)$ holds for a certain α containing only p. If $\alpha = p$ then $C\alpha\beta = Cp\beta$, and $S(C\alpha\beta)$ holds because $CCppCCp\beta$ $Cp\beta$ is a substitution of Syl. If $\alpha \neq p$, we note that $CC\alpha\alpha CC\alpha\beta C\alpha\beta$ is a substitution of Syl and obtain, by means of $CCppC\alpha\alpha$ (provable on the basis of $S(\alpha)$) and Syl p/Cpp, $q/C\alpha\alpha$, $r/CC\alpha\beta C\alpha\beta$, the proposition $CCppCC\alpha\beta C\alpha\beta$. But this is $S(C\alpha\beta)$, as was to be proved.

LEMMA 2. Syl $\rightarrow C\beta CCpp\beta$, *if β contains only p, but is distinct from p.*

Proof. β is not a variable and is hence of the form $C\gamma\alpha$; if $\alpha = p$, $C\beta CCpp\beta = CC\gamma pCCppC\gamma p$, which is a substitution of Syl. We may therefore assume that $\alpha \neq p$. But then according to the previous lemma Syl $\rightarrow CCppC\alpha\alpha$ and hence also Syl $\rightarrow$ (1) $CCC\alpha\alpha\beta CCpp\beta$. Furthermore, the proposition $CC\gamma\alpha CC\alpha\alpha C\gamma\alpha$ = (2) $C\beta CC\alpha\alpha\beta$ is a substitution of Syl, and we obtain finally from (1) and (2) by means of Syl the desired proposition $C\beta CCpp\beta$.

LEMMA 3. Syl & $CpC\beta p$ & $\beta \rightarrow CpCCppp$, *if β contains only p.*

Proof. Our formula plainly holds if β is a variable. Otherwise Syl & $\beta \rightarrow CCpp\beta$ holds (lemma 2), consequently also Syl & $\beta \rightarrow$ $CC\beta pCCppp$, and finally, by means of Syl $q/C\beta p$, $r/CCppp$, the desired formula.

This completes the proof of Theorem 35, if one substitutes p for all variables in α.

THEOREM 36. Ax_C (Syl, Peir, α, $CpC\alpha p$), *if α is a valid C-proposition* (Ths. 34, 35).

THEOREM 37. Ax_C (Syl, Peir, $CpC\alpha p$), *if α is a consequence of this axiom-triple* (Th. 36).

THEOREM 38. Ax_C (Syl, Peir, $Cq\alpha$), *if α is a valid C-proposition not containing q.*

Proof. If we substitute for all variables in α the proposition $CCppCpp$ we obtain a proposition δ, for which it is shown below that $C\delta CCppCpp$ is a consequence of Syl and Peir. Hence

$CqCCppCpp$ is a consequence of Syl, Peir, and $Cq\alpha$, and consequently this triple is an axiom system of the C-calculus by Th. 16.[1] We now prove the formula Syl & Peir $\rightarrow CSCCppCpp$, or in fact the following:

THEOREM 39. *Let* $\epsilon = CCppCpp$ *and for every proposition* α *let* $\phi(\alpha)$ *denote the property that* α *results from a certain* C-*proposition by substituting* ϵ *for all its variables. If* $\phi(\delta)$, *then* Syl & Peir $\rightarrow \delta$ & $C\epsilon\delta$ & $C\delta\epsilon$.

We denote this theorem by $S(\delta)$ and prove it by means of the following lemmas.

LEMMA 1. Syl & Peir $\rightarrow CC\epsilon\epsilon\epsilon$.

Proof.

(a) $CCpqCCCprsCCqrs$	(Syl)
(b) $CCpCpqCpq$	(Syl, Peir)
(c) $CCCpp\epsilon\epsilon$	(b. p, q/Cpp)
(d) $CC\epsilon\epsilon$	(a. p/Cpp, q, r, s/ϵ; Syl q, r/p; c)

LEMMA 2. Syl $\rightarrow C\epsilon C\epsilon\epsilon$ (Syl p, q, r/Cpp)

LEMMA 3. *If* $\phi(\delta)$, *then* Syl & $\delta \rightarrow C\epsilon\delta$.

Proof. If $\delta = \epsilon$, this formula holds.

By lemma 2 of Th. 35 (with the remark that in it p should be replaced by ϵ), the formula Syl $\rightarrow CSCC\epsilon\epsilon\delta$ holds, and consequently Syl & $\delta \rightarrow CC\epsilon\epsilon\delta$, whence by lemma 2 Syl & $\delta \rightarrow C\epsilon\delta$. This proves the lemma.

LEMMA 4. Syl & $C\epsilon\alpha$ & $C\beta\epsilon$ & $CC\epsilon\epsilon\epsilon \rightarrow CC\alpha\beta\epsilon$.

Proof.

(a) $CCpqCCCprsCCqrs$	(Syl)
(b) $CCqpCCsqCCprCsr$	(Syl, cf. Th. 34, b)
(c) $CC\epsilon\beta CC\epsilon\epsilon C\epsilon\epsilon$	(b. q/β, p, r, s/ϵ; $C\beta\epsilon$)
(d) $CCC\epsilon\epsilon C\epsilon\epsilon C\epsilon\epsilon$	($CC\epsilon\epsilon\epsilon$, p/Cpp)
(e) $CCC\epsilon\epsilon C\epsilon\epsilon\epsilon$	(d, $CC\epsilon\epsilon\epsilon$; Syl[2])
(f) $CC\epsilon\beta\epsilon$	(c, e; Syl[2])
(g) $CC\alpha\beta\epsilon$	(a. p/ϵ, q/α, r/β, s/ϵ; $C\epsilon\alpha$, f)

---

[1] By means of the same substitution, using Ths. 38 and 39, it can in general be shown that Ax_C (Syl, Peir, $C\alpha_1 \ldots \alpha_n Cq\alpha_{n+1}$), if q does not occur in $\alpha_1 \ldots \alpha_{n+1}$ and if the third axiom is valid.

Lemma 5. Syl & $CC\epsilon\epsilon\epsilon \to \epsilon$ & $C\epsilon\epsilon$.

Proof. By Syl^2 we obtain from $C\epsilon C\epsilon\epsilon$ (lemma 2) and $CC\epsilon\epsilon\epsilon$ the proposition $C\epsilon\epsilon$, and then by $CC\epsilon\epsilon\epsilon$ the proposition ϵ.

Lemma 6. *If* $\phi(\delta)$, *then* Syl & $CC\epsilon\epsilon\epsilon \to \delta$ & $C\epsilon\delta$ & $C\delta\epsilon$.

Proof. If we denote this statement by $S(\delta)$, then by lemma 5 $S(\delta)$ holds for $\delta = \epsilon$, i.e. for the shortest proposition with the property $\phi(\delta)$. Hence it suffices to show that, for $C\alpha\beta \neq \epsilon$, $S(C\alpha\beta)$ follows from $S(\alpha)$ and $S(\beta)$. Let $C\alpha\beta \neq \epsilon$ be a proposition with the property $\phi(C\alpha\beta)$; plainly $\phi(\alpha)$ and $\phi(\beta)$ hold, and it follows from $S(\alpha)$ and $S(\beta)$ that

(1) Syl & $CC\epsilon\epsilon\epsilon \to C\alpha\epsilon$ & $C\epsilon\beta$ & $C\epsilon\alpha$ & $C\beta\epsilon$.

But (2) Syl & $C\alpha\epsilon$ & $C\epsilon\beta \to C\alpha\beta$, and further by Lemma 3,

(3) Syl & $C\alpha\beta \to C\epsilon C\alpha\beta$,

and by lemma 4

(4) Syl & $C\epsilon\alpha$ & $C\beta\epsilon$ & $CC\epsilon\epsilon\epsilon \to CC\alpha\beta\epsilon$.

It follows that

(5) Syl & $CC\epsilon\epsilon\epsilon \to C\alpha\beta$ & $C\epsilon C\alpha\beta$ & $CC\alpha\beta\epsilon$,

as was to be shown.

Finally we obtain the desired theorem: if $\phi(\delta)$, then Syl & Peir $\to \delta$ & $C\epsilon\delta$ & $C\delta\epsilon$ (lemmas 6 and 1).

Theorem 41.† *Let* $p, q, r\ldots$ *be all the variables of our vocabulary and let* $p_1, p_2, q_1, q_2\ldots$ *be certain mutually distinct variables. For every proposition* α, *let* α' *be the proposition obtained from* α *by substituting* $Cp_1p_2, Cq_1q_2\ldots$ *for the variables* $p, q\ldots$ *respectively. If a C-proposition* β *is a consequence of the C-propositions* $\alpha_1, \alpha_2,\ldots$, *then* β' *is a consequence of* $\alpha_1'\alpha_2'\ldots$.

Proof. If β is a substitution of one of the α_i, e.g. $\beta = \alpha_i p/\gamma, q/\delta$, $\ldots$, where $p, q\ldots$ are all the variables of α_i, then $\beta' = \alpha_i p/\gamma', q/\delta'$, $\ldots$, and also $\beta' = \alpha_i' p_1/\gamma_1', p_2/\gamma_2', q_1/\delta_1', q_2/\delta_2'\ldots$, where $\gamma_1', \delta_1'\ldots$ and $\gamma_2', \delta_2'\ldots$ are respectively the antecedents and the consequents of $\gamma', \delta'\ldots$. Hence β' is a substitution of α_i' and the theorem holds for the case where β is a substitution of $\alpha_1, \alpha_2\ldots$. Conse-

---

† [*Ed. note.* Wajsberg appears to have misnumbered this and subsequent theorems, omitting 40.]

quently it is sufficient to show that if $(C\gamma\delta)'$ and γ' are consequences of α'_1, α'_2..., then so is δ'. But, since $(C\gamma\delta)' = C\gamma'\delta'$, this assertion is obvious and our proof is ended.

THEOREM 42. Ax_C (Syl, Peir, $CpCpp$ (A)).

Proof. (a, b, d–g) as in the proof of Th. 34.

(c) $CCpqCqCpq$	(a. p, r/Cpq, q/Cqq, s/q; Syl r/q; A p/q)
(h) $CCpqCpCpq$	(g. p/Cpq, $q/CqCpq$, $r/CpCpq$;
	Syl r/Cpq; c)
(i) $CCpqCCCpqCpqCpCpq$	(h; Syl q, r/Cpq; Syl^2)
(j) $CCsCpCpqCCCpqCpqCsCpq$	(b. $q/CpCpq$, p, r/Cpq; d)
(k) $CCpqCCCpqCpqCCCpqCpqCpq$	(i, j. $s/CCpqCpq$; Syl^2)
(l) $CCpqCCCpqCpqCpq$	(k; d. $p/CCpqCpq$, q/Cpq; Syl^2)
(m) Cpp	(d. q/p; A)

If we note that line l results from the proposition $A = CpCCpppp$ of Theorem 34 by substituting Cpq for p, we may conclude, with the help of the preceding theorem and Th. 34, that

(n) Syl & Peir & $l \to CCp_1p_2CCq_1q_2Cp_1p_2$ (let $\beta = CpCqp$).

From n and m we obtain $CCqqCpp$, and finally by A p/q and Syl^2 the proposition $CqCpp$, which, together with Syl and Peir, constitutes an axiom system of the C-calculus by Theorem 11.

§2. Axiom systems of the C-N and C-0-calculi

THEOREM 43.

(a) Ax_{CN}(Syl, $CCNppp$, $CpCNpq$)	(Łukasiewicz)
(b) Ax_{CN} (Simpl, Freg, Transp)	(Łukasiewicz)

THEOREM 44. *The proposition COp, together with any axiom system of the C-calculus, forms an axiom system of the C-0-calculus.* (MB, § 5)

THEOREM 45. Ax_{CN} (Simpl, Freg^1, Transp).

Proof. It suffices, by Ths. 33 and 43, to deduce Peir from the above formulae.

(a) $CNqCqp$	(Syl^2 (Th. 32b), Simpl p/Nq, q/Np; Transp)
(b) $CCNqqCNqNp$	(Freg^1 p/Nq, r/Np; a. p/Np)
(c) $CCNqqCpq$	(Syl^1, b, Transp p/q, q/p)

(*d*) $CpCCNqqq$ (Comm[1] = Th. 32 *f*; *c*)

(*e*) $CCNppp$ (*d. p/d., q/p*)

(*f*) $CCCpqpp$ = Peir   (Th. 32 *m. p/Np, q/Cpq, r, s/p*;

$\qquad\qquad\qquad\qquad\qquad\qquad\qquad$ *a. p/q, q/p*; *e*)

Note. In the above theorem Transp can be replaced by
Transp*, for Comm[1] is a consequence of Simpl and Freg[1] (see
Th. 32*f*).

THEOREM 46. Ax_{CN} (Simpl, Freg[1], $CCNppp$, $CpCNpq$).

As proof we deduce from these formulae the proposition Peir
(see the derivation of line *f*, Th. 45). Taking note of Ths. 45*a*
and 33, we see that $CpCNpq$ can be replaced by $CNpCpq$.

THEOREM 47. *Adding to an axiom system of the C-calculus
the rule Transp[1] does not yield an axiom system of the C-N-calculus.*

Proof. We see without difficulty that from a set M of C-
propositions only those C-N-propositions are derivable which
are substitutions of pure C consequences of M. A proposition
of the form $CN\alpha N\beta$ can be a substitution of a valid C-proposi-
tion only if $\alpha = \beta$. But if we use on $CN\alpha N\alpha$ the rule Transp[1], we
obtain $C\alpha\alpha$, a consequence of Cpp. Therefore no new consequences are derivable by Transp[1] from a set of C-propositions
yielding Cpp, which completes our proof.

THEOREM 48. Ax_{C0} (Simpl, Freg[1], $CCCp0Cq0Cqp$ (*A*)).

Proof. If we write $N\alpha$ for $C\alpha0$, then A = Transp. By Th. 45
all valid C-propositions are derivable from the above propositions. Using $CqCpp$ (32*g*) we obtain from A *q*/0 the proposition
$C0p$, which by Th. 44 gives the required result.

THEOREM 49. Ax_{C0} (Simpl, Freg[1], Peir[1], $C0p$). (Th. 33, Th.
44).

THEOREM 50. Ax_{C0} (Simpl, Freg[1], $CCCp00p$ (*A*)).

Proof. By the previous theorem it suffices to derive $C0p$ and
Peir.

(*a*) $Cqr \to CCpqCpr$ (Th. 32*b*)

(*b*) $CCCprqCrq$ (Syl[1], Simpl; Th. 32)

(*c*) Cpp (Th. 32*h*)

(d) $Cpq \to CCCprsCCqrs$ (Th. 32 *m*)

(e) $C0p$ (*b.* $p/Cp0$, $r/0$, q/p; *A*)

(f) $CCCp0pCCp00$ (Freg¹ $p/Cp0$, q/p, $r/0$; *c.* $p/Cp0$)

(g) $CCCp0pp$ (*a.* $q/CCp00$, r/p, $p/CCp0p$; *A, f*)

(h) $CCp0Cpq$ (*a.* $q/0$, r/q; *e.* p/q)

(i) $CCCpqpp$ = Peir   (*d.* $p/Cp0$, q/Cpq, r, s/p; *h, g*)

Note. In the above theorem $CCCp00p$ may obviously be replaced by $CCCpq0p$.

Theorem 51. Ax_{C0} (Simpl, Freg¹, $CCCp0qCCq0p$ (*A*)).

Proof. It suffices, by the previous theorem, to deduce $CCCp00p$.

(a) $CpCqr \to CqCpr$ (Th. 32 *f*)

(b) Cpp (Th. 32*h*)

(c) $CC00CCCp00p$ (*a.* $p/CCp00$, $q/C00$, r/p; *A* $q/0$)

(d) $CCCp00p$ (*c, b.* $p/0$)

Note. A may be replaced above by any of the following propositions (as may be shown by putting 0 for *q*):

1. $CCq0CCCp0qp$
2. $CCCpqqCCq0p$
3. $CCq0CCCpqqp$.

Theorem 52. Ax_{C0} (Simpl, Freg¹, $C0p$ (*A*), $CCp0p \to p$ (*B*)).

Proof. By Th. 49 it is sufficient to deduce the rule Peir¹ = $CCpqp \to p$ from the above formulae.

(a) $Cqr \to CCpqCpr$ (Th. 32*b*)

(b) $CCp0Cpq$ (*a, A* p/q)

(c) Syl¹   (Th. 32)

(d) $CCCpqpCCp0p$ (*c.* $p/Cp0$, q/Cpq, r/p; *b*)

(e) $CCpqp \to p$ (*d, B*)

Theorem 53. Ax_{C0} (Simpl, Freg¹, $C0p$, $CCp00 \to$ p).

Proof. It is sufficient, by the above theorem, to deduce the rule $CCp0p \to p$ from the above formulae. To this end we derive $CCCp0pCCp00$ from Simpl and Freg¹ (as in the proof of Th. 50) and then use $CCp00 \to p$.

Addenda

THEOREM 54. Transp & Transp$_1$ & Syl^1 → Syl^{*1}.

Proof.

(a) $Cpq \to CNqNp$	(Transp$_1$)
(b) $CNqNp \to CCNpNrCNqNr$	(Syl^1 $p/Nq, q/Np, r/Nr$)
(c) $Cpq \to CCNpNrCNqNr$	(a, b)
(d) $CCNpNrCNqNr \to CCrpCNqNr$	(Transp$_1$ $p/r, q/p$; Syl^1)
(e) $Cpq \to CCrpCNqNr$	(c, d)
(f) $CCrpCNqNr \to CCrpCrq$	(Transp $f/q, q/r$; Syl^1)
(g) $Cpq \to CCrpCrq = $ Syl^{*1}	(e, f)

Similarly we may prove

THEOREM 55. Transp & Transp$_1$ & Syl^{*1} → Syl^1.

Correction. In footnote 2, p. 45 of my work, 'Untersuchungen über den Aussagenkalkül von A. Heyting', we should read: 'According to . . . J. Johansson if one replaces V2 (p. 46) by the deductively equivalent and independence-preserving formula V2′.$CCpNqCqNp$.'

Łomża, 19. xi. 1938

16

THE FULL THREE-VALUED
PROPOSITIONAL CALCULUS†

JERZY SŁUPECKI

A SYSTEM of propositional logic, defined by a two- or many-valued[1] matrix, is called a *full* system when every possible propositional function of finitely many arguments may be defined in it by means of the primitive concepts of the system. This property is not possessed by the three-valued propositional calculus whose sole primitive concepts are implication and negation and which is defined by the following matrix:

C	0	1	2	N
0	1	1	1	1
*1	0	1	2	0
2	2	1	1	2

It is easy to prove by induction that it is not possible to define, by means of the functions 'C' of implication and 'N' of negation, any function of one argument which takes the value 2 for the values 0 or 1 of its argument. Such a function, which is denoted by Łukasiewicz as the 'tertium' and represented symbolically by Tp, makes it possible to define, with the help of implication and negation, every function of the three-valued calculus with finitely many arguments.

† This paper appeared originally under the title, 'Der volle dreiwertige Aussagenkalkül', in *Comptes rendus des séances de la Société des Sciences et des Lettres de Varsovie*, Cl. iii, 29 (1936), pp. 9–11. Translated by S. McCall.
[1] The reader who wishes more ample information concerning many-valued logic is referred to the following works: J. Łukasiewicz and A. Tarski, 'Untersuchungen über den Aussagenkalkül', *Comptes rendus des séances de la Société des Sciences et des Lettres de Varsovie*, Cl. iii, 23 (1930); J. Łukasiewicz, 'Philosophische Bemerkungen zu mehrwertigen Systemen des Aussagenkalküls', ibid. [Translated as paper 3 of this volume—Ed.]

The system defined by means of the matrix

	C	0	1	2	N	T
	0	1	1	1	1	2
I.	1	0	1	2	0	2
	2	2	1	1	2	2

is a full system of three-valued logic.

One obtains the axiomatization of this system by adding to the four axioms of the three-valued calculus,[1] whose primitive concepts are implication and negation, two new axioms containing the primitive concept T:

$$CTpNTp$$
$$CNTpTp.$$

In these two axioms is contained one of the principal intuitions of three-valued logic: propositions of the value 2 are equivalent to their own negations.

The full three-valued calculus is a complete system in the sense that every proposition constructed out of the symbols C, N, T and propositional variables either is a consequence of the axioms of this logic or results in a contradiction when added to the axioms. It is also plain that the system is consistent, for not every meaningful proposition of the full three-valued calculus satisfies matrix I. Thirdly, it can be easily shown, using the matrix method of Łukasiewicz, that these axioms are independent of one another.

We see therefore that the foregoing propositional calculus possesses all the fundamental properties that we meet with in two-valued logic (for example, that which has implication and negation as primitive concepts): it is a full, complete, and con-

---

[1] These axioms are:

1. $CqCpq$
2. $CCpqCCqrCpr$
3. $CCCpNppp$
4. $CCNqNpCpq$.

See M. Wajsberg, 'Aksjomatyzacja trójwartościowego rachunku zdań', *Comptes rendus des séances de la Société des Sciences et des Lettres de Varsovie*, Cl. iii, **24** (1931). [Translated as paper 13 of this volume—Ed.]

sistent system, possessing a set of independent axioms. By contrast the three-valued calculus whose sole primitive concepts are implication and negation is not a full system, as was remarked above; it is also not a complete system in the above-mentioned sense.

17

ANTINOMIES OF FORMAL LOGIC†

LEON CHWISTEK

WHEN Bertrand Russell published his theory of logical types,[1] it might have appeared that the difficulties which had since ancient times been inherent in the foundations of logic were finally resolved. Nevertheless some additional assumptions of Russell's theory, known collectively as the principle of reducibility, could have at once inspired a certain distrust in view of their utterly arbitrary character. The form of the principle of reducibility, which in Russell's notation states

$$(\exists \, \phi) \, . \, \phi! x \equiv_x \psi x,$$

already provides an indication that this is a typical existence axiom, clearly similar in this respect to the much discussed axiom of choice. Henri Poincaré, the first critic of the theory of types, spoke against the axiom of reducibility, making the objection that Russell introduced a synthetic *a priori* statement into the axioms of logic.[2] A few years later I drew attention to some paradoxical consequences which follow from the acceptance of the axiom of reducibility.[3] It became clear that if we accept this axiom as an integral part of the system of formal logic, we have to give up once and for all the logic of constructive propositions and functions, that is, of such functions as may be constructed by means of letters or symbols given beforehand

† This paper appeared originally under the title 'Antynomje logiki formalnej' in *Przegląd Filozoficzny* 24 (1921), pp. 164–71. Translated by Z. Jordan, with certain small changes in notation.

[1] 'Mathematical logic as based on the theory of types', *American Journal of Mathematics*, 30 (1908), pp. 222–62.

[2] H. Poincaré, *Dernières Pensées*, ch. iv.

[3] 'Zasada sprzeczności w świetle nowszych badań Bertranda Russella' (The principle of non-contradiction in the light of Bertrand Russell's recent investigations), *Rozprawy Akademii Umiejętności* 30, Cracow, 1912, pp. 270–334.

by enumeration, and we have to resort to functions which cannot be thus constructed. This seems to be inconsistent with the basic conception of logic, which apparently should be independent of metaphysical assumptions such as the existence of non-constructive functions. In particular, this is clearly incompatible with the starting-point of Russell's system, which tries to avoid being involved in problems of this sort; for instance, it does not resolve the question as to the existence of infinite classes.

Considerations of this kind led me to the conclusion that the principle of reducibility cannot be an axiom of formal logic. This was the starting-point of my efforts to construct a system of formal logic based on Bertrand Russell's theory of types with the exclusion of the principle of reducibility. At the same time I continued to investigate the consequences of the axiom of reducibility; in particular, the possibility of the appearance of new antinomies. A reconstruction of Richard's paradox in the theory of types is a result of these investigations. To present this antinomy is the task of the first part of this paper. In the second part I deal with the reconstruction of the old paradox of Eubulides and with the assumptions necessary for its resolution.

1. Richard's antinomy

Let us assume that there are infinitely many inductive (natural) numbers of a given type, and let the letter R denote any relation holding between classes of the same type as that of the given inductive numbers $(0, 1, 2,...)$ and classes of these classes.

We shall now construct the function

$$\Theta n \underset{df}{.} = : n \in D'R \,.\, n \notin \breve{R}'n$$

where $D'R$ stands for the domain of the relation R, and $\breve{R}'n$ for the only class that satisfies the condition $nR(\breve{R}'n)$.

It is clear that according to the theory of classes and relations of Whitehead and Russell[1] the type of the function $\Theta\hat{n}$ is fully

---

[1] *Principia Mathematica*, Cambridge, 1910, vol. i. Henceforth, for the rest

determined, if its argument n is of the same type as the given inductive numbers 0, 1, 2,.... .

Let us now consider *all the propositional functions* of the same type as the function $\Theta\hat{n}$. These functions can be divided into those which may be represented by a set of letters or symbols given beforehand by enumeration, and the remaining functions. The functions of the first kind shall be called constructive. It follows from Richard's investigations that we can enumerate all the expressions of the form '$\hat{n}(\phi n)$'—in words: 'the class of all classes satisfying the function $\phi\hat{n}$'—where the function $\phi\hat{n}$ is of the same type as $\Theta\hat{n}$. Therefore, we can assign a numeral to every class $\hat{n}(\phi n)$ comprising the elements which satisfy a certain constructive function $\phi\hat{n}$ of the same type as the function $\Theta\hat{n}$.[1] This implies that we can construct a certain many-one relation R_0, whose domain $D`R_0$ is the class of inductive numbers 0, 1, 2,..., and whose converse domain $\varPi`R_0$ is the class of classes whose elements satisfy a certain constructive function of the type $\Theta\hat{n}$.

If we now construct the function:

$$\varPhi_0 n \mathbin{.} \underset{df}{=} : n \in D`R_0 \mathbin{.} n \notin \check{R}_0`n,$$

we see that this function is (*a*) constructive, (*b*) of the same type as $\Theta\hat{n}$. It follows from the definition of the relation R_0 that if we adopt the definition:

$$\bar{\omega} \underset{df}{=} \mathbin{.} \hat{n}(\varPhi_0 n),$$

the class $\bar{\omega}$ must correspond by means of the relation R_0 to at least one of the inductive numbers: 0, 1, 2,.... . This means that the proposition

(I) $(\exists\, n) \mathbin{.} n R_0 \bar{\omega}$

must hold.

Now it is easy to see that this proposition is false.

Let us ask whether the number n satisfying the condition $n R_0 \bar{\omega}$ is or is not contained in $\bar{\omega}$.

of this paper, the figures preceded by an asterisk refer to the theorems to be found in this work.

[1] Of course it is not excluded that the same class may be assigned different numerals, but one and only one class corresponds to a given numeral.

First, let us assume that

$$nR_0\bar\omega \,.\, n \in \bar\omega.$$

The definition of classes and of the relation which holds between an element satisfying a predicative function and this function itself (*20.01, *20.02) imply that

(1) $\qquad nR_0\bar\omega \,.\, n \in \bar\omega \,.\, \supset \,.\, n \in D'R_0 \,.\, n \notin \check{R}_0'n.$

Moreover, and irrespective of the assumption that $n \in \bar\omega$, by the definition of the symbol $\check{R}'n$ and in virtue of the fact that R_0 is a many-one relation the following thesis is established:

(2) $\qquad nR_0\bar\omega \,.\, \supset \,.\, \bar\omega = \check{R}_0'n.$

It is clear that

(3) $\qquad nR_0\bar\omega \,.\, n \in \bar\omega \,.\, \supset \,.\, n \notin \bar\omega$

can be derived from theses (1) and (2) by a simple deduction.

If we now accept the hypothesis that

$$nR_0\bar\omega \,.\, n \notin \bar\omega,$$

by means of de Morgan's law (*4.51) we obtain:

$$nR_0\bar\omega \,.\, n \notin \bar\omega \,.\, \supset \,.\, n \notin D'R_0 \,.\, \lor \,.\, n \in \check{R}_0'n.$$

In view of the thesis

$$nR_0\bar\omega \,.\, \supset \,.\, n \in D'R_0,$$

which is a simple application of the definition of the domain $D'R_0$ and of the principle *9.1, we obtain

$$nR_0\bar\omega \,.\, n \notin \bar\omega \,.\, \supset \,.\, n \in \check{R}_0'n,$$

from which, together with thesis (2), we deduce

(4) $\qquad nR_0\bar\omega \,.\, n \notin \bar\omega \,.\, \supset \,.\, n \in \bar\omega.$

Applying the transformation rule based on *3.3 and the principle of the *reductio ad absurdum* (*2.01) to theses (3) and (4),

(5) $\qquad nR_0\bar\omega \,.\, \supset \,.\, n \in \bar\omega$

(6) $\qquad nR_0\bar\omega \,.\, \supset \,.\, n \notin \bar\omega$

can be derived.

Applying *4.82 to theses (5) and (6), we obtain

$$-nR_0\bar\omega.$$

Since n is arbitrary, we reach the conclusion

(II) $(n) \cdot -nR_0\bar{\omega},$

which is the flat negation of (I).

It is advisable to observe that the proof of (II) is closely connected with the proof of *102.71 (vol. ii) and that it is based solely on the assumptions of *Principia Mathematica* and the definition of the relation R_0. On the other hand, the proof of (I) makes use of the concept of 'expression' or that of 'set of symbols or letters', which is alien to the system of Whitehead and Russell. The proof does not, however, make any assumption which could not be found in Richard's classical antinomy. Therefore we have to reject, first of all, the assertion of Whitehead and Russell that the theory of types resolves Richard's antinomy.

If a conclusion concerning the merits of Whitehead's and Russell's system is to be drawn from what has been said above, it should be pointed out first of all that the antinomy could have been established only by means of the principle of reducibility. It is in virtue of this principle that the function $\Phi_0\hat{n}$ is of the same type as the function $\Theta\hat{n}$. If we assume for the moment that the principle of reducibility is false, we are unable to assert that the function $\Phi_0\hat{n}$ is of the same type as $\Theta\hat{n}$, and thus theorem (I) cannot be obtained. In view of this fact the conclusion is inevitable that we have to reject the principle of reducibility, if we wish to resolve Richard's paradox without altering the basic structure of the theory of deduction contained in *Principia Mathematica*.

It is possible, however, to take a different line. Richard's paradox makes use of concepts which constitute no part of the system of logic. Some logicians, for instance Peano and Zermelo, regard these concepts as not clear enough and do not want to consider them. In this case only one antinomy remains, namely that which can be derived from the permissible assumption of the function

$$-\Phi(\Phi\hat{x}).$$

To eliminate this antinomy, it is sufficient to accept the simple theory of types which distinguishes individuals, functions of individuals, functions of these functions, and so on. The distinction of orders of functions of a given argument, thereby the introduction of predicative functions, and the consequent resort to the axiom of reducibility, is from this point of view a superfluous complication of the system. It should be noted that the elimination of these elements from the theory of types of Whitehead and Russell would render this theory exceedingly simple and manifest. If Whitehead and Russell decided not to carry out this simplification, they undoubtedly did so because of the conviction that a system of logic admitting Richard's antinomy could not be regarded as the final word on what it is possible to attain in this field.

Leaving this question aside, let us restrict ourselves to the statement that the theory of types combined with the principle of reducibility cannot be maintained. For either the principle is false or else it represents in an involved form what is fundamentally simple.[1]

2. The antinomy of Eubulides

Let us now turn to another problem. Let us assume that the theory of types has been freed from the principle of reducibility and that we have managed to construct a system of formal logic based on this theory.[2] Could such a system be regarded as a reliable scheme, on which every deductive science should be modelled? The following antinomy seems to provide evidence to the contrary.

Let us assume that *I consider as interesting those and only those elementary propositions which are equivalent to the proposition*:

---

[1] It should be observed that Richard's antinomy could be resolved by the assumption that there cannot be infinitely many inductive numbers of a given type. This hypothesis would, however, be completely arbitrary, and there is no doubt that a system of logic based on this hypothesis could be regarded only as a particular case to be accommodated within a comprehensive system of formal logic, which is independent of how the question of actual infinity is solved.

[2] See 'The theory of constructive types', so far in manuscript. [*Translator's note*: 'The theory of constructive types' was published in *Annales de la Société Polonaise de Mathématique*, 2, 1924, pp. 9–48; 3, 1925, pp. 92–141.]

'*I consider as interesting an elementary proposition and this proposition is false.*' It appears to me that such an assumption can be made, for I may consider anything I like as interesting. Yet this supposition leads to an antinomy.

If the function $M(p)$ denotes the expression 'I consider the elementary proposition p as interesting' and the proposition A is determined by the definition:

$$A \underset{df}{=} . (\exists\, p) . M(p) . -p,$$

the assumption I wish to examine can be expressed by the formula:

(I) $M(p) . \equiv . p \equiv A.$

From (I) first follows:

$$p \equiv A . \supset . M(p),$$

and applying *10.23 we obtain:

(1) $(\exists\, p) . p \equiv A . \supset . (\exists\, p) . M(p).$

Now if we take advantage of the typical ambiguity of the symbol $\vee$ and substitute A for p in *5.15, we obtain:

$$A \equiv q . \vee . A \equiv -q,$$

where q stands for any elementary proposition. Since each member of the disjunction entails $(\exists\, p) . p \equiv A$, the antecedent of the implication (1) is true, and thus also the consequent:

(2) $(\exists\, p) . M(p).$

Let us now assume that A is true. We obtain:

$$A . \supset . (\exists\, p) . p \equiv A . -p$$

$$[*4.11] . \supset . (\exists\, p) . -p \equiv -A . -p$$

(3) $[*5.36] . \supset . -A.$

On the other hand, we have:

$$-A . \supset . M(p) \supset_p p.$$

Moreover, it follows from (I) that

$$M(p) . \supset_p . p \equiv A.$$

By these two last theses we obtain:

$$-A . \supset : M(p) . \supset_p . p . p \equiv A$$
$$[*10.23] . \supset : (\exists p) . M(p) . \supset . (\exists p) . p . p \equiv A$$
$$[(2)] . \supset : (\exists p) . p . p \equiv A$$

(4) $\quad [*5.36] . \supset . A.$

It is clear from (3) and (4) that the function $M(p)$, satisfying the condition laid down in (I), cannot exist. I cannot, therefore, consider as interesting those and only those propositions which I wish so to consider.

The above antinomy, which, as can be seen, is independent of the principle of reducibility, appears to indicate that even the pure theory of types does not free us from antinomies. However, the statement (I) is not a simple statement of fact, for we can regard as a fact only what is directly given, and what is directly given must be stated by means of *elementary propositions* or predicative functions such as 'I feel pain', 'x is red', and the like, *from the formulation of which the use of apparent variables is excluded.* We cannot regard theorem (1) as a result of the substitution of A in the predicative function

$$M(p) . \equiv . p \equiv \hat{P},$$

for A is by itself neither true nor false and can become either true or false only by theorem (1). As long as theorem (1) is not established, we cannot substitute A for P in the above function. In view of that we have to regard (I) either as the definition of the function $M(p)$ or as an axiom. The first possibility is disposed of by the simple observation that $M(p)$ is a statement of a different type from that of $p \equiv A$; we cannot therefore consider

$$M(p) . \underset{df}{=} . p \equiv A$$

to be meaningful.

Moreover, it seems to be right to exclude *a limine* any definition in which a symbol is defined by means of itself. Should we regard (I) as an axiom, there would arise the theoretical necessity of proving its consistency with the axioms of logic. In this case the antinomy under discussion could be considered to provide a proof that axiom (I) cannot be added to the axioms of logic.

18

THE DEVELOPMENT OF MATHEMATICAL LOGIC IN POLAND BETWEEN THE TWO WARS†

ZBIGNIEW A. JORDAN

I. *Historical Notes on the Lwów–Warsaw School*

IN the first few years after the 1914–18 war a group of Polish logicians, with names hard to remember and still harder to pronounce, began to make a name for themselves in the circle of specialists all over the world. In the course of time the group increased in number and its reputation was established by the importance of its contributions. It became known under the collective name of the Lwów–Warsaw School. The sudden appearance of Polish logicians, of whom nothing had been heard before, taking a leading part in the development of logic, may be regarded as an unaccountable event. It does not hold, however, any mystery, though its history is rather long, going back to the end of the last century. It begins in 1895, when Casimir Twardowski, after a few years of lecturing at Vienna University, came to Lwów and was appointed to the chair of philosophy. Twardowski was neither a logician nor a man whose mind was influenced by a close acquaintance with mathematical logic. His papers, which are few in number and small in volume, deal mostly with problems of 'philosophical logic', with the problems of percept, concept, proposition, etc. They provided a solid basis for future semantic investigations, at the same time being the starting-

† This work comprises the first six sections and the relevant parts of the bibliography of the author's booklet *The Development of Mathematical Logic and of Logical Positivism in Poland between the Two Wars*, which appeared as no. 6 in the series 'Polish Science and Learning', Oxford University Press, 1945. Certain minor changes have been made by the author.

point of some interesting psychological analysis.[1] But Twardowski prepared the ground for future development by training his numerous pupils in the use of scientific method in philosophy, by inculcating their minds with a great respect for clear thinking and precision of expression, and in general by fertilizing philosophic thought with the results of the scientific habit of mind. The influence of his teaching was assuredly extended by the particular circumstances under which at that time the exact sciences were developing. After the analytic work of Cauchy, Weierstrass, Dedekind, and Cantor, mathematics had just reached the exactitude of method and the conceptual clearness for which it was prematurely famous in past centuries; the emerging *Grundlagenproblem* showed no limit to the control we can exercise over the methods adopted, every step forward being rewarded by ample and important results. In logic Frege was rediscovered, Peano was at his work, and the new knowledge was extended and systematized in *Principia Mathematica*. Physics was beginning to enter into the era of an intense urge for neatness, lucidity, and clearness of structure, which marks its present development. The study of the procedure of obtaining observational knowledge and a logical scrutiny of this procedure were being successfully applied, with the result that some revolutionary changes had to be introduced, the theory of relativity being the first of them.[2] From this and other results came the recognition that 'it is actually an aid to the search for knowledge to understand the nature of the knowledge which we seek' (Eddington). Mathematics, logic, physics were showing how concentration on methods and the careful analysis of procedure yield far-reaching results and lead to new discoveries.

In his teaching and writing Twardowski himself set the example of great analytical acumen and of eminent lucidity of thought and statement. His influence went far beyond the circle

[1] I am thinking first of all of numerous and fine descriptions of different kinds of mental images given by a young Lwów psychologist, L. Blaustein.

[2] To fix more exactly the change referred to it should be stated that the common concepts of time and space had to undergo a radical correction as the result of the logical examination to which Einstein subjected the observational procedures intended to ascertain distant simultaneity. The examination showed that instructions for these procedures always end in a vicious circle.

of 'professional' philosophers. W. Witwicki, a distinguished
Warsaw psychologist; J. Kleiner, an historian of literature; R.
Ganszyniec, a classical scholar, claimed equally the right of
being his pupils. More than this, Twardowski was one of those
men who, though little known outside the walls of colleges and
universities, exercise a wide and powerful influence on the life
of the nation to which they belong. We can say about him,
with some measure of justice, what J. S. Mill said about his
father: 'He did not revolutionize or create one of the great
departments of human thought. But in the power of influencing
by mere force of mind and character the convictions and purposes of others, he left few equals among men.'

One of the oldest and perhaps the most eminent of Twardowski's pupils was J. Łukasiewicz. Łukasiewicz was the first
logician in Poland who fully mastered and appreciated the new
ideas and methods in logic. Beginning in 1906 as a lecturer at
Lwów University, he taught and spread the knowledge of
mathematical logic. Under Twardowski's and his leadership a
new generation of logicians and philosophers grew up, permeated
with the new ideas and fully prepared for work on their development. To this generation belong Z. Zawirski, T. Kotarbiński,
S. Kaczorowski, T. Czeżowski, and K. Ajdukiewicz. They were
joined by S. Leśniewski, who had begun his studies under H.
Cornelius in Germany, but completed them in Lwów, where,
influenced by Łukasiewicz, he made his first acquaintance with
mathematical logic.

In independent Poland Twardowski's and Łukasiewicz's pupils
scattered all over the country, being appointed to the chairs of
logic and philosophy in the three reopened universities, Warsaw,
Poznań, and Wilno. Łukasiewicz himself, Leśniewski, and
Kotarbiński went to Warsaw, Czeżowski to Wilno, Zawirski to
Poznań. Ajdukiewicz remained ultimately in Lwów. Round
them gathered quickly large groups of students, of whom many
later on made numerous and important contributions to the
development of logic. To this second generation belong A.
Lindenbaum, S. Jaśkowski, S. Presburger, J. Słupecki, B.
Sobociński, A. Tarski, and M. Wajsberg. Among them Tarski

swiftly won a prominent position and joined his teachers, Łukasiewicz and Leśniewski, as a leader of what in the course of time became known as the Lwów–Warsaw School.

Besides Warsaw, which was by far the biggest centre of logical research—philosophers, logicians, and mathematicians joining hands in a fruitful co-operation—and Lwów, there was still a third centre playing an important role in the development of logic in Poland. This was Cracow University. Its development was independent of activities originated in Lwów by Twardowski and Łukasiewicz, though some incentive might have come from there. From then on, Cracow jealously guarded its independence, remaining outside the Lwów–Warsaw School. The man who in Cracow did the same thing as Łukasiewicz in Lwów was J. Śleszyński. Śleszyński shares with Łukasiewicz the honour of having done most to promote and to spread the knowledge of modern logic in Poland. He was assisted by an eminent scientist, S. Zaremba, who gained a world-wide reputation for his work in physics, but was interested also in the logical foundations of physics and mathematics. L. Chwistek is Śleszyński's and Zaremba's pupil and is of the same age as the first generation of the Lwów–Warsaw School. He was for many years a lecturer in Cracow and only shortly before the war was appointed to the chair of logic in Lwów. W. Hetper, J. Herzberg, and J. Skarzeński were Chwistek's pupils. Contemporaneous with Chwistek were W. Wilkosz and O. Nikodym, known for their work on the theory of sets. On the whole the distinctive feature of the Cracow centre was perhaps a closer association of logic with mathematics, logic being considered not for its own sake, but only in so far as it provides means and methods for solving the foundational problems of mathematics and of some branches of physics.

There was no special periodical on logic in Poland and logical papers were published either in one of the three main philosophical periodicals—*Ruch Filozoficzny*, *Kwartalnik Filozoficzny*, *Przegląd Filozoficzny*, published in Lwów, Cracow, and Warsaw respectively—or in *Fundamenta Mathematicae*, or in *Comptes Rendus* of different scientific associations (*Comptes rendus des*

séances de la Société des Sciences et des Lettres de Varsovie being
the most important from this point of view), or in different
foreign mathematical periodicals. To remove the language
barrier and to enable foreign specialists to follow the development of logic and philosophy in Poland a new periodical, *Studia
Philosophica*, was founded in Lwów in 1935. The articles published there were written only in one of the principal world
languages. Eventually, for the same purpose, in 1937 the Institute for the Encouragement of Science in Warsaw began to
publish *Organon*, an international periodical devoted to the
psychology, sociology, and methodology of sciences. Logical
papers were published in both of these periodicals. A partial
bibliography of the Polish logical and philosophical literature was
published in *Erkenntnis*, v, pp. 186, 189, 190, 194–5, 199–203. It
may be found also in Carnap's *Logical Syntax of Language* and
Quine's *Mathematical Logic*.[1] The full list of anything of importance is contained in the bibliography which is being published
by *The Journal of Symbolic Logic*. The bibliographical references
and annotations below do not lay any claim to completeness.

It remains to say a few words about textbooks of symbolic
logic and the way logic was taught at schools and universities.
The most widely known and used textbooks were Ajdukiewicz's
mimeographed lectures, *The Main Principles of Methodology of
Science and of Formal Logic*, and Łukasiewicz's *Elements of
Mathematical Logic*.[2] They treat of the calculus of propositions,
the theory of quantifiers, the Aristotelian syllogistic in an
axiomatic presentation, and the fundamental notions of metalogic. I must say I do not know textbooks of symbolic logic of
equal value in English, French, or German logical literature.
A few years later Tarski published a handbook, *Mathematical
Logic and Methodology of Deductive Sciences*, to provide those
reading for mathematics with an elementary introduction to
mathematical logic. It was successively extended in its German
and English translations.[3] Its high merits are known to all inter-

---

[1] I used these sources in compiling the bibliography appended to this paper.
[2] Ajdukiewicz (5) and Łukasiewicz (7) (see Bibliography).
[3] Tarski (19), (20), (21).

ested in the subject. The more advanced students used Carnap's *Abriss der Logistik* and Hilbert and Ackermann's *Grundzüge der theoretischen Logik*. A detailed summary of *Principia Mathematica* in Polish was given in Z. Zawirski's *The Relation of Logic and Mathematics in the Light of Modern Research*.[1]

The courses of lectures on logic were organized differently at different universities. As a rule there was a general course in the faculty of Arts and a special one in the faculty of Science. Generally speaking the special course covered in about three years the three volumes of *Principia*, including metalogical research as far as it was developed. In the general course the calculus of propositions, including metalogic, was treated at length, and then the calculus of classes and of relations, though (the students lacking the necessary mathematical knowledge) not as fully as in the special course.

There were some attempts to introduce mathematical logic into secondary schools. The curriculum prescribed two hours a week in the last form for the teaching of the fundamental notions of logic and psychology. It was planned to replace the teaching of an outline of Aristotelian syllogistic by the teaching of preliminaries to mathematical logic.

II. *The First Steps: Critical Examination of 'Principia'; Łukasiewicz's Improvements; the Theory of Deduction*

Although—if my memory serves me right—the first references to and a short exposition of the main ideas of modern logic, of its symbolism, of the way in which its theorems are mathematically proved, etc., may be found in Łukasiewicz's *The Principle of Non-Contradiction in Aristotle* (1910), yet the proper start of the actual research work on mathematical logic falls in the first years after the 1914–18 war. They form a period of transition, in which much energy had to be diverted to the rebuilding of universities, to the reorganization of teaching, to catching up with the developments of logic in war-time. As everywhere else, *Principia Mathematica* provided at first the starting-point for further research.

[1] Zawirski (4).

Among the first well-known improvements of Whitehead's and Russell's system made by the Polish logicians are those of Chwistek.[1] Like many others Chwistek could not agree with the *Principia* definition of classes and he found out that on the basis of Russell's theory of types and with the aid of the reducibility axiom Richard's antinomy can be re-formulated.[2] This discovery pushed him in the direction of rebuilding the system in such a way as to do without the much discussed axiom of reducibility.

Chwistek's solution of the problem is known by the name of the theory of constructive types. It rejects the distinction of orders of propositions and propositional functions and retains only that of types. The reason for it is that the logical antinomies of the type of Russell's antinomy disappear if the existence of classes of different types is recognized. The logical antinomies are the only ones with which we are likely to meet in mathematics. Secondly, Chwistek saw for the first time that the notion of class may be replaced by the notion of function; that is to say, that the expression 'the set of all things x such that O' (where 'O' stands for a definite condition such as '$x > o$', '$x < \sqrt{2}$', etc.) and '$x \epsilon C$' (where 'C' denotes the class in question) are equivalent.[3] We now recognize this equivalence by writing $C = \underset{f(x)}{E} [\ldots]$, where the blank in the square brackets is left for the condition in question. Consequently Chwistek considered restricting and making more precise the notion of a propositional function—in such a way that it ensures, so to say automatically, the absence of circularity—as the proper way out of difficulties. The theory of constructive types can be regarded as a system of rules concerning the ways in which signs and letters are put

---

[1] Chwistek (2), (3).

[2] Chwistek (2). In (5) Chwistek recognized that the reconstruction of Richard's antinomy was made possible by the introduction of some semantic expressions which do not belong to logic, and that consequently the system of Whitehead and Russell is free from Richard's antinomy.

[3] This substitution reflects a procedure which is very common in mathematics, where the use of the expression 'the set of all numbers (points) such that O' is frequently used in order to obtain a well determined set of numbers or a new kind of geometrical configuration. A clear and comprehensive account of the whole question of antinomies is to be found in A. Fraenkel, *Einleitung in die Mengenlehre*, 3rd edition, Berlin, 1928.

together and made into valid formulae of logic and mathematics. He thought at the same time that in this way metaphysics is removed from logic and mathematics, the reader's intuition being exclusively confined to distinguishing different signs from each other and to comprehending the ways in which they are put together.

Chwistek's simple theory of types is the final result of a thorough and detailed examination of *Principia Mathematica* in which he attempted to remedy some technical defects, remove some inconsistencies of symbolism, banish some metaphysical assumptions, and put in their proper place the extra-logical axioms of infinity and reducibility, as well as the axiom of choice and the hypothesis of the continuum which are considered by Chwistek as existential hypotheses.

In the 'Introduction to the Second Edition'[1] Russell wrote: 'One point in regard to which improvement is obviously desirable is the axiom of reducibility. This axiom has a purely pragmatic justification: it leads to the desired results, and to no others. But clearly it is not the sort of axiom with which we can rest content. On this subject, however, it cannot be said that a satisfactory solution is as yet obtainable. Dr. L. Chwistek took the heroic course of dispensing with the axiom without adopting any substitute; from his work it is clear that this course compels us to sacrifice a great deal of ordinary mathematics.' Chwistek agreed to Russell's criticism only in so far as the theory of constructive types curtails to some extent the theory of sets, or rather the Cantorean theory of sets, since it makes it impossible to prove the existence of the continuum. It goes without saying that the theory of alephs, the continuum hypothesis, etc., go overboard too. It is, however, doubtful whether it entails an actual curtailment of classical mathematics. Poincaré suggested long ago that classical mathematics is independent of the notion of the continuum as a set (*Acta Mathematica*, xxxii, p. 199). We can add also that in this respect Chwistek is in the good company of a whole school of celebrated mathematicians, the Parisian

---

[1] Whitehead and Russell, *Principia Mathematica*, vol. i, p. xvi, Cambridge, 1925.

Empirical School, which does not recognize Cantor's proof of the existence of the continuum and instead simply accepts sets which can be proved not to be denumerable.[1] At the time of the publication of *The Theory of Constructive Types* Chwistek was convinced that on the basis of his system such a complicated notion as Lebesgue's measure could be introduced. In a paper published later[2] he showed that the whole body of analysis proper may be easily reconstructed in his systems.

Concerning the theory of sets Chwistek observed that the controversy marks the point where different theories of sets diverge, some assuming the existence of the continuum, others only the existence of a certain type of denumerable sets, the continuum being denoted by a typically undetermined symbol and providing denumerable sets of a particular type. Whitehead's and Russell's system cannot be regarded as an absolute measure by which other systems are valued. Their investigations showed that there are in the theory of sets theorems which cannot be proved by purely logical means: they depend on the axioms of multiplication and infinity. Whitehead's and Russell's investigations belong to the same category as those of Sierpiński (4): they are concerned with the determination of the decision domain (*Entscheidungsbereich*—see below p. 368) of different sets of axioms. There is so far no axiomatic theory of sets which could be considered as complete, not even that of Zermelo.[3] Chwistek put forward the suggestion that we cannot speak yet of *the* theory of sets. Rather, there are many equally justifiable theories of sets ('Non-Euclidean theories of sets'). In (4) Chwistek analysed the 'range' of some of them, when he investigated different propositions which can be adopted instead

---

[1] See especially E. Borel, *Leçons sur la théorie des fonctions*, 3rd edition, Paris, 1928. (Notes): '. . . nos connaissances précises sur les puissances diverses n'excédent guère la remarque suivante: il y a des ensembles dénombrables et des ensembles non dénombrables, cette dernière notion étant surtout négative.' The question whether the continuum is a set and, if so, what meaning is to be attached to it, is still among the Intuitionists a controversial problem. See also A. Fraenkel, 'Zur Diagonalverfahrung Cantors', *Fund. Math.* 25 (1935), and the literature given there.　　　　　　　　　　　　　　　　[2] Chwistek (6).

[3] The proposition '*a* is a transfinite number' is the simplest example of an irresoluble problem. It is obvious that it can be proved only for at most a denumerable set of *a*'s.

of the axioms of choice and infinity ('the axiom of denumerability', 'the axiom of transcendence', 'the axiom of affinity').

The theory of constructive types is an inappropriate tool for the study of the theory of sets. For this purpose Chwistek gave an analogue of it, known as the simplified theory of types.[1] In this theory the notion of class is no longer reduced to that of propositional function. Its main idea can be summarized in the following statements. We assume certain objects, let us say x, y, z, . . . which cannot be exemplified in the theory of sets. x, y, z, . . . are called individuals. A class X consists of individuals. The class of sub-classes of X is a class of the first type (X_1). The sub-classes of X_1 form classes of the second type (X_2). In this way we can construct successively classes of the 3rd, 4th . . . nth type. The expression 'all classes' is an abbreviation for 'all classes of a certain type'. No other restriction is required to assure the elimination from the theory of sets (and from mathematics in general) of 'illegitimate totalities'.

Łukasiewicz's first discoveries,[2] though perhaps by their implications of a more important nature, were less ambitious than those of Chwistek. At the beginning Łukasiewicz devoted almost his whole attention to the calculus of propositions, which he considered as the simplest and the most perfect example of a deductive theory, in which all its peculiarities and characteristics may be easily studied. It must be added that Łukasiewicz had a great share in bringing the calculus of propositions to its present condition of perfection. He devised a symbolism of his own (the CN symbolism), making it possible to dispense with dots and brackets, an essential step forward in the strict formalization of logical proof. The use of dots and brackets requires special rules, awkward if not impossible to formalize. Łukasiewicz's symbolism is not limited to his own calculus or to a calculus of propositions, but lends itself to a wide application in any logical system.[3] Łukasiewicz discovered simultaneously with P. Bernays[4] that one of the five axioms in Whitehead's and Russell's

---

[1] Chwistek (4). It was first suggested in Chwistek (1a).

[2] Łukasiewicz (4), (6), (10), and Łukasiewicz–Tarski (8).

[3] See footnote 4, p. 357; Tarski (11); Chwistek (5).

[4] Bernays's results are of a more general nature. He proved not only that

calculus of propositions is not independent. He constructed many systems of this calculus which should be regarded as masterpieces of mathematical precision. The best known of them assumes negation and implication as the primitive terms and only three axioms—as compared with six in Frege's and four (after the reduction) in Whitehead's and Russell's system. Written in Russell's notation these axioms are:

1. $p \supset . \sim p \supset q$
2. $\sim p \supset p . \supset . p$
3. $p \supset q : \supset . q \supset r . \supset . p \supset r$

This set of axioms of the theory of deduction is far from being final. In 1921 Tarski proved that given the rules of substitution and detachment all theorems of the theory of deduction, involving the sign of implication as the only logical constant (the restricted propositional calculus), can be derived from the following set of axioms:

4. $p \supset . q \supset p$
5. $p \supset q . \supset : q \supset r . \supset . p \supset r$
6. $p \supset q . \supset . r : \supset : p \supset r . \supset . r$

If we allow in the inferences from the axioms (4)–(6) the introduction of definitions and the use of the traditional rules concerning quantifiers, the axioms (4)–(6) suffice to construct the whole propositional calculus. At any rate the 'completeness' of the axioms (4)–(6), that is to say the coincidence of all meaningful implicational propositions with the decision domain of the axioms (4)–(6), using the rules of substitution and of detachment, was proved by Tarski in 1926.[1] In 1925 Tarski worked out a method which makes it possible to reduce the axioms of the restricted propositional calculus, involving the rules of detach-

the axiom of association is deducible from the remaining axioms, but also that there are three other systems, each consisting of four formulae, which may be assumed as axioms of Whitehead's and Russell's propositional calculus. See P. Bernays, 'Axiomatische Untersuchungen des Aussagenkalküls der "Principia Mathematica" ', *Math. Zeitschr.* 25 (1926).

[1] See Leśniewski (8) and Łukasiewicz–Tarski (8), Th. 29. Tarski's proof does not seem to have been published. Another proof is to be found in Wajsberg (7). Hilbert and Bernays point out in *Grundlagen der Mathematik*, vol. i, Berlin, 1934, that 6 can be replaced by the formula: $p \supset q . \supset p : \supset p$.

ment, substitution, distribution of quantifiers, 'extensionality', and formative rules of definition, to one axiom only.[1] This method, as Tarski showed, can be carried over to systems including implication among their primitive terms. The reduction of the axioms of the complete propositional calculus involving implication and negation as its primitive terms (axioms (1)–(3)) to one axiom only, was actually carried out by Łukasiewicz and Wajsberg and later simplified by Sobociński and Łukasiewicz.[2]

Another single axiom of the theory of deduction, including, besides propositional variables, universal quantifiers and functions taking propositional arguments (variable functors) had been already given by Leśniewski, Tarski, and Wajsberg in the years 1923–6.[3] Leśniewski took advantage of Tarski's (1) results and derived his Protothetic from one axiom involving the sign of equivalence as the only primitive sign.

At the same time Łukasiewicz, Tarski, Wajsberg, and Sobociński, in close mutual collaboration, investigated the Nicod–Sheffer axiom, their researches resulting in finding many equivalent formulae and in simplifying it considerably. In particular Łukasiewicz, without changing Nicod's rules of inference, reduced the stroke-function axiom to a formula involving four instead of five variables, which is by no means a trivial simplification.[4]

---

[1] See Leśniewski (8) and Łukasiewicz–Tarski (8), Ths. 8 and 25.

[2] Unfortunately I am not at the moment in a position to substantiate the information with all the relevant formulae and successive simplifications. They are to be found in Łukasiewicz–Tarski (8), Sobociński (1), and Łukasiewicz (16).

[3] Leśniewski (8). The axiom which for typographical reasons I omit to state here is to be found on page 59 of Leśniewski's paper.

[4] Leśniewski (8); see the explanatory note in footnote 2 above. The following formula was given by Wajsberg (2). Let us replace '$p|q$' by 'Dpq'. Then Nicod's axiom takes the form:

(A) $DDpDqrDDtDttDDsqDDpsDps$.

Wajsberg proved that (A) can be deduced from:

(B) $DDpDqrDDDsrDDpsDpsDpDpq$.

(B) is an improvement on (A) because of the above-mentioned reduction in the number of variables and because (B) is an organic axiom, that is, no proper part of (B) is a thesis like $DtDtt$ in (A).

Łukasiewicz proved that (B) is equivalent to:

(C) $DDpDqrDDpDrpDDsqDDpsDps$.

I cannot say whether (C) is identical with the formula mentioned in Leśniewski (8).

It is to be emphasized that the calculus of propositions based on one axiom only, involving implication and negation as its primitive terms, is a higher achievement than the propositional calculus based on a single axiom involving the stroke sign. This is because of the fact that Nicod's rule of detachment is a somewhat stronger means of inference than the usual rule of detachment.

After 1920 Łukasiewicz widely applied truth-tables or truth-matrices and connected with them the graphical method, the so-called 'zero-one method', of determining the truth-value of a function or formula. Subsequently truth-matrices have played an important part in logic. A logical matrix was defined by Łukasiewicz as an ordered quadruple

$$\mathfrak{M} = [A, B, f, g]$$

that consists of two mutually exclusive classes A and B, of a function f of two variables and of a function g of one variable. Both functions are defined for all elements of $A + B$ and admit only these as their values. Let S be the set of all meaningful expressions of the calculus of propositions, which beside propositional variables involve only the signs of implication and negation. If x and y are propositions, let $c(x, y)$ and $n(x)$ be names denoting the propositions 'Cxy' and 'Nx'. The function h is a *value function* of the matrix

$$\mathfrak{M} = [A, B, f, g]$$

if it fulfils the following conditions: (1) the function h is defined for every $x \in S$; (2) if x is a propositional variable, then $h(x) \in A + B$; (3) if $x \in S$ and $y \in S$, then $h(c(x, y)) = f(h(x), h(y))$; (4) if $x \in S$, then $h(n(x)) = g(h(x))$. We say that a proposition x is satisfied by the matrix

$$\mathfrak{M} = [A, B, f, g],$$

symbolically: $x \in \mathfrak{E}(\mathfrak{M})$,

if for every value function of this matrix the formula $h(x) \in B$ holds. The elements of the set B are called, with Bernays, the *designated* elements of the matrix

$$\mathfrak{M} = [A, B, f, g].$$

It should be noted that by the use of matrices we can dispense entirely with proofs of theorems in the usual sense. Whether a formula, expressed in terms of the system, holds or not, can be determined by the 'zero-one method' applied for all combinations of the truth values of the elements involved in the given formula. Accordingly we can say, for instance, that by the complete propositional calculus is understood the set of all propositions which satisfy the matrix

$$\mathfrak{M} = [A, B, f, g],$$

in which $A = \{0\}$, $B = \{1\}$, and the functions f and g are determined by the equations:

$$f(0, 0) = f(0, 1) = f(1, 1) = 1, \quad f(1, 0) = 0,$$

$$g(0) = 1, \quad g(1) = 0.$$

Into the same period fall Kuratowski's and Tarski's first contributions to mathematical logic. Kuratowski's name[1] is associated with the so-called Wiener–Kuratowski theory of relations, now almost universally accepted and widely applied. The Wiener–Kuratowski theory consists in defining ordered couples by means of class theory and consequently in reducing relation theory to class theory. Kuratowski, utilizing Hessenberg's and Hartogs' ideas concerning the way in which the notion of order is introduced into the theory of sets, and using Janiszewski's notion of a set saturated with respect to a given property, gave a new definition of the expression, 'the class of classes of individuals M establishes order in the class m'. This definition permits us to establish a one-one correspondence between 'ordering classes' in a given set and the different ways of ordering this set.[2] The notion of an ordered couple can be then determined in the following way. Let A be a set consisting

---

[1] Kuratowski (1); see also Sierpiński (3); A. Fraenkel, 'Bemerkungen zum Begriff der geordneten Menge', *Fund. Math.* 7 (1925); W. V. Quine, *Mathematical Logic*, pp. 197–202, New York, 1941.

[2] That is not the case with Hessenberg's definition of the above expression, which allows, given an 'ordering class', to determine the ordered set univocally, but not conversely, since for a given ordered set there are many 'ordering classes'.

of two elements, a and b. There are then two ordering classes establishing order in the set A:

$$((a, b), (a)) \text{ and } ((a, b), (b)).$$

Kuratowski defines an ordered couple whose first element is a and the second b as the class $((a, b), (a))$. The relation R taken in extension is then to be considered as the class of ordered couples. To say 'xRy' is logically equivalent to saying 'x paired with y is a member of R'. It is not necessary to assume that A be a set of individuals. The notion of ordered couples assumes only that its elements are of the same logical type (homogeneous relations).

In his doctor's thesis (1) Tarski found the striking and somewhat paradoxical fact that, granted the use of variable functions with propositional arguments and of the universal quantifier, all logical constants of the propositional calculus can be defined in terms of the equivalence relation. To prove this the following theorem is established:

$$(p, q) : : p \cdot q \equiv : . (f) : . p \equiv : (r) \cdot p \equiv f(r) \cdot \equiv . (r) \cdot q \equiv f(r).$$

The right-hand side of the equivalence can be regarded as the definition of the sign of conjunction by means of the equivalence sign and the universal quantifier. By adding the sign of conjunction to those just mentioned, the signs of the remaining logical constants can be defined by means of the theorems:

$$(p) : . \sim p . \equiv : p \equiv . (q) \cdot q$$
$$(p, q) : . p \supset q . \equiv : p \equiv . p \cdot q$$
$$(p, q) : p \vee q . \equiv . \sim p \supset q$$

Tarski's discovery is of importance for all systems involving one logical constant as the primitive term. If such a system is to be correctly constructed, it is necessary to avoid, in the enunciation of definitions, the introduction of any special constant, distinct on the one hand from the primitive term, and on the other from the terms previously defined or to be defined. According to Leśniewski Whitehead's and Russell's system does not involve only '$\sim$' and '$\vee$' as primitive terms, but beside these two the sign involved in the expression of the type '$p = q \; Df$'. Similarly

in Nicod's system '|' as well as the sign involved in expressions of
the type '$p = q\ Df$' (Nicod) or '$p = q$' (Sheffer) is to be counted
among the primitive terms of the system. The sign of equivalence, if accepted as the only primitive term, offers from this
point of view a special advantage. It allows a strict adherence
to the explicitly mentioned rule and provides for definitions the
most natural and handy form, that is to say the form of an
equivalence. In his system of propositional calculus (Protothetic) Leśniewski took advantage of Tarski's discovery and
adopted the equivalence sign as the only primitive term.

In the second part of his thesis, Tarski (2) gave the necessary
and sufficient conditions that a function f (with a propositional
variable) must satisfy in order to be a truth-function in Whitehead's and Russell's sense.[1] These necessary and sufficient conditions are stated in the formula:

$$(p) : f(p) \equiv . \ p \lor \sim p \ : \lor . \ (p) f(p) \equiv .p. \ \lor . \ (p) f(p) \equiv \ \sim p.$$
$$\lor \ : (p) f(p) \equiv p. \sim p$$

Tarski gave at the same time some formulae equivalent to the
'law of substitution', that is to say to the theorem which asserts
that the function f satisfies the condition of being a truth-
function (in Whitehead's and Russell's sense):

$$(p, q, f) : p \equiv q. f(p) \supset f(q).$$

The equivalent formulae are of importance in that the above-
mentioned theorem can be proved to be independent of the
Principia system of axioms. The results for the function f of one
variable may be extended to a function f of many propositional
variables.

Łukasiewicz's above-mentioned reduction of the number of
axioms in Whitehead's and Russell's calculus of propositions (now
only of an historical significance), Leśniewski's investigations
of different systems of Protothetic, together with similar work
on other axiomatic systems and on the axiomatization of different logical theories, drew attention to the importance of investigations concerning the general properties of formalized deductive

---

[1] *Principia Mathematica*, vol. i, p. 115, Cambridge, 1925.

systems. This last branch of logical research, first known by the name of the theory of proof, developed in Poland very widely.

As is well known, the theory of proof dealt at the beginning with the methods of constructing a deductive system, including such problems as the consistency, completeness, and independence of a set of axioms, the method of proving that a given axiomatic system fulfils the above-mentioned conditions, the function of definition and of rules of definition within a deductive system, the notion of formal proof, etc. (the investigations concerning the consistency and completeness of a set of axioms being the most important factor in the further extension of the domain of methodological research). It is known also that this new kind of investigation required the creation of a whole system of new notions and precise definitions. The logicians in Poland took a very active part in this work.[1] Without paying much attention to chronological order[2] we shall mention some of the most outstanding contributions. Łukasiewicz found, independently of Post and Bernays, the method of proving the consistency of the axioms of a logical system, and later he succeeded in proving the completeness and independence of the axioms of his calculus of propositions. Leśniewski's contributions, about which more will be said later on, are of outstanding importance. He did pioneer work in applying the new methods to the theory of definition, and his work resulted in the rejection of the view that definitions are conventional abbreviations for economy and

---

[1] See first of all Łukasiewicz–Tarski (8).

[2] The neglect of chronological order is unavoidable since many members of the Lwów–Warsaw School did not bother to publish even their important results as soon as they were found. They were, however, as personal communication was taking place constantly, known at once to all interested, and used and developed by them in their further research (see, for instance, Leśniewski (8), Tarski (1), (2), and Sobociński (2)). Some of the papers leave a vivid impression of being written in haste, under the high pressure of exploiting new ideas and discoveries just made, leaving hardly any time for writing. (See, for instance, Tarski (11).) 'The aim of this paper', writes Leśniewski (8), 'is to liquidate a painful situation, in which I have found myself for many years. This situation consists in having in my possession a great number of unpublished scientific results concerning different investigations on the foundations of mathematics. The number of unpublished results increases from one year to another. They stand in a relation of mutual dependence to those of other research workers in this domain, and they pile up still more new technical difficulties which are bound up with the preparation for publication of the results.'

convenience in the presentation of a formal system, without which the system can do as well. Far from being an 'expedient in presentation' or a 'mere assignment of names', definitions are an act of choosing the complex and various ideas which are to become the object of study.[1] As early as 1921 Leśniewski contended that the necessary condition for a system, making use of definitions, to involve one and only one sign as a primitive term, is that definitions be stated in terms of the accepted primitive sign, without introducing a special definition sign, which is itself a new primitive term. He showed in the case of Nicod's system how this should be done.[2] The use of definitions requires, therefore, special formative rules, carefully and exactly formulated. In his system of Protothetic Leśniewski gave the first completely formalized definitions obeying the above-stated rule. The same care should be taken in respect to the operations used in the construction of axiomatic systems. Leśniewski emphasized the importance of rules of inference in every system. The set of consequences of a given set of propositions does not depend exclusively upon the choice of axioms, but also upon the rules of inference actually used.[3] Which rules are actually used can be ascertained only from that formulation which leaves no doubt as to the nature of the accomplished transformation. In this respect every existing system fails to come up to the required standard. 'As a matter of fact', writes Leśniewski, 'I do not know of a single work purporting to provide the foundations of mathematics which states, in a way excluding any doubt as to their interpretation, the set of rules sufficient to infer all the theorems actually deducible in this system, and which at the same time would not lead in some way or other to contradictions.'[4] To avoid misunder-

[1] Leśniewski's theory of definition was adopted by Tarski, so that those who are anxious to spare themselves the valuable though somewhat excruciating experience of going through Leśniewski's writings, may find it applied in numerous papers of Tarski's.

[2] For instance, the definition of negation should take in Nicod's notation the following form:

$$\sim p.|:p|p.|.p|p\therefore|\therefore p|p.|:\sim p.|.\sim p::|::\sim p.|:p|p.|.p|p\therefore p|\therefore p.|:\sim p.|.\sim p$$

[3] In Chwistek (5) may be found a striking example of the extent to which axioms may be simplified by strengthening the rules of inference.

[4] Leśniewski (8); compare Lindenbaum (3) and Leśniewski (9).

standings and misuse of rules of inference the following suggestion is implicitly made in Leśniewski (8). Since the inference
rules do not belong to the system and cannot, therefore, be
formulated in its terms, ordinary speech must be used. For this
purpose a number of 'terminological explanations', concerning the
words involved in the enunciation of the rules, should be added
(there are not less than forty-nine 'terminological explanations'
concerning five inference rules in Leśniewski's Protothetic). For
the sake of convenience and abbreviation a symbolic language
can be introduced, in terms of which all 'terminological explanations' and rules are formulated. Thus, e.g., the rule of detachment is enunciated in Leśniewski (8) in the following way:

$$[\exists\ C,\ D].\ (C\ \epsilon\ thp\ (A)\ .\ D\ \epsilon\ thp\ (A)\ .\ B\ \epsilon\ cnsqeqvl\ (C,\ D)).$$

To decipher the preceding sequence of signs it would be necessary to go successively through all 'terminological explanations'
—which shows with what thoroughness and exactitude the
formative and transformative rules are formulated.

Finally, Wajsberg's research should be mentioned here too.
Although its results were published later and consequently made
ample use of metalogical methods, some of his results were reached
and known earlier. Wajsberg found the first proof ever given that
the system of Strict Implication is not reducible to Material
Implication.[1] He gave for Hilbert's and Ackermann's 'restricted'
functional calculus an exactly k-numerically identical formula,
from which every k-numerically identical formula is deducible:

$$(Ax_k)\ \underset{l\leqslant k}{\overset{l}{D}}[F_l\ (x_l){\rightarrow}\underset{l<m\leqslant k+1}{\overset{m}{D}}\ [F_l\ (x_m)]\,]^2$$

[1] C. I. Lewis and C. H. Langford, *Symbolic Logic*, Appendix ii, New York,
1932.

[2] A formula of the predicate calculus is k-numerically identical (for every
finite k, different from 0), if, applied to a range of individuals consisting of
k things, it is a universally valid formula of the calculus. A formula which is
a k-numerically but not $(k+1)$-numerically identical formula is called an exactly
k-numerically identical formula. Let the figures $1, 2 \ldots, k$ denote the individuals of a k-numerical range. Then the application of a predicate calculus
formula to a k-numerical range means that (a) $A(1).A(2)\ldots A(k)$ and
(b) $A(1)\vee A(2)\vee \ldots \vee A(k)$ can be substituted for $(x) A(x)$ and $(\exists x)A(x)$
respectively. For the sake of abbreviation (b) is written:

$$\underset{l\leqslant k}{\overset{l}{D}}[A\ (l)\ ]$$

From (Ax_k) we obtain, for example, for $k = 1, 2$

(Ax_1) $F_1(x_1) \to F_1(x_1)$

(Ax_2) $[F_1(x_1) \to F_1(x_2) \vee F_1(x_3)] \vee [F_2(x_2) \to F_2(x_3)]$

He proved generally that from any (Ax_k) every k-numerically identical formula can be inferred.[1] He axiomatized a special case of the 'extended' Hilbert–Ackermann calculus of predicates which is originated by the interpretation of propositional variables as variable predicates.[2] The proof of completeness of the system is given and it is shown that by an appropriate interchange of logical constants and variables and the addition of a new axiom[3] the extended 'calculus of predicates' passes into Lewis's system of Strict Implication.[4] Applying Tarski's notion of the degree of completeness ($Vollständigkeitsgrad$) of a deductive system, Wajsberg proved that the degree of completeness of the 'restricted' and 'extended' functional calculi of Hilbert and Ackermann, as well as of $Principia\ Mathematica$, is equal to the continuum.[5] These results, however, were reached with the aid of metalogical investigations, made possible by the introduction of metalogical symbols, and it brings us into the middle of the stream described in the following chapter.

III. *Metalogic*

A rather misty and somewhat loose relation between logic and the theory of proof gained in precision when Hilbert introduced the fundamentally important distinction between mathematics

[1] Wajsberg (5). See also D. Hilbert and P. Bernays, *Grundlagen der Mathematik*, vol. i, Berlin, 1934, § 4. According to Hilbert and Bernays the totality of k-numerically identical formulae is 'deductively closed'; that is to say, that by the addition to the axioms of the predicate calculus of k-numerically identical formulae, only k-numerically identical formulae can be inferred. Wajsberg proved that the set of k-numerically identical formulae is 'deductively closed' if any exactly k-numerically identical formula is added to the axioms of the predicate calculus.

[2] Wajsberg (3). See also D. Hilbert and W. Ackermann, *Grundzüge der theoretischen Logik*, Berlin, 1928, chaps. 2, 3, 4.

[3] The axiom $C11$ of Appendix ii in Lewis, and Langford's *Symbolic Logic*.

[4] C. I. Lewis, *A Survey of Symbolic Logic*, University of California Press, 1918, chap. v. Cf. E. V. Huntington, 'The Mathematical Structure of Lewis's Theory of Strict Implication', *Fund. Math.* 25 (1935).

[5] Tarski (10); Wajsberg (6).

and metamathematics, meaning by the latter a system of rules having mathematical formulae for their object. Simultaneously, in close relation to the Hilbertean idea, the Lwów–Warsaw School formulated the distinction between logic and metalogic. Perhaps we could say that the distinction between mathematics and metamathematics was at first made for the limited purpose of proving the self-consistency of the axioms of mathematics, and that the distinction between logic and metalogic, and analogously and more generally between a theory and a meta-theory, was brought about by the recognition of the necessity to observe carefully the distinction between using a language and speaking about a language. The latter is a matter of necessity if we wish to escape the antinomies which are unavoidable in a language which, like ordinary discourse, is semantically closed. Tarski showed[1] that in a semantically closed language, as the result of the use of 'duplicated' terms, the number of potential antinomies is unlimited. This shows that some principles and rules specifying and limiting the kinds of expressions to be used in a language are required. A distinction must, therefore, be adopted between language that speaks of objects and facts, and language that speaks of terms and sentences.[2] In the case of logic this leads first of all to fulfilling very strictly Frege's demand to distinguish a designation from a designated expression, which in turn makes it necessary to introduce special metalogical (or syntactical, as they are called nowadays) symbols, such as $\mathfrak{E}(\mathfrak{M})$, $Cn(X)$, $\mathfrak{S}$ (the set of all those propositions which satisfy a certain matrix, the set of all propositions deducible from X,

---

[1] See Tarski (13).

[2] The alternative to this procedure is the arithmetization of language in the way in which it was done for the first time by Gödel. By establishing a one-one correspondence between certain numbers and expressions of a language the formulae consisting of sequences of signs pass into sequences of numbers. Since syntactical propositions are then propositions concerning arithmetical properties and relations of number sequences, syntax becomes a part of Arithmetic. Moreover, if the language in question contains an Arithmetic, it follows that the syntax is formulated in terms of the object language itself. In a syntax of this kind self-referring sentences must occur. They do not entail a reappearance of antinomies. But the inclusion of the arithmetic of natural numbers implies that in every such system there are unprovable or irresoluble problems (K. Gödel, 'Über formal unentscheidbare Sätze der Principia Mathematica und verwandter Systeme', *Monatsh. f. Math. u. Phys.* 38 (1931)).

the set of all deductive or closed systems), etc. The introduction of special metalogical notation is advisable for reasons of convenience, as it enables us to avoid an excessive use of quotation marks and guarantees a certain conciseness and elegance in speech and writing. Not only this, it was justified later on by its fruitfulness and the numerous important results it yielded. Consequently it became an established practice in the Lwów–Warsaw School.

Historically speaking the incentive in this direction was given by Leśniewski, who in his *Grundzüge eines neuen Systems der Grundlagen der Mathematik* gave the first definitions expressed in terms of a completely symbolized language. Yet most of the important improvements and discoveries in this domain, made in Poland, are bound up indissolubly with Tarski's name. He took the lead and his investigations showed that it may be most desirable to completely symbolize the metalogical (syntactical) language and thus reduce to a minimum, or even get rid of, ordinary speech.

Tarski made his reputation by numerous contributions concerning different particular problems, which belong more or less directly to the domain of metalogic. He subjected its concepts to a close scrutiny, and, making his way with apparently no effort through the unexploited land of metalogic, improved what he found, framed new notions, discovered new methods and ideas, and gathered a rich harvest of 'incidental' results from the application of his analysis to a variety of subjects belonging to logic and to pure and applied mathematics. The two papers by Tarski, *Fundamentale Begriffe der Methodologie der deduktiven Wissenschaften* and *Einige methodologische Unter-suchungen über die Definierbarkeit der Begriffe*,[1] are very typical of the kind of work he was doing. The first of them especially is a brilliant achievement, striking in its precision, which goes hand in hand with matchless clearness, and in its skilfulness and neatness; almost dangerous because it leaves the feeling that nothing more can be said on the subject.

In the methodology of deductive sciences we come again and

[1] Tarski (10), (15).

again upon two groups of notions: one that of axiom, consequence, rule of inference, proof; and the other that of primitive term (sign), derived term (sign), rule of definition, definition—which though materially different show a certain analogy when their function in the construction of a deductive system and their mutual relations are considered. Tarski showed first that the notions of deductive system, of logical equivalence of two sets of propositions, of a set of propositions capable of axiomatization,[1] of a set of axioms, of the independence, completeness, and self-consistency of a set of axioms, etc., are reducible to those of a meaningful statement ('S') and of the consequence relation ('Cn'). These two are accepted as the primitive notions and their meaning exhibited by four axioms. Among them there is one very characteristic of Tarski's general attitude, the one which states that the power of S (the set of all meaningful statements) is less than or equal to aleph zero, that is, that the set S is at most denumerable. The reduction is carried out in a strictly deductive way, formal proofs following one other. The new definitions of consistency and completeness of a system of propositions (A) should be emphasized. Let '$\mathfrak{Aq}(S)$' represent the class of classes equivalent to S, '$\mathfrak{P}(A)$' the power set of A and '$\mathfrak{W}$' the class of all consistent sets of propositions. Then:

$$\mathfrak{W} = \mathfrak{P}(S) - \mathfrak{Aq}(S)$$

The meaning of this definition is made clearer in the following theorem:

In order that $A \in \mathfrak{W}$ it is necessary and sufficient that $A \subseteq S$ and $Cn\,(A) \neq S$.

In ordinary speech, the theorem says that a set of propositions A is consistent if it is not equivalent to the class of all meaningful statements.

To define completeness Tarski introduces the notion of the decision domain of a set of propositions A. The decision domain of A is the set of all propositions either deducible from A or which added to A gives an inconsistent set of propositions.

---

[1] Lindenbaum (1) found a somewhat paradoxical result. Although in practice we deal almost without exception with axiomatic systems, yet deductive systems are as a rule incapable of axiomatization.

A set of propositions A is complete if its decision domain comprises all meaningful propositions. Let '$\mathfrak{Ent}(A)$' stand for the decision domain of A and '$\mathfrak{V}$' for the class of all complete sets of propositions. Then symbolically:

$$\mathfrak{Ent}(A) = Cn(A) + \underset{x}{S}.\,E[A + \{x\} \in \mathfrak{W}], \;\; \mathfrak{V} = \mathfrak{P}(S).\underset{X}{E}[\mathfrak{Ent}(X) = S].$$

Tarski's definitions of consistency and completeness depart from the familiar way of defining these notions and are applicable also to cases in which the old ones had no definite meaning.[1] They make it, moreover, possible to throw a new light on the dependence existing between them. A theorem established by Lindenbaum states that every consistent set of propositions may be made complete. Unfortunately the theorem does not give any hint how in a particular case this completion may be accomplished.

Tarski showed further that with reference to the notions of primitive sign, of equivalence of two sets of signs, etc., the notion of definability of a sign 'a' plays the same role as that of the consequence relation with reference to the formerly considered group of notions. A sign 'a' is definable by means of a set of signs B and with respect to a set of propositions X if 'a', as well as all the signs of B, occur in the propositions of the set X and at least one definition of the sign 'a' by means of the signs of B is deducible from the set of propositions X. A definition of 'a' is a function of the form:

$$(x) : x = a. \equiv .\, \phi\,(x;\,b',\,b'',\,\ldots),$$

where $\phi\,(x;\,b',\,b'',\ldots)$ stands for any propositional function with x as the only real variable and in which no other non-logical constants occur but b', b'', $\ldots$ of B. The relativization of the definability of a sign 'a' to a set of propositions X becomes obvious if we realize that we cannot consider the reducibility of a sign to one or more others before we know the meaning of the

---

[1] The usual way of defining the notions of completeness and consistency uses the notion of a negated proposition. The usual definition of consistency cannot therefore be applied to a system that does not involve the notion of negation at all, for instance, to the restricted calculus of propositions. For a similar reason the usual definition of completeness is inappropriate for any system involving free variables.

sign in question. In a deductive theory there is no other way of doing this but to describe the propositions, accepted as true, in which the sign occurs. The theorem that states precisely the necessary and sufficient conditions of the definability of a sign 'a' makes it clear that the notion of definability may be reduced to that of the consequence relation.[1]

Tarski found that the problems of definability and independence of signs[2] correspond exactly to the problems of deducibility and independence of propositions. The problem of the completeness of a system of signs,[3] though closely related to the problems of completeness and categoricity[4] of a system of axioms, does not

---

[1] In order that a sign 'a' be definable by means of a set of signs B and with respect to a set of propositions X it is necessary and sufficient that the formula:

$$(x) : x = a. \equiv. (\exists z', z'', \ldots). \psi (x; b', b'', \ldots; z', z'', \ldots)$$

be deducible from the set of propositions X. The function '$\psi (x; b', b'', \ldots, z', z'' \ldots)$' stands for the conjunction of all elements of X, and 'z'', 'z'''', $\ldots$ are variables substituted for the non-logical constants (if any) occurring in the propositions of X.

An equivalent theorem provides a theoretical foundation for Padoa's method, which allows us in certain cases to prove the independence of a sign from a given set of signs.

[2] From Lindenbaum (Tarski and Lindenbaum (5)) come many striking and astonishing examples of the independence of mathematical notions. For instance, the metric four-termed relation of the congruence of two pairs of points and the descriptive three-termed relation of betweenness are independent in one-dimensional geometry. But each of them is definable in terms of the other in a more than one-dimensional geometry.

[3] To define the notion of the completeness of a set of signs we have to introduce first the following definition: A set of propositions Y is essentially richer in its specific signs than a set of propositions X, if (a) each proposition of X belongs to Y; and (b) in the propositions of Y there occur signs which do not occur in X and which are indefinable in the set Y by means of the specific signs of the set X only.

A set of propositions X is complete with respect to its specific signs, if there is no categorical set of propositions Y which is essentially richer in its specific signs than X.

According to this definition, various sets of axioms of geometry are incomplete, and various axiom systems of arithmetic complete with respect to their specific terms. A theorem establishes that if a set of propositions is monotransformable, then it is also complete with respect to its specific terms.

[4] The term 'categorical', as a qualification of a set of axioms, is used here in a somewhat stronger sense than that given to it by O. Veblen. A set of propositions is categorical in Tarski's sense, if for any two interpretations of it there is at least one relation which establishes their isomorphism. A set of propositions is monotransformable if there is at most one such relation. A set of propositions is strictly categorical if it is categorical and monotransformable.

It may be shown that the various axiom systems of arithmetic (of natural, real, complex, hyper-complex numbers) are strictly categorical, while the various allow us to establish a similar correspondence. As a side-result it turns out that the problem of reducing the notions of mechanics to those of geometry amounts to establishing a categorical system of geometry (certain modern physical theories assume a geometrical system which is not categorical). If, however, a certain particular system of four-dimensional Euclidean geometry—which being formally equivalent to the arithmetic of hypercomplex numbers is categorical as well as strictly categorical—is accepted, the above-mentioned reduction is equivalent to constructing a strictly categorical system of mechanics. Since the last-mentioned problem meets with some hopeless difficulties, Tarski's side-result does not further its solution, though it clears the ground for a successful tackling of the problem and points to the direction in which the solution is to be sought.

The ideas laid down in Tarski (10) were developed and used for the foundation of a Calculus of Systems in Tarski (17). For this purpose a set of axioms analogous to that in Tarski (10) was assumed, the most important modification consisting in replacing 'Cn (X)' by 'Cn (0)', also denoted by 'L'; that is to say, in making the set of consequences of the empty class of propositions (the set of all logically true propositions) the subject-matter of further investigations. On the ground of this modified system the postulates of Boolean Algebra, as formulated in Tarski (23), can be reconstructed. The Algorithm of Propositions, which will be left out of account, and the Calculus of Systems, are respectively a complete and a partial interpretation of Boolean Algebra. In particular, if we replace in all postulates of the usual Algebra of Logic the variables 'x', 'y', 'z' by 'X', 'Y', 'Z', which represent deductive systems, and the symbols 'B', '$<$' '$+$' '0', '1', by '$\mathfrak{S}$', ' $\subseteq$ ', '$\dotplus$', 'L', 'S', respectively, then all postulates, with the exception of VII(c)[1] are satisfied. In the Calculus of Systems founded in this way, what constitute the subject-

axiom systems of geometry (projective, descriptive, metric, and topology) are categorical but not strictly categorical.
[1] However, the consequence of VII(c) still holds:

$$\text{If } X, Y \in \mathfrak{S} \text{ and } X.Y = L, \text{ then } Y \subseteq X.$$

matter of discourse are not the elements of S, their mutual relations and the operations which are carried out on them, but
certain chosen sets of these elements, namely deductive systems.
Consequently the propositions which belong to the Calculus of
Systems involve only the three following kinds of symbols and
terms: (1) terms of a general, logical character; (2) constants
involved in axioms as well as symbols defined by means of these
constants; (3) variables representing deductive systems, classes
of systems or relations between them, classes of these classes and
relations between them, etc. The Calculus can be extended by
introducing infinite operations (the product of a class of deductive systems and the sum of the elements of this product) and
the notion of an axiomatizable system. A set X is axiomatizable
if and only if there is an $x \in S$ such that $X = Cn(\{x\})$, where x may
be a conjunction of several propositions. A theorem can be then
established (Tarski (17), theorem 17), which asserts that the
axiomatizable systems are the only ones satisfying the principle
of excluded middle. The theorem 17 can be regarded as the definition of an axiomatizable system, in accordance with which X is
axiomatizable if and only if $X \dotplus \bar{X} = S$; in other words, if the
principle of excluded middle is applicable to it.

Since the principle of excluded middle fails in the case of non-
axiomatizable systems and can be applied to them only in its
'weakened' form, it follows that the double negation theorem
holds merely in one direction and only the triple negation
theorem takes the form of an equivalence:

$$\text{If } X \in \mathfrak{S}, \text{ then } X \subseteq \bar{\bar{X}} \text{ and } \bar{X} = \bar{\bar{\bar{X}}}.$$

Owing to these circumstances the Calculus of Systems reveals
a striking formal likeness to Heyting's propositional calculus.[1]
We can sum it up in the following sentence: if on the ground of
Intuitionist Logic a calculus of classes were founded, it would
not differ formally from the Calculus of Systems.

By collaboration with mathematicians—Banach, Kuratowski, Lindenbaum—on various mathematical problems (the de-

---

[1] A. Heyting, *Die formalen Regeln der intuitionistischen Mathematik*,
Sitzungsberichte der Preuss. Ak. d. Wiss., Phys.-math. Klasse, 1930.

finition of a finite set, the axiomatization of geometry with the notion of sphere instead of that of point as the primitive notion),[1] Tarski not only achieved some valuable results, but also demonstrated the fruitfulness of the application of logic to mathematical problems, with regard to which many a mathematician used to show great reluctance. The results found in collaboration with Kuratowski deserve special mention.[2] Tarski discovered a method of solving on the ground of 'metatheory' problems belonging to the theory itself, the solution of which in the ordinary way had so far met with setbacks or was successful only in a very complicated manner. The method consists in pointing out that if certain logical procedures in defining mathematical notions are applied, they enable us by using the structural or morphological properties of the definitions to draw some inferences about the properties of the defined notions themselves.

Although Tarski demonstrated his method on a particular example, viz. that of the definability of a set of real numbers, yet it is capable of a wide, though probably not general, application. Roughly speaking the leading idea of Tarski's method consists in a metalogical construction of a mathematical notion (in this case of the expression 'the set X of real numbers is definable') and then in repeating it within mathematics proper, taking as a starting-point the mathematical substratum of the investigated metalogical expression (in this case the notion of propositional function on the one hand and of the set of finite sequences of real numbers on the other) and in retracing all the steps of the metalogical investigations by means familiar to every mathematician, enriched only by a specified logical language. The guidance provided by the metamathematical investigations is only one factor in the successful mathematical reconstruction.[3] The second, equally important, is brought out by the necessity of using in the reconstruction a logical notation

---

[1] Tarski (3), (4), (9).   [2] Tarski (11), Kuratowski–Tarski (2).

[3] It is to be recognized that the reconstruction may be often only partial. In the case in question the family of actually definable sets of real numbers is at most denumerable. By the application of the diagonal procedure it is always possible to define a set of real numbers which is not contained in the family

enabling us to make the above-mentioned inferences from the morphology of the mathematical definitions.

We could say also that Tarski's method consists in finding a mathematical interpretation of the metamathematical construction, or in the case of certain mathematical theories in establishing a logico- or metalogico-mathematical dualism. Such dualism can be easily established in the case of projective sets of points (in Lousin's sense). Using Lebesgue's notation a propositional function $\phi(x_1,...,x_n)$ of n real variables is called projective if the set

$$\underset{x_1...x_n}{E}\phi(x_1,...,x_n),$$

that is to say the set of points of an n-dimensional space which satisfy the function ϕ, is projective. Since on the one hand equivalences between the five logical operations used by Tarski, namely, the operations of negation, logical addition and multiplication, universal and existential quantification,[1] and the operations proper to the theory of sets (complementary set, sum and product of two sets) can be established; and on the other hand by means of the formula defining the operation E

$$t \in \underset{x}{E}\phi(x) \equiv \phi(t)$$

it may be shown that the operation E is permutable with each of the five logical operations, each of them changing its logical meaning into a mathematical one (or vice versa); it is obvious that if a propositional function $a(x)$ is derivable from the functions ϕ_1, $\phi_{11},...,\phi_k$ by the application of the five logical operations, then the set $\underset{x}{E}a(x)$ is derivable in the same way from the sets $\underset{x}{E}\phi_1,..., \underset{x}{E}\phi_k$. Hence we can prove the following theorem:

The five logical operations carried out on the projective propositional functions lead always to projective propositional functions.

from which we started. For a similar reason it is impossible to define the general notion of a definable set of the nth order. It is, however, possible by purely mathematical means to reconstruct the notion of a definable set of the 1st, 2nd, 3rd, . . ., nth order, where 'n' stands for a fixed natural number.

[1] From the two pairs named in the second and third place one member of each can be defined in terms of the other and thus eliminated.

From a theorem, which correlates to the logical operation Σ (of existential quantification) a geometrical continuous operation, comes a wide range of applications of logical calculus to topology. Kuratowski gave in (3) many examples showing how the use of logical notation enables us to evaluate in many cases the Borelean or projective class of a given set of points and to prove directly many theorems found by means of different methods, often excessively long and complicated. It is hardly possible to find a better instance of the usefulness of logic and metalogic for mathematics.

The rapid development of metalogical investigations made it eventually possible to carry into effect the idea of an axiomatized metalogic. Whatever the objections of speculative Micawberists, the axiomatization of metalogic does not mean simply a shifting or a relegation to somewhere else of the actually insoluble problem of constructing a logical system dispensing with the appeal to intuition or to the feeling of self-evidence. It is an actual step forward towards the elimination, in the construction and analysis of a logical system, of any mental operations entering into the consideration of the meaning of otherwise extremely complicated and unmanageable logical notions and expressions (Hilbert's *formale Betrachtung* as opposed to the *inhaltliche*). The axiomatic way of dealing with metalogical problems dates from 1931. It was introduced by Gödel for a limited purpose, and later used by Tarski,[1] who presented a formalized metalogical system having a symbolic notation as strict and systematic as the notation of the logical system of which it treats.

In the present state of research it is hardly possible to speak of a single metalogic having as the object of its investigations the language of any deductive system. It is true that there are a number of important metalogical notions, common to all investigations of this kind; but they can be applied only to deductive systems which have reached a certain degree of formal development making such application possible; for instance,

---

[1] Tarski (13), (14). It should, however, be emphasized that Tarski had already to some extent applied the axiomatic method to metalogic in 1930—see Tarski (10)—or even in 1928, when in a paper read at a meeting of the Polish Mathematical Society he laid down the main ideas of (10). See Tarski (6), (7).

the calculus of propositions, some parts of arithmetic, etc. As a rule every theory requires a metatheory of its own. Tarski's axiomatized metalogical system applies to a language based on *Principia Mathematica*, with respect to which exact formalization and far-reaching simplifications are carried out, without prejudicing its capacity to formulate any idea comprised by *Principia*.[1] In Tarski's axiomatized metalogic there occur two kinds of notions:

1. Notions of a general, logical nature, as they are used in any sufficiently extensive system of logic (for instance in that of Whitehead and Russell), especially those which occur in the calculus of classes and in the arithmetic of natural numbers.

2. Specific notions of a structural and descriptive nature, which represent the structural properties and relations between the expressions of the object language, such as: 'the class of all expressions'—'A'; 'negation'—'$-$'; 'implication'—'$<$'; 'the sign of set membership'—'$\in$'; 'universal quantifier'—'Π'; 'variable with a stroke above and below—'ϕ_k^1'; and 'expression that consists of two expressions following immediately each other'—'$\zeta^\frown \eta$'. Since there are two kinds of notions, there are also two kinds of axioms: (1) axioms of mathematical logic; (2) formulae showing the characteristics of the notions belonging specifically to the metalanguage. On such a basis and with the aid of some definitions introduced for the sake of simpler exposition, it is possible to define exactly and morphologically the notions of a meaningful statement and of the consequence relation. Since, as Tarski showed in *Fundamentale Begriffe der Methodologie der deduktiven Wissenschaften*, many metalogical notions may be reduced to those two, the system, as it is described so far, provides a secure basis for research in a strictly formal way on different metalogical problems. By the addition

---

[1] The alterations concern four main points: (*a*) the ramified theory of types is replaced by the simplified theory and the reducibility axiom is rejected; (*b*) the law of extensionality is retained; (*c*) all logical constants, except the primitive terms, are suppressed; (*d*) the variables admit as their values only individuals (objects of the 1st order)—symbolically: '$x_|{}^|$', '$x_|{}^{||}$', . . . '$x_|{}^\kappa$', . . .; classes of individuals (objects of the 2nd order)—symbolically: '$x_{||}{}^|$', '$x_{||}{}^{||}$', . . ., '$x_{||}{}^\kappa$' . . .; families of classes (objects of the 3rd order)—symbolically: '$x_{|||}{}^|$', '$x_{|||}{}^{||}$', . . . , '$x_{|||}{}^\kappa$' . . . , etc.

of some auxiliary symbols we can extend the system in such a way that it includes other concepts and, in particular, the notions of ω-consistency and ω-completeness, which cannot be reduced to the notions previously referred to.[1]

Apart from its own virtue the formalization of metalogic enables us to define quite correctly and precisely what is meant by formal proof, its definition marking a further stage reached on the way towards the formalization of logic. Besides, axiomatized metalogic has opened up new prospects and laid bare new possibilities of constructing systems of logical calculi, of which full use was soon made in the semantics of the Hilbert–Bernays or Chwistek type, or in the still more general attempt of Carnap to provide a secure foundation for the logical syntax of language.[2]

Last, though not least, is the construction, which but for the formalization of metalogic would have been impossible, of a language which on the one hand is not ω-complete, and on the other is consistent in the ordinary sense but not ω-consistent. This sensational discovery was made first by Gödel and published in the now famous paper, 'Über formal unentscheidbare Sätze der Principia Mathematica und verwandter Systeme' (*Monatsh f. Math. u. Phys.* xxxviii, 1931). Tarski constructed another comparatively simple example of a ω-incomplete language which may be proved to be consistent but not ω-consistent. It brings

---

[1] The two above-mentioned notions are defined by Tarski formally and precisely. For our purpose some explanations should suffice. Let '*E*' be the name of a property which classes of individuals may have and which may be formulated in the language in question. Let $\xi_n(E)$, where n is a natural number, and $\xi_\omega(E)$ be propositions of the language, which say that every class of at most n or ω individuals possesses the property E. A set of propositions X is called ω-consistent, if for no E are all the propositions of the infinite sequence $\xi_n(E)$ and the negation of $\xi_\omega(E)$ simultaneously deducible from X. The set X is ω-complete if, whenever all the propositions $\xi_n(E)$ are deducible from X, then the proposition $\xi_\omega(E)$ is also deducible from X.

[2] Carnap characterizes the relation between metalogic and metamathematics on the one hand, and the syntax of language on the other, in the following way: 'We call a (logical) syntax a purely formal investigation of a calculus, that is to say one not considering the meanings of signs. If this calculus is a mathematical system, then its syntax is metamathematics in Hilbert's sense; if this calculus is a purely logical system, then its syntax is metalogic in the Warsaw logicians' sense' (R. Carnap, 'Die Antinomien und die Unvollständigkeit der Mathematik', *Monatsh f. Math. u. Phys.* xli, 1934).

to the surface a rather unexpected and unwelcome fact, whose consequences were discussed by Tarski in his *The Concept of Truth in Formalized Languages*. It was previously thought that all purely structural operations which always lead from true to true propositions might be completely reduced to the rules of inference used in deductive sciences. Secondly it was believed that the consistency of a deductive system is a guarantee that certain pairs of propositions would not, on account of a definite structural relation in which they stand to each other, be simultaneously true. Since there are languages highly formalized, ω-incomplete, consistent in the ordinary sense but not ω-consistent, the two above-mentioned assumptions are deprived of their justification. The difficulty could be removed by the addition of the so-called rule of infinite induction, by which every deductive system would become *ex definitione* ω-complete, and ω-consistency would not be different from consistency in the ordinary sense.[1] But the use of the rule of infinite induction is very problematic in the actual construction of a deductive system, and would require a new conception of the deductive method itself, working on a 'finitary basis' in accordance with which at every stage of construction only a finite number of signs and propositions is actually given.

There does not seem to be any escape from the conclusion that the formalized concept of the consequence relation, and the ordinary one, that used in everyday language, are actually different; and that the consistency of a system does not preclude the possibility of 'structural falsehood'. If we consider as the ideal deductive method the one making use only of general logical and structural-descriptive notions in the actual construction of a system and in the formulation of its rules of inference, then such an ideal deductive method is unfortunately incompatible with an ideal deductive science, by which is meant a system comprehending all true propositions formulated in terms of the given language. If disciplines of an elementary logical structure are excluded, there remains a comprehensive class of formalized deductive systems which, being consistent

[1] For the definition of the so-called rule of infinite induction see Tarski (14).

and incomplete, contain pairs of undecidable contradictory propositions. In such systems the notions of truth and provability
fail to coincide. While all provable propositions are true, not all
true propositions are provable.

Finally we owe to Tarski the conception of semantics[1] conceived as the whole body of investigations concerning the notions
in which we state correlations between the expressions of a
language and certain objects, namely, those which are denoted
by the respective expressions. In this we are able to remove
some well-known difficulties concerning the impossibility of
expressing in metalogical notation the correlations between the
names of the logical expressions and the expressions themselves,
as well as difficulties of an allied nature connected with the
formally correct and materially valid definition of the Aristotelian notion of truth, shown by Tarski himself in the first part
of his *The Concept of Truth in Formalized Languages*.[2]

In this important book Tarski deals with the definition,
materially adequate and formally correct, of a true statement.
He states expressly and precisely what he means by the expression 'materially adequate definition'. He then shows that it is
possible to give such a definition if, and only if, the language in
terms of which the problem is handled is a 'richer' one, that is
to say, if it comprises expressions of a higher logical type than
the language for which the definition of a true statement is to
be given. He then shows in a general way that a 'true statement'
is definable for any formal language of a finite order with the
help of morphological terms only, that is of terms which refer
only to the shape of expressions and to the relations between
them. A still more general result is achieved when he shows
that we can construct an axiomatized semantics of any formalized language of any order, whose axioms and primitive
notions refer only to the morphological characteristics of the
language.

Considered as a whole Tarski's work contributed greatly to
the extension of the domain of metalogical studies and caused
a change in their fundamental character. Tarski describes the

[1] Tarski (18).   [2] Tarski (13).

change in his *Introduction to Logic* (21) in the following way: 'The conception of the methodology of deductive sciences confined to methods of constructing a deductive system, during the historical development of the subject, turned out to be too narrow. The analysis and critical evaluation of methods applied in practice in the construction of deductive sciences ceased to be the exclusive or even the main task of methodology. The methodology of the deductive sciences became a general science of deductive sciences in an analogous sense as arithmetic is the science of numbers and geometry is the science of geometrical configurations. In contemporary methodology we investigate deductive theories as wholes as well as the sentences which constitute them; we consider the symbols and expressions of which these sentences are composed, properties and sets of expressions and sentences, relations holding between them (such as the consequence relation) and even relations between expressions and the things which the expressions "talk about" (such as the relation of designation); we establish general laws concerning the concepts.'

IV. *Leśniewski's and Chwistek's Systems*

Owing to the development of metalogical research the theory of types did not, except for Chwistek's simplification, arouse later on much interest. If it were practicable to prove the self-consistency of the axioms of a given deductive system, the system could dispense with Russell's or any other theory of types. The rapid development of metalogic produced the impression that with the progress of time it would be possible to accomplish almost anything. Metalogic gave rise to great expectations and held out hopes which might never be fulfilled.

F. P. Ramsey's criticism of the reducibility axiom was almost universally accepted. The rejection of this axiom together with the simultaneous retention of the theory of types was—as it entailed the curtailment of a large part of mathematics—an untenable position; perhaps more so in Poland than in any other country. The long acquaintance of Polish mathematicians with pure logic could only have accentuated their naturally formalistic attitude. W. Sierpiński, who after Janiszewski's premature death held with Mazurkiewicz much the same place among the mathematicians as Łukasiewicz and Leśniewski among the logicians, is beside E. Landau the most pure representative of practical formalism in mathematics I can think of.

On this account those whose interest went beyond the limits of theoretical logic were seeking for the solution of the difficulties of the theory of classes in the way shown by Zermelo, which was the more attractive because it had some indubitable successes to its credit. Zermelo's method consisted in restricting the number of elements accepted in his *Mengenlehre* by specifying in a set of axioms the procedures by means of which classes may be constructed, and giving in their formulation some kind of guarantee that the paradoxes, at least those which were known so far, would not recur in it. Special axioms were adopted providing for the existence of the null-class, the class of all subclasses of a given class, the class of all members of members of a given class, the class having one or two given things as sole members. Besides, the axiom of choice was adopted to allow the generation of further classes. Fraenkel, who perfected Zermelo's axiomatic *Mengenlehre*, hoped also that one day it would be possible to give a proof of self-consistency of the whole set or at least of a part of the set of axioms in question (excluding the axiom of choice which may prove to be similar to Euclid's axiom of parallels, marking the point where different *Mengenlehren* would branch off). Fraenkel's hope was never fulfilled, but the general idea of eliminating paradoxes was taken up and developed.

The most thorough, original, and philosophically significant attempt to provide a logically secure foundation for the whole of mathematics comes from Leśniewski.[1] Leśniewski's system consists of three parts. The first of them, called Protothetic, corresponds to what is known as the calculus of equivalent statements, *Aussagenkalkül*, or the theory of deduction, together with the theory of the apparent variable.[2] It makes use of one

---

[1] Leśniewski (7), (8), (9), (10), (11), (12).
[2] Compare Whitehead and Russell, *Principia Mathematica*, vol. i, Cambridge, 1925, pp. 90–126.

axiom and of one logical constant only.[1] The tautological character of all theorems of Protothetic is assured by the mechanical
application of a curious metatheorem established by Łukasiewicz
and Leśniewski (8): any iterated equivalence is tautologous
provided that each of its components occurs an even number of
times, from which it also follows at once that two iterated
equivalences are equivalent, if the same components appear an
odd number of times in each. Protothetic is followed by what is
called Ontology. This name is due to the fact that its only axiom
gives the precise meaning of the copula 'is' in its existential
sense (this axiomatic definition of the copula 'is' has greatly
influenced the formation of the purely philosophical views of the
logistically-minded Warsaw philosophers, especially those of
Kotarbiński, his view growing slowly into something like a
philosophical system, known by the name of Reism), and according to Leśniewski's intentions covers the ground assigned
by Aristotle to the πρώτη φιλοσοφία, the general theory of
objects (*Gegenstandstheorie*). If we think of *Principia Mathematica*, Ontology would correspond, roughly speaking, to the
theory of classes and relations. But it differs from it in some
most essential respects, as will become apparent from further
remarks. Ontology is derived from one axiom only. This single
axiom was at first long, and consisted of three apparently
independent parts. It was enormously simplified later on, by
Leśniewski, Tarski, and Sobociński, to a very short formula.
Sobociński gave an account of the way in which it was done.[2] It
is a delight for every one to watch the skill and the resourceful-
ness with which the axiom in question was handed on from one
man to another till the final, simple result was reached.

The third part of Leśniewski's system consists of Mereology,
the theory of the part-whole relation, which was to establish a
general theory of sets free from Russell's paradoxes.[3] To the
solution of these paradoxes Leśniewski devoted much time and
energy and managed to find a way of avoiding them in a manner
which, considering its theoretical consequences, shows some

---

[1] See Section II and note 3, p. 357.
[2] Sobociński (2).   [3] Compare Leśniewski (6).

formal affinities with the already known theories of logical types, but in so far as its intuitive aspect is concerned goes back to the tradition of Aristotle's *Categories* and to Husserl's *Bedeutungskategorien*. The theory of semantic categories was to replace the 'hierarchy of types', which for Leśniewski was lacking any intuitive justification. As he says, he would feel bound to accept the theory of semantic categories, if he wished to speak sense, were there no antinomies whatever in the world. The theory of semantic categories is responsible for the peculiar way in which, in Protothetic, Ontology, and Mereology, the rules of inference and the formative rules of definition are enumerated. In concrete application the signs and expressions belonging to different semantic categories are made apparent by being enclosed in different kinds of brackets, whose use is as much restricted by exact rules as other logical operations, to which so far attention was almost exclusively directed.

Leśniewski was very critical about the attempts to remove the antinomies undertaken by modern mathematicians and logicians (Frege, Zermelo, and Russell). According to his opinion they went further and further from the historical and intuitive basis out of which the antinomies had grown. Moreover, they fostered the disappearance of the sense of difference between mathematics, considered as a group of deductive theories intended to provide an exact language for the formulation of the laws governing the world, and such consistent deductive theories as enable us to prove an ever increasing number of irrelevant theorems, un-related to reality. The only way of solving antinomies is provided by an 'intuitive undermining' of such consequence relations and assumptions as together lead to contradictions. A non-intuitive mathematics cannot remedy the shortcomings of intuition.[1]

Since Leśniewski excludes the notion of a null-class, Mereology lends itself to a much wider application than merely to the

---

[1] The theory of semantic categories was established in 1921 (Leśniewski (8)). The foundations for it were already laid in a series of articles published in the years 1912–14; see Leśniewski (1), (2), (3), (5). It is of some interest to note that the articles referred to date from Leśniewski's prelogistic period and that in them he made no use of symbolic methods. I now realize that they show a certain affinity with the Cambridge Analytical School.

theory of sets, viz. to application as a calculus of individuals, meaning by individuals whatever is represented by signs of the lowest logical type within a given system. In this role it might prove to be a powerful instrument in logical constructions and in the analysis of some philosophical problems (for instance, in the dispute between the nominalists and realists about the problem of universals). There is a close formal affinity between Mereology and the extended Boolean Algebra, where by the extended Boolean Algebra is to be understood the usual Algebra of Logic supplemented by two operations, addition and multiplication, both of an infinite character. The differences may be reduced to exactly one point: in Mereology there is nothing corresponding to the symbol '0' of Boolean Algebra. The wide applicability of Mereology was demonstrated by Tarski when he used it in the axiomatization of Woodger's 'P' and 'T' relations.[1]

These desultory remarks do not pay tribute to Leśniewski's work (he died unexpectedly in 1939). He was the forerunner and originator of many ideas, which taken up by logicians in Poland and abroad proved to be of the utmost importance. His ideas concerning the definition of extensionality, worked out with a matchless precision, belong to the highest achievements of logical semantics, and his investigations of the function of definitions within a formalized system, and of the exact and precise way in which the rules of definition should be formulated, yielded results of great importance for the development of logic and philosophy. His extremely high demands of precision (for which he was famous among some people, notorious among others), combined with deep philosophical intuitions, which he tried to symbolize and express in a mathematical language, made him a formidable personality and a mind to be admired and dreaded at the same time. There was hardly anybody of distinction who escaped his influence, and no one able to stand his ground against his combined power of criticism and of invention. He was anything but a formalist, though in his papers of a system-

[1] See Tarski (23); J. H. Woodger, *Axiomatic Method in Biology*, Appendix E, Cambridge, 1937.

atic character he made hardly any use of ordinary speech. Tarski calls his basic attitude 'intuistic formalism' and there are good reasons for attaching this label to his work. 'I should not have taken pains', writes Leśniewski (8), 'over systematizing and repeatedly checking the rules in my system, if I had not attached to its theorems a certain quite definite, just this and not different, meaning, by virtue of which the axioms of the system and the methods of definition and inference codified in the rules possess for me an intuitive, irresistible validity. I do not see any contradiction in saying that just for this reason I practise a rather radical "Formalism". I do not know any effectual method of conveying my "logical intuitions" to the reader but the method of formalization of the deductive theory in question, which, however, under the influence of formalization does not cease by any means to consist of clearly meaningful propositions possessing for me an intuitive validity.' Leśniewski did not consider a formalized system as a play with signs, a game of chess, or anything of this kind. He believed in the absolute truth of some assumptions and formed his opinions about different systems from this point of view. He did not wish to add another system to the large number of already existing ones. He thought that some logical calculi were of no philosophical interest if, instead of their technical correctness, their material validity were considered.

Another logical system, as ambitious as that of Leśniewski in its attempt to give formal foundations to the whole of mathematics, comes from Chwistek.[1] His system, as he says himself, was constructed in opposition to 'verbal' philosophy, as distinguished from the philosophy of common sense dealing only with the use of objects, in this case with the construction of symbolic expressions. In his basic attitude Chwistek was influenced by what we might call *la règle de Poincaré*: 'Ne jamais envisager que des objets susceptibles d'être définis en un nombre fini de mots.'[2] 'I contend', says Chwistek, 'that I am not going to consider in

---

[1] Chwistek (4), (5), (6), (7); Chwistek–Hetper–Herzberg (8); Chwistek–Hetper (9); Chwistek (10). Chwistek (10) is now available in an English translation which differs, however, from the original Polish edition of 1935.

[2] H. Poincaré, *Dernières Pensées* (Flammarion).

my system any objects which are not capable of construction'
(6). Some important consequences are then inescapable. Chwistek's system does not comprise, for example, the theory of
alephs and in general the whole realm of Cantorean numbers.
But it is possible to construct sets of cardinal numbers which
may somehow replace the set of alephs. At any rate it is possible
to construct logically the whole of analysis proper in such a way
that it forms a part of logic or 'pure mathematics' or of metamathematics based on semantics.[1] As the terms 'logic', 'pure
mathematics', 'metamathematics', 'semantics', etc., suggest,
Chwistek evolved the idea of a stratified system or of hier-
archically ordered systems. The idea was originated in 1929[2]
and brought to completion in his *Limits of Science*.

I cannot say I quite understand Chwistek's system. His idea
of a stratified system appears to me to spring from two main
considerations: (*a*) Hilbert's metamathematics cannot do entirely without *inhaltliche Überlegungen*; they go beyond the
proper field of logic and they constitute, so to speak, a plunge
into darkness. The new metamathematical investigations require systematization and formalization. This consideration
would then correspond to the motive that led Gödel to the arithmetization and Tarski to the axiomatization of metalogic. (*b*)
The passage from a lower to a higher language system enables
us to speak in terms of a higher type and thus to solve specific
problems which Chwistek intended to solve. In an outline of a
stratified system (6) Chwistek says, for example, that in 'pure
mathematics' only measurable sets *B* can be defined. To reach
non-measurable sets *L* we have to pass to the metamathematical
system based on semantics.

By semantics Chwistek understands the set of rules and
formulae by means of which the expressions of logic and mathematics can be constructed mechanically. Chwistek emphasized

---

[1] Chwistek uses sometimes a terminology of his own. By 'pure mathematics'
he understood mathematical notions and theorems expressed in terms of a
logical system.

[2] Chwistek (5), (6). Chwistek's idea of a stratified system of logic has nothing
to do with W. V. Quine's theory (see M. H. A. Newman, 'Stratified systems of
Logic', *Proceedings of the Cambridge Philosophical Society* 39, 1943).

very strongly the mechanical character of all operations carried out within semantics. The semantical axioms are simple descriptions of essential characteristics of expressions in terms of a number of fundamental semantical notions and others constructed out of them. Semantical propositions are extensionally equivalent to those accepted and used in every science, for example, as I should imagine, to those of which use is made to introduce an axiomatic metalogic or metamathematics. 'F comes after E', 'H is the result of the substitution of G for F in E', 'E is semantically identical with F', etc., are examples of semantical propositions. 'E is a logical expression', 'E is in H a real variable', 'H is a consequence of G with respect to the type sign', etc., are propositions in which semantics is applied to logic. Chwistek symbolized the propositions of pure and applied semantics and, since these symbolical expressions are sometimes of considerable length, he introduced a system of symbols abbreviating the semantical expressions. He emphasized that a table of abbreviations is, fundamentally speaking, needless. It should be noticed, however, that an extensive system of abbreviations amounts to a double formalization, with results which cannot be foreseen without special investigations. Chwistek's semantics is not presented in the form of an axiomatic system in the strict meaning of this word. It indicates exactly the 'universe of discourse' and enumerates a considerable number of formal and informal postulates, determining the boundaries within which semantics is allowed to move.

Chwistek starts by introducing a very restricted preliminary language, which is the first one in a hierarchy of languages, each constituting the medium in terms of which the propositions of the consecutive languages are formulated. The preliminary language consists of different kinds of directions applied to letters and ciphers, and somehow fulfils the function of a machine for producing symbolic expressions (they may be brought into correspondence with Carnap's rules of formation). The construction of significant expressions by the exclusive use of specified directives leaves no room for any doubt or ambiguity (except that caused by the possible vagueness of the shape of the letters and ciphers). In this way the domain left to intuition and to intuitive logic has been restricted as much as possible. The preliminary language enables us to construct the theorems of the auxiliary system which contains no variables at all. It is used to build up elementary systems containing real and apparent variables, which in turn lead to the elementary semantical calculus used in the arithmetic of integers and of rational numbers, in the theory of classes, in the metamathematical calculus, and so on. It is to be noticed that in Chwistek's system the proof of Gödel's theorem is entirely formal and does not differ from an ordinary arithmetical proof.

Chwistek, whose interests were wide and very varied—he was a painter, and an instigator of new tendencies in painting; an art critic, courageous, outspoken, and full of invention; a philosopher of great culture, wide knowledge, and deep convictions, expressed in a forcible language, perhaps a little at variance with the carefulness and exactness characterizing his logical activities —considered that to lay the formal foundations of mathematics is one of the most urgent tasks of a philosopher, in order that he may counteract the rising wave of irrationalism, originating in the misinterpretations of some difficulties met with in logic and mathematics, which by its social implications threatens a catastrophe. It is up to the philosopher to attack the evil within his own boundaries and at its very source. Hence Chwistek's programme: *Die Metaphysik aus den Grundlagen der Mathematik auszuschalten*—a programme on which he had worked from the time of his examination of *Principia Mathematica* up to the day when he could see it completed in his *Limits of Science*—possessed for him a more general purpose than *un tour de force d'un expert desillusionné*.[1]

---

[1] While I am writing these words Chwistek's tall and bulky figure still floats before my eyes. He was full of fun, eager to discuss for hours as well as enjoy life, and he knew perfectly how to do it. Even for us, the Poles, born legend-makers, he possessed some qualities lending themselves to legend. The younger of us used to refer to him as 'this Renaissance man', probably intending to convey something similar to what G. K. Chesterton said of Dickens: 'he is not a person, he is a crowd'. He did not shun the limelight, nor avoid a fight whenever his pugnacious spirits saw an opportunity for it. Numberless anecdotes were told about him. When he was a young man, says one of them, he was asked

V. *The Discovery of Many-Valued Systems of Logic*

Whatever value may be attached to the above-mentioned results, the discovery by Łukasiewicz of many-valued systems of logic stands out against all of them.[1] Without any doubt it is a discovery of the first order, eclipsing everything done in the field of logical research in Poland.

It is well known from some remarks made by Łukasiewicz in his book, *The Principle of Non-Contradiction in Aristotle* (namely, when he comes to the examination of the Aristotelian doctrine that statements concerning future events are neither true nor false), as well as from some personal statements, that *l'ideé-force* leading him to the discovery of many-valued systems of logic was the conviction of the indeterminism of future events. The idea of many-valued systems of logic, however, took shape only in 1920 with the introduction of truth matrices. In the same year the three-valued system was originated; that is to say, the truth matrices determined, the definitions of primitive ideas formulated, and the directions of interpretation indicated. The generalized many-valued systems (of any denumerable set of values) came two years later (1922). In the course of time it became apparent that they form a general scheme under which there fall, as particular cases, Lewis's strict-implication theory, Brouwer's logic as formalized by Heyting, O. Becker's calculus with six and ten truth-values (if they are different from Heyting's logic), etc. At first Łukasiewicz was most interested in the three-valued system in connexion with the possibility which it offers of reducing the logic of modality to extensional logic. By a three-valued propositional calculus is to be understood the set of all propositions determined by a three-valued matrix and comprising all 3^3 functions with one argument and all 3^9 functions with two arguments. The three-valued system was

by his mother what he would like to do. He replied that he was anxious to become a painter. But his mother thought that besides being a painter he must have a profession to secure his living. Chwistek was reported to have agreed with his mother's opinion and to be anxious to act in accordance with her wishes. Consequently he matriculated to read mathematics.

[1] Łukasiewicz (5), (9), (13), (15); Łukasiewicz–Tarski (8); Wajsberg (4); Slupecki (1), (2), (3); see also Lewis and Langford, *Symbolic Logic*, chap. vii.

axiomatized by M. Wajsberg and J. Słupecki, the latter's axiomatization making use of the function T in addition to Łukasiewicz's C and N. The three primitive terms are negation ('Np'), implication ('Cpq'), and 'tertium' ('Tp'). Logical sum, product, and equivalence are defined in terms of negation and implication and correspond to $p \vee q$, $p.q$, $p \equiv q$ of the two-valued system respectively. The truth matrices give the following equations:

$$C00 = C01 = C02 = C11 = C21 = C22 = 1;$$
$$C10 = 0;$$
$$C12 = C20 = 2;$$
$$N0 = 1, N1 = 0, N2 = 2;$$
$$T0 = T1 = T2 = 2;$$

where '0' means false, '1'—true, '2'—tertium. The value of the remaining functions of two arguments is determined as follows: (a) $p.q = 1$ if and only if $p = q = 1$; $p.q = 2$ if $p \neq 0$ and $q \neq 0$; $p.q = 0$ in all other cases; (b) $p \vee q = 0$ if and only if $p = q = 0$; $p \vee q = 1$ if at least one of the arguments has the value 1; $p \vee q = 2$ if at least one argument has the value 2; (c) $p \equiv q = 1$ if and only if $p = q$ (p and q have the same truth-value), $p \equiv q = 2$ if at least one argument has the value 2; $p \equiv q = 0$ in the remaining cases.

The system consists of six axioms:

1. $CpCqp$
2. $CCpqCCqrCpr$
3. $CCCpNppp$
4. $CCNpNqCqp$
5. $CTpNTp$
6. $CNTpTp$

and of two rules of inference, taken from the ordinary calculus, the rule of detachment and the rule of substitution.

The three-valued matrix shows that the relation 'Cpq' has a meaning different from that in the two-valued system. The rule of detachment, however, can be safely applied. The matrix method never proves the validity of a theorem by establishing the validity of the implication relation between two propositions

p and q. But it provides a simple means of ascertaining whether a given inference is valid. The three-valued matrix shows that the axioms (1)–(6) are identically true propositions. Hence we are assured that only tautological implication relations are included in the system and therefore the result of any inference from the axioms or from already proved propositions is an identically true proposition of the system.

The system of axioms (1)–(6) is self-consistent, independent, and complete, in the same sense as the two-valued system. It is, however, impossible to give an interpretation of the six axioms, retaining the two rules of inference, in terms of the two-valued system without falling into contradiction. In the three-valued system it is possible to introduce the traditional modal functions into the propositional calculus. For this purpose let us define a new function Mp ('p is possible') as follows:

$$Mp \equiv CNpp$$

The three other modal functions can be then introduced, namely:

$$NMp \ (NCNpp), \ MNp \ (CpNp), \ NMNp \ (NCpNp)$$

which are interpreted: 'p is impossible', 'p is possibly false', 'p is necessary', respectively. The matrix determines the modal functions in the following way:

$$M1 = NMN1 = NM0 = MN0 = M2 = MN2 = 1$$
$$NM1 = MN1 = M0 = NMN0 = NM2 = NMN2 = 0.$$

As we see, the matrix determines the functions Mp, NMp, MNp, $NMNp$ categorically; they have under all circumstances either the value 1 or 0. They differ in this respect from the function of one variable Np, which admits the value 2 also. But considering this case, namely $Np = 2$, and the interpretations of the modal functions, we come to the conclusion that their categorical determination for $p = 2$ is consistent with $Np = 2$ for $p = 2$.

Unfortunately the new functions coincide only partially with the traditional modal functions, a fact for which the ambiguity in ordinary speech of the terms 'possible', 'impossible', 'necessary' is largely responsible. The function Mp admits a univocal meaning, if 'possibility' is interpreted relatively, as in the calculus of probability, Mp being interpreted as 'p is not certainly false'. With an absolute interpretation of 'possibility', as is found in Lewis's system of Strict Implication, the ambiguity arises. An example of it was provided by Łukasiewicz himself in the formula:

$$(A) \qquad CNpNMp$$

('not p implies that p is impossible'). If 'p is impossible' means 'it is not possible that p be true', then (A) is certainly a valid proposition. But it does not hold, if 'p is impossible' means 'p is inconceivable'. The falsity of p does not imply its inconceivability.

The many-valued systems are established by the following fundamental law. Let p and q designate certain numbers of the interval $(0, 1)$. Then the following equations hold:

$$Cpq = 1 \qquad \text{for } p \leqslant q$$
$$Cpq = 1-p+q \quad \text{for } p > q$$
$$Np = 1-p$$

From this set of equations n-valued systems can be established in a way analogous to the two- or three-valued systems; that is to say, by excluding from the interval $(0, 1)$ all values except the n intended. An n-valued system would include all n^n functions of one variable and all n^{n^2} functions of two variables. As in the case of the three-valued system only the nature of the axioms would decide whether a given n-valued calculus is a system of tautologies. For any value of n the rule of detachment would be then applicable in the system.

The question, whether in the n-valued calculus the 'truth-value' of a proposition may be identified with its probability, was answered negatively.[1] Assuming the axiom of extension-

---

[1] Ajdukiewicz (5), Mazurkiewicz (1), (2); Tarski (16). The view in question was held, for instance, by H. Reichenbach. He concluded that only a many-valued propositional calculus, in which each proposition is merely probable, can be usefully employed in science. Strictly speaking there are in science no

ality, such an idea is not feasible. Mazurkiewicz pointed out that in an extensional many-valued calculus the 'truth-value' of the logical sum or product is univocally determined by the 'truth-values' of their arguments. That is not the case with the probability of a logical sum or product, their probability not being a univocally determined function of their factors. This *a priori* reason was combined with an argument concerning the notion of probability itself. There was a tendency in the Lwów–Warsaw School not to consider the arguments of the probability functor as propositions or propositional variables. In accordance with Keynes[1] the notion of probability should be relativized to a set of propositions. The symbol

$$'P\ (x,\ Y)'$$

denotes the probability of the proposition x with respect to the set of propositions Y. If x is deducible from Y, then $P(x, Y) = 1$; if the negation of x is deducible from Y, then $P\ (x,\ Y) = 0$; if, however, neither x nor its negation is deducible from Y, then a probability, lying between 0 and 1, may be assigned to the proposition x. Such a conception of the notion of probability is not appropriate to the interpretation of n-valued calculi in terms of the probability calculus.

An extensive research into n-valued systems was carried out by Łukasiewicz, Wajsberg, Sobociński, and others. The results however remained, so far as I know, unpublished.[2] In particular it was known that Wajsberg axiomatized some n-valued calculi; five-, seven-, and eleven-valued calculi, if my memory serves me right.†

The difficulty with the n-valued systems does not consist so much in technical problems, considerable as these are, as in finding an interpretation of the n 'truth-values' involved in the system. Without an interpretation assigning a definite logical

---

p's and q's such that p implies q, but only such that p implies q to such and such a degree, which is determined by the 'truth-value' of the implication.

[1] J. M. Keynes, *A Treatise on Probability*, London, 1921.

[2] See Łukasiewicz–Tarski (8), § 3.

† [*Ed. note.* The author can find no reference in the literature to these axiomatizations by Wajsberg, and relies here only on his vivid memory of their having been reported in discussions in the 1930s.]

meaning to the n 'truth-values' any given n-valued calculus remains an abstract structure.[1] The importance of studying such structures cannot be denied. But according to the accepted opinion, coming ultimately from Wittgenstein, the value of a logical system consists in providing a set of rules (possessing definite properties) for transforming expressions of a given meaning into other expressions, and thus in revealing their hidden properties and relations. This requirement is not satisfied by an abstract n-valued calculus.

It seems safe to say that many-valued logic has grown out of the stage during which it was just a new field of research, its results leaving now no room for doubts as to its validity. It seems also very likely that the implications of the discovery concern such widely different problems as the theory of probability, Brouwer's intuitionist logic, quantum mechanics, the discussion about universal strict determinism, etc. But it is difficult to fully realize now all the consequences of Łukasiewicz's discovery. Investigations in this direction have gone little beyond the first shocking conclusion that logical truth has got a multi-form character and that there is a variety of ways in which it may be considered. In some respects the discovery and the foundation of many-valued logic makes us think of the shattering blow dealt by the discovery of the Non-Euclidean geometries to the deeply-rooted conviction that there is one and only one way of constructing the spatial reference-frame of our experiences. Similarly it was supposed that a consistent deductive system must follow the Aristotelian pattern, in accordance with his most general 'laws of thought'. In particular it was believed that a statement must be either true or false. In some circles there arose doubts concerning this principle when Brouwer constructed definite examples of mathematical theorems—dealing with 'the infinite' as it occurs in analysis—which are neither true nor false. The construction by Łukasiewicz of a self-consistent deductive system in which the proposition 'a statement is

---

[1] To give an example of a logical abstract structure we could point to E. V. Huntington, 'The Mathematical Structure of Lewis's Theory of Strict Implication', *Fund. Math.* 25 (1935).

either true or false' no longer holds turned the balance definitely against the Aristotelian assumption.

VI. *History of Logic*

This sketch of the development of logic in Poland would not be complete without mentioning the awakening of a lively interest in the history of logic. Łukasiewicz once again was the energetic and influential protagonist. He himself for the first time drew attention to many so far neglected or misunderstood discoveries made in ancient Greece (the School of Megara, the Stoics) and in the Middle Ages (Duns Scotus, Petrus Hispanus).[1] Łukasiewicz combined the expert knowledge of a logician with a thorough acquaintance with ancient sources. This combination made it possible for him to rehabilitate the logic of the Stoics which for a long time had been scornfully treated.

It was till lately the accepted opinion to regard the fundamental distinction between the calculus of propositions and the calculus of propositional functions—the first being logically prior to the second and no less distinct from it than arithmetic from geometry—as a result due to modern logic. Łukasiewicz found out that the first to see it clearly were the Stoics. They created, in conscious opposition to the Aristotelian logic, a theory very similar to the present calculus of propositions; that is to say, a theory consisting of formulae admitting only propositions as the values of their arguments. They considered these formulae as rules of inference—in this they differed from the modern point of view—accepting five of them as indemonstrable and deriving all the others from five axiomatic rules. But they knew a simple procedure which allowed them to transform any rule of inference into a logical assertion. They realized that their theory of deduction permits the Aristotelian syllogistic, the tiny part of the functional calculus at that time regarded as the whole of logic, to be based on a secure foundation.

The Stoics used four functions: negation, implication, conjunction, and disjunction. The first three they defined as truth

---

[1] Łukasiewicz (14). This paper contains results achieved already by Łukasiewicz in 1923.

functions, exactly as it is done nowadays. The definition of implication as a truth-function comes from Philo of Megara; Philo's definition gave rise to a long and famous dispute in antiquity, in which Diodoros Kronos is known as his chief opponent. Diodoros Kronos, who wanted to define implication with the aid of modality, defended an opinion which comes very near to C. I. Lewis's views. The Stoics accepted clearly Philo's definition. The only function that was defined by the Stoics in a way different from the modern is disjunction. They interpreted it in the exclusive meaning of 'or'. That seems at least to have been the case with Chrysippos, though a little later the conviction is also shared that '$p \vee q$' is equivalent to '$\sim p \supset q$'. Only Petrus Hispanus accepted plainly, though unwillingly, a disjunction to be true when 'p' and 'q' are both true. He defined '$p \vee q$' to be false if and only if 'p' and 'q' are both false, and true otherwise.

Petrus Hispanus added to the Stoics' formulae one law of simplification, and those two laws which since De Morgan have come to be known as De Morgan's laws. Philo's definition of implication seems to have been forgotten in the Middle Ages; but in medieval research about *consequentiae*—especially in the form that Duns Scotus gave to it—everything necessary to the rediscovery of this definition is present.

In his researches on the logic of the ancient world Łukasiewicz benefited by the collaboration of several classical scholars. Among them A. Krokiewicz should be especially mentioned; and it may be added that the help he gave to Łukasiewicz influenced in turn his own work. In general Łukasiewicz gave a new impetus to historico-systematic studies. Under the combined influence of the work of Łukasiewicz and O. Toeplitz (Bonn), the present writer, making use of modern logical ideas, made some contributions concerning the theory and practical applications of the axiomatic method from Plato and Eudoxus down to Euclid's *Elements*.[1] He showed how out of the crisis in

---

[1] Jordan (1). Toeplitz was not only a distinguished mathematician, but also an original historian of mathematics. His criticism of Spengler's *Der Untergang des Abendlandes*, so far as it lies in a mathematician's competence, said the final word on the subject. He was one of the founders of *Quellen und Studien*

Greek mathematics, caused by the discovery of irrational numbers (incommensurable magnitudes), had grown the 'foundation problem' of mathematics—bearing some essential resemblance both to the 'technical' crisis of the time of Weierstrass, Dedekind, and Cantor, and to the foundation crisis of the present day. He also showed how in the solution of this problem the germs of the modern dispute between the intuitionists and the formalists are contained. The Revd. J. Salamucha[1] and the Revd. I. M. Bochenski, O.P., encouraged by Łukasiewicz, made important contributions to the history of ancient and medieval logic.

Łukasiewicz, whose own studies provided a crushing criticism of Prantl's *Geschichte der Logik*, demanded that the whole history of logic, till now distorted by involuntary ignorance of the subject and consequently limited to the Aristotelian syllogistic, should be rewritten by someone who had mastered equally modern logic and the technique of historical studies. H. Scholz's *Geschichte der Logik* (Berlin, 1931), the first step in this direction, was made thanks to the lead of, and in collaboration with, Łukasiewicz and his pupils.

Polish Armoured Division
    Easter, 1944.

zur Geschichte der Mathematik, Physik, und Naturwissenschaften and he remained its editor even when deprived of his chair by the German racial regulations. He did not die in a concentration camp, but his premature death was closely connected with the painful circumstances under which he had to live.

[1] He was first deported, as were other professors of Cracow University, to the Sachsenhausen–Oranienburg concentration camp. He was later shot by the Germans together with the wounded whom he refused to leave alone in a Warsaw hospital during the rising of 1944 (Added 1965).

BIBLIOGRAPHY

Note. Items that have been translated into English are marked with an asterisk.

K. AJDUKIEWICZ:

(1) 'O odwracalności stosunku wynikania' (On the Reversibility of the Consequence Relation), *Przegląd Filozoficzny* 16 (1913).

(2) 'Czas absolutny i relatywny' (Absolute and Relative Time), *Przegląd Filozoficzny* 24 (1921).

(3) 'Założenia logiki tradycyjnej' (The Presuppositions of Traditional Logic), *Przegląd Filozoficzny* 29 (1926).

(4) *Z metodologii nauk dedukcyjnych* (Contributions to the Methodology of Deductive Science), Lwów, 1921.

(5) *Główne zasady metodologii i logiki formalnej* (Fundamental Principles of Methodology of Science and of Formal Logic), Warszawa, 1928 (mimeographed).

(6) 'Sprache und Sinn', *Erkenntnis* 4 (1934).

(7) 'Das Weltbild und die Begriffsapparatur', *Erkenntnis* 4 (1934).

*(8) 'Die syntaktische Konnexität', *Studia Philosophica* 1 (1936). [Eng. trans. this volume.]

(9) *Logiczne podstawy nauczania* (Logical Foundations of Teaching), Warszawa, 1934.

(10) 'W sprawie powszechników' (On Universals), *Przegląd Filozoficzny* 38 (1935).

(11) 'Der logistische Antiirrationalismus in Polen', *Erkenntnis* 5 (1935).

I. M. BOCHEŃSKI:

(1) 'Notes historiques sur les propositions modales', *Revue des sciences philos. et théol.* 26 (1937).

(2) *Elementa logicae graecae*, Rome, 1937.

(3) 'De consequentiis scholasticorum earumque origine', *Angelicum* 15 (1938).

L. CHWISTEK:

(1) 'Zasada sprzeczności w świetle nowszych badań Bertranda Russella' (The Principle of Contradiction in the Light of Recent Investigations of Bertrand Russell), *Rozprawy Akademii Umiejętności* 30, Kraków, 1912.

*(1a) 'Antynomie logiki formalnej' (Antinomies of Formal Logic), *Przegląd Filozoficzny* 24 (1921). [Eng. trans. this volume.]

(1b) 'Miara Lebesgue'a' (Lebesgue's Measure), *Archiwum Tow. Naukowego we Lwowie*, Lwów, 1922.

(1c) *Wielość rzeczywistości* (The Plurality of Reality), Kraków, 1921.

(2) 'Über die Antinomien der Prinzipien der Mathematik', *Math. Zeitschr.* 14 (1922).

(3) 'The Theory of Constructive Types', *Annales de la Société Polonaise de Mathématique* 2 (1924); 3 (1925).

(4) 'Über die Hypothesen der Mengenlehre', *Math. Zeitschr.* 25 (1926).

(5) 'Neue Grundlagen der Logik und Mathematik', *Math. Zeitschr.* 30 (1929).

(6) 'Une méthode métamathématique d'analyse', *Comptes rendus du Premier Congrès des Mathématiciens des Pays Slaves*, Warszawa, 1929.

(7) 'Die nominalistische Grundlegung der Mathematik', *Erkenntnis* 3 (1932–3).

(8) (with Hetper and Herzberg) 'Fondements de la métamathématique rationelle', *Bull. de l'Acad. Pol. des Sc. et des Lett.*, 1933.

(9) (with Hetper) 'New Foundations of Formal Metamathematics', *Journal of Symbolic Logic* 3 (1938).

*(10) *Granice nauki* (The Limits of Science), Lwów–Warszawa, 1935. [Eng. trans. London 1948.]

(11) *Zagadnienia kultury duchowej w Polsce* (Problems of Spiritual Culture in Poland), Kraków, 1933.

T. Czeżowski:

(1) *Teoria klas* (Theory of Classes), Lwów, 1918.

(2) 'Imiona i zdania' (Names and Sentences), *Przegląd Filozoficzny* 21 (1918).

(3) *O pewnych stosunkach logicznych* (On Some Logical Relations), Lwów, 1921.

(4) *Klasyczna nauka o sądzie i wniosku w świetle logiki współczesnej* (The Classical Theory of Proposition and Inference in the Light of Modern Logic), Wilno, 1927.

(5) *O pewnym uogólnieniu logiki klasycznej* (On a Certain Generalization of Classical Logic), Lwów, 1931.

(6) *Jak powstało zagadnienie przyczynowości* (How Did the Problem of Causality Arise ?), Wilno, 1933.

(7) 'O zależności między treściami i zakresami pojęć' (On the Relation between Intension and Extension of Concepts), *Przegląd Filozoficzny* 35 (1932).

(8) 'Arystotelesa teoria zdań modalnych' (Aristotle's Theory of Modal Propositions), *Przegląd Filozoficzny* 39 (1936).

J. Herzberg:

(1) See Chwistek (8).

W. Hetper:

(1) See Chwistek (8).

(2) See Chwistek (9).

(3) 'Semantische Arithmetik', *Comptes rendus de la Société des Sciences et des Lettres de Varsovie*, Cl. iii, 27 (1934).

400 BIBLIOGRAPHY

Z. Janiszewski:

(1) 'Podstawy geometrii' (Foundations of Geometry), *Poradnik dla Samouków*, Warszawa, 1915.

(2) 'Idealizm i realizm w matematyce' (Realism and Idealism in Mathematics), *Przegląd Filozoficzny* 19 (1916).

(3) 'Nowe kierunki w geometrii (New Tendencies in Geometry), *Wiadomości Matematyczne* 14 (1910).

S. Jaśkowski:

(1) 'On the Rules of Suppositions in Formal Logic', *Studia Logica* 1 (1934).

Z. Jordan:

(1) *O matematycznych podstawach systemu Platona* (Mathematical Foundations of Plato's System), Poznań, 1937.

M. Kokoszyńska:

(1) 'Z semantyki funkcji zdaniowej' (Semantics of Propositional Functions), *Księga Pamiątkowa Towarzystwa Filozoficznego we Lwowie*, Lwów, 1931.

(2) 'Nauka o *suppositiones termini* u Piotra Hiszpana' (The Theory of *Suppositiones Termini* of Petrus Hispanus), *Przegląd Filozoficzny* 37 (1934).

(3) 'Über den absoluten Wahrheitsbegriff und einige andere semantische Begriffe', *Erkenntnis* 6 (1936).

T. Kotarbiński:

(1) *Szkice praktyczne* (Practical Essays), Warszawa, 1913.

(2) *Utylitaryzm w etyce Milla i Spencera* (Utilitarianism in Mill's and Spencer's Ethics), Kraków, 1915.

(3) 'O istocie doświadczenia wewnętrznego' (On the Nature of Internal Experience), *Przegląd Filozoficzny* 25 (1922).

(4) 'Sprawa istnienia przedmiotów idealnych' (The Problem of the Existence of Ideal Objects), ibid. 24 (1921).

*(5) *Elementy teorii poznania, logiki formalnej i metodologii nauk* (Elements of the Theory of Knowledge, Formal Logic and Methodology of Science), Lwów, 1929. [English translation, *Gnosiology—The Scientific Approach to the Theory of Knowledge*, London 1966.]

(6) 'Le réalisme radical', *Proceedings of the 7th Internal. Congr. of Philos*, held at Oxford, 1930. Oxford, 1931.

(7) 'The Development of the Main Problem in the Methodology of Francis Bacon', *Studia Philosophica* 2 (1937).

(8) 'Grundlinien und Tendenzen der Philosophie in Polen', *Slavische Rundschau*, Prag, 1933.

C. Kuratowski:

(1) 'Sur la notion de l'ordre dans la théorie des ensembles', *Fund. Math.* 2 (1921).

*(2) (with Tarski) 'Les opérations logiques et les ensembles projectifs', ibid. 17 (1931). [English translation in Tarski, *Logic, Semantics, Metamathematics*, Oxford 1956.]

(3) 'Évaluation de la classe borélienne ou projective d'un ensemble de points à l'aide des symboles logiques', ibid. 17 (1931).

S. Leśniewski:

(1) 'Przyczynek do analizy zdań egzystencjalnych' (A Contribution to the Analysis of Existential Propositions), *Przegląd Filozoficzny* 14 (1911).

(2) 'Próba dowodu ontologicznej zasady sprzeczności' (An Attempt to Prove the Ontological Principle of Contradiction), ibid. 15 (1912).

(3) 'Krytyka logicznej zasady wyłączonego środka' (Criticism of the Logical Principle of the Excluded Middle), ibid. 16 (1913).

(4) *Logical Studies* (in Russian), St. Petersburg, 1913.

(5) 'Czy klasa klas nie podporządkowanych sobie jest podporządkowana sobie' (Is the Class of Classes not Subordinate to Themselves Subordinate to itself ?), *Przegląd Filozoficzny* 17 (1914).

(6) *Zasady ogólnej teorii mnogości.* I (Foundations of the General Theory of Sets I), Moscow, 1916.

(7) 'O podstawach matematyki' (Foundations of Mathematics), *Przegląd Filozoficzny* 30 (1927); 31 (1928); 32 (1929); 33 (1930); 34 (1931).

(8) 'Grundzüge eines neuen Systems der Grundlagen der Mathematik', *Fund. Math.* 14 (1929); *Collectanea Logica* (1939).

*(9) 'Einleitende Bemerkungen zur Fortsetzung meiner Mitteilung u. d. T. "Grundzüge . . ."' *Collectanea Logica* (1939). [Eng. trans. this volume.]

(10) 'Über Funktionen deren Felder Abelsche Gruppen in Bezug auf diese Funktionen sind', *Fund. Math.* 14 (1929).

(11) 'Über die Grundlagen der Ontologie', *Comptes rendus des Séances de la Société des Sciences et des Lettres de Varsovie* Cl. iii, 23 (1930).

*(12) 'Über Definitionen in der sogenannten Theorie der Deduktion', ibid. Cl. iii, 24 (1931). [Eng. trans. this volume.]

A. Lindenbaum:

(1) See Tarski (5) and (5a).

(2) 'Remarques sur une question de la méthode axiomatique', *Fund. Math.* 15 (1930).

(3) 'Bemerkungen zu den vorhergehenden Bemerkungen des Herrn J. v. Neumann', ibid. 17 (1931).

J. Łukasiewicz:

(1) *O zasadzie sprzeczności u Arystotelesa* (The Principle of Non-Contradiction in Aristotle), Kraków, 1910.

(2) 'Über den Satz des Widerspruches bei Aristoteles', *Bull. de l'Acad. de Sc. à Cracovie* (1910).

(3) *Die logischen Grundlagen der Wahrscheinlichkeitsrechnung*, Kraków, 1913.

(4) 'Logika dwuwartościowa' (Two-valued Logic), *Przegląd Filozoficzny* 23 (1921).

*(5) 'O logice trójwartościowej' (On Three-valued Logic), *Ruch Filozoficzny* 5 (1920). [Eng. trans. this volume.]

(6) 'Démonstration de la compatibilité des axiomes de la théorie de la déduction', *Ann. de la Soc. Pol. de Math.* 3 (1925).

*(7) *Elementy logiki matematycznej* (Elements of Mathematical Logic), Warszawa, 1929 (mimeographed). [Eng. trans., London 1963.]

*(8) (with Tarski) 'Untersuchungen über den Aussagenkalkül', *Comptes rendus des Séances de la Société des Sciences et des Lettres de Varsovie*, Cl. iii, 23 (1930). [Eng. trans. in Tarski, *Logic, Semantics, Metamathematics*, Oxford 1956.]

*(9) 'Philosophische Bemerkungen zu mehrwertigen Systemen des Aussangenkalküls, ibid. [Eng. trans. this volume.]

(10) 'Ein Vollständigkeitsbeweis des zweiwertigen Aussagenkalküls', ibid., Cl. iii, 24 (1931).

(11) 'Uwagi o aksjomacie Nicoda i o "uogólniającej dedukcji" ' (Remarks about Nicod's Axiom and 'Generalizing Deduction'), *Księga Pamiątkowa Polskiego Towarzystwa Filozoficznego we Lwowie*, Lwów, 1931.

(12) *O nauce* (On Science), Lwów, 1934. (*Second impression.*)

(13) 'Zur vollen dreiwertigen Aussagenlogik', *Erkenntnis* 5 (1935–6).

*(14) 'Zur Geschichte der Aussagenlogik', ibid. [Eng. trans. this volume.]

(15) 'Die Logik und das Grundlagen-Problem', *Les Entretiens de Zürich sur les fondements et la méthode des sciences mathématiques*, 6–9 décembre, 1938, Zürich, 1941.

(16) 'Logistyka a filozofia' (Logistic and Philosophy), *Przegląd Filozoficzny* 39 (1936).

S. MAZURKIEWICZ:

(1) 'Zur Axiomatik der Wahrscheinlichkeitslehre', *Comptes rendus des Séances de la Société des Sciences et des Lettres de Varsovie*, Cl. iii, 25 (1932).

(2) 'Über die Grundlagen der Wahrscheinlichkeitsrechnung', *Monatsh. f. Math. u. Phys.* 41 (1934).

M. PRESBURGER:

(1) 'Über die Vollständigkeit eines gewissen Systems der Arithmetik ganzer Zahlen, in welchem die Addition als einzige Operation hervortritt', *Comptes rendus du Premier Congrès des Mathématiciens des Pays Slaves*, Warszawa, 1929.

J. SALAMUCHA:

(1) *Pojęcie dedukcji u Arystotelesa i św. Tomasza* (The Notion of Deduction in Aristotle and St. Thomas), Warszawa, 1930.

*(2) 'Dowód *ex motu* na istnienie Boga' (The Proof *ex motu* for the

Existence of God), *Collectanea Theologica* 15 (1934). [Eng. trans. in *The New Scholasticism* 32 (1958).]

(3) 'Logika zdań u Wilhelma Ockhama' (Logic of Propositions in William Ockham), *Przegląd Filozoficzny* 38 (1935).

(4) 'Pojawienie się zagadnień antynomialnych na gruncie logiki średniowiecznej' (The Appearance of the Problems of Antinomies in Medieval Logic), *Przegląd Filozoficzny* 40 (1937).

W. SIERPIŃSKI:

(1) *Zarys teorii mnogości* (Introduction to the Theory of Sets), Warszawa, 1923.

(2) *Leçons sur les nombres transfinis*, Paris, 1930.

(3) 'Une remarque sur la notion de l'ordre', *Fund. Math.* 2 (1921).

(4) 'L'axiome de M. Zermelo et son rôle dans la théorie des ensembles et l'analyse', *Bull. de l'Acad. des Sciences*, Cracovie, 1919.

(5) *Hypothèse du continu*, Monografie Matematyczne 4, Warszawa, 1934.

J. SŁUPECKI:

*(1) 'Der volle dreiwertige Aussagenkalkül', *Comptes rendus des Séances de la Société des Sciences et des Lettres de Varsovie*, Cl. iii, 29 (1936). [Eng. trans. this volume.]

(2) 'Dowód aksjomatyzowalności pełnych systemów wielowartościowych rachunku zdań' (A Proof of the Axiomatizability of Full Systems of Polyvalent Propositional Calculus), ibid. Cl. iii, 32 (1939).

(3) 'Kryterium pełności wielowartościowych systemów logiki zdań' (A Criterion of the Definability of Functors in Polyvalent Systems of Propositional Calculus), ibid.

B. SOBOCIŃSKI:

(1) 'Z badań nad teorią dedukcji' (Investigations on the Theory of Deduction), *Przegląd Filozoficzny* 35 (1932).

*(2) 'O kolejnych uproszczeniach aksjomatyki "ontologji" Prof. St. Leśniewskiego' (On the Successive Simplifications of the Axioms of Leśniewski's Ontology), *Fragmenty Filozoficzne*, Warszawa, 1934. [Eng. trans. this volume.]

(3) 'Aksjomatyzacja implikacyjno-konjukcyjnej teorii dedukcji' (Axiomatization of the Implicational-Conjunctive Propositional Calculus), *Przegląd Filozoficzny* 37 (1935).

(4) 'Aksomatyzacja pewnych wielowartościowych systemów teorii dedukcji' (Axiomatization of Certain Polyvalent Systems of the Theory of Deduction), *Roczniki Prac Naukowych Zrzeszenia Asystentów Uniwersytetu Józefa Piłsudskiego w Warszawie* 1 (1936).

J. ŚLESZYŃSKI:

(1) *Teoria dowodu* (Theory of Proof), Kraków, 1925–9.

A. TARSKI:

*(1) 'Sur le terme primitif de la logistique', *Fund. Math.* 4 (1923). [English translation in Tarski, *Logic, Semantics, Metamathematics,* Oxford 1956.]

*(2) 'Sur les truth-functions au sens de MM. Russell et Whitehead', ibid. 5 (1924). [Eng. trans. in Tarski, *op. cit.*]

(3) 'Sur les ensembles finis', ibid. 6 (1924).

(4) (with S. Banach) 'Sur la décomposition des ensembles de points en parties respectivement congruentes', ibid. 6 (1924).

(5) (with Lindenbaum) 'Communication sur les récherches de la théorie des ensembles', *Comptes rendus des Séances de la Société des Sciences et des Lettres de Varsovie,* Cl. iii, 19 (1927).

(5a) (with Lindenbaum) 'Sur l'indépendance des notions primitives dans les systèmes mathématiques', *Rocznik Polskiego Towarzystwa Matematycznego* 5 (1926).

*(6) 'Über einige fundamentale Begriffe der Metamathematik', *Comptes rendus des Séances de la Société des Sciences et des Lettres de Varsovie,* Cl. iii, 23 (1930). [Eng. trans. in Tarski, *op. cit.*]

(7) 'Remarques sur les notions fondamentales de la méthodologie des mathématiques', *Ann. de la Soc. Pol. de Math.* 7 (1929).

*(8) See ŁUKASIEWICZ (8).

*(9) 'Les fondements de la géométrie des corps, *Księga Pamiątkowa Pierwszego Polskiego Zjazdu Matematycznego,* Kraków, 1929. [Eng. trans. in Tarski, *op. cit.*]

*(10) 'Fundamentale Begriffe der Methodologie der deduktiven Wissenschaften I', *Monatsh. f. Math. u. Phys.* 37 (1930). [Eng. trans. in Tarski, *op. cit.*]

*(11) 'Sur des ensembles définissables de nombres réels I', *Fund. Math.* 17 (1931). [Eng. trans. in Tarski, *op. cit.*]

*(12) See KURATOWSKI (2).

*(13) *Pojęcie prawdy w językach nauk dedukcyjnych* (The Concept of Truth in Formalized Languages), Warszawa, 1933. German translation: 'Der Wahrheitsbegriff in den formalisierten Sprachen', *Studia Philosophica* 1 (1936) (reprint dated 1935). [Eng. trans. in Tarski, *op. cit.*]

*(14) 'Einige Betrachtungen über die Begriffe der ω-Widerspruchsfreiheit und der ω-Vollständigkeit', *Monatsh. f. Math. u. Phys.* 40 (1933). [Eng. trans. in Tarski, *op. cit.*]

*(15) 'Einige methodologische Untersuchungen über die Definierbarkeit der Begriffe', *Erkenntnis* 5 (1935–6). [Eng. trans. in Tarski, *op. cit.*]

(16) 'Wahrscheinlichkeitslehre und mehrwertige Logik', ibid.

*(17) 'Grundzüge des Systemenkalküls', *Fund. Math.* 25 (1935). [Eng. trans. in Tarski, *op. cit.*]

*(18) 'Grundlegung der wissenschaftlichen Semantik', *Actes du Congrès International de Philosophie Scientifique,* Paris, 1936. [Eng. trans. in Tarski, *op. cit.*]

BIBLIOGRAPHY 405

(19) *O logice matematycznej i metodzie dedukcyjnej* (Mathematical Logic and Methodology of Deductive Sciences), Warszawa, 1935.

(20) *Einführung in die mathematische Logik und die Methodologie der Mathematik*, Wien, 1937.

(21) *Introduction to Logic*, New York, 1941.

(22) 'On Undecidable Statements in Enlarged Systems of Logic and the Concept of Truth', *Journal of Symbolic Logic* 4 (1939).

*(23) 'Zur Grundlegung der Booleschen Algebra I', *Fund. Math.* 24 (1935). [Eng. trans. in Tarski, op. cit.]

K. TWARDOWSKI:

(1) *Zur Lehre vom Inhalt und Gegenstand der Vorstellungen*, Wien, 1894.

(2) *Über begriffliche Vorstellungen*, 16. Jahresbericht der Philosophischen Gesellschaft an der Universität zu Wien, 1903.

(3) *O istocie pojęć* (The Nature of Concepts), Lwów, 1924.

(4) *Rozprawy i artykuły filozoficzne* (Philosophical Essays and Articles), Warszawa, 1927.

M. WAJSBERG:

(1) 'Über Axiomensysteme des Aussagenkalküls', *Monatsh. f. Math. u. Phys.* 39 (1932).

(2) 'Ein neues Axiom des Aussagenkalküls in der Symbolik von Sheffer', ibid.

(3) 'Ein erweiterter Klassenkalkül', ibid. 40 (1933).

*(4) 'Aksjomatyzacja trójwartościowego rachunku zdań' (Axiomatization of the Three-valued Propositional Calculus), *Comptes rendus des Séances de la Société des Sciences et des Lettres de Varsovie*, Cl. iii, 24 (1931). [Eng. trans. this volume.]

(5) Untersuchungen über den Funktionenkalkül für endliche Individuenbereiche', *Math. Annalen* 108 (1933).

(6) 'Beitrag zur Metamathematik', ibid. 109 (1933).

*(7) 'Metalogische Beiträge', *Wiadomości Matematyczne* 43 (1937). [Eng. trans. this volume.]

N. WILKOSZ:

(1) *Podstawy ogólnej teorii mnogości* (Foundations of the General Theory of Sets), Kraków, 1925.

Z. ZAWIRSKI:

(1) 'Związek przyczynowy i funkcyjny' (Causal and Functional Connection), *Przegląd Filozoficzny* 15 (1912).

(2) 'Metoda aksjomatyczna i przyrodoznawstwo' (The Axiomatic Method and Natural Science), *Kwartalnik Filozoficzny* 1 (1924).

(3) 'O wiecznym powrocie światów' (The Everlasting Return of the Worlds), ibid. 5 (1927).

(4) *Stosunek logiki i matematyki w świetle badań współczesnych* (The

Relation of Logic to Mathematics in the Light of Modern Research), Kraków, 1927.

(5) 'W sprawie indeterminizmu fizyki kwantowej' (A Contribution to the Problem of Indeterminism in Quantum Physics), *Księga Pamiątkowa Polskiego Towarzystwa Filozoficznego we Lwowie*, Lwów, 1931.

(6) *L'Évolution de la notion du temps*, Cracovie, 1936.

(7) *Stosunek logiki wielowartościowej do rachunku prawdopodobieństwa* (Relation of the Many-valued Systems of Logic to the Calculus of Probability), Poznań, 1934.

E. ŻYLIŃSKI:

(1) 'O przedstawialności funkcji prawdziwościowych' (The Representation of a Truth Function by another Truth Function), *Przegląd Filozoficzny*, 30 (1927).

(2) 'Some Remarks Concerning the Theory of Deduction', *Fund. Math.* 7 (1925).

PRINTED IN GREAT BRITAIN
AT THE UNIVERSITY PRESS, OXFORD
BY VIVIAN RIDLER
PRINTER TO THE UNIVERSITY